D1179563

The Zebra Finch

Oxford Ornithology Series
Edited by C. M. Perrins

The Zebra Finch

A SYNTHESIS OF FIELD AND LABORATORY STUDIES

RICHARD A. ZANN
School of Zoology,
La Trobe University, Australia

Line drawings by Michael Bamford

Oxford   New York   Melbourne
OXFORD UNIVERSITY PRESS
1996

Oxford University Press, Walton Street, Oxford OX2 6DP

Oxford   New York
Athens   Auckland   Bangkok   Bombay
Calcutta   Cape Town   Dar es Salaam   Delhi
Florence   Hong Kong   Istanbul   Karachi
Kuala Lumpur   Madras   Madrid   Melbourne
Mexico City   Nairobi   Paris   Singapore
Taipei   Tokyo   Toronto
and associated companies in
Berlin   Ibadan

Oxford is a trade mark of Oxford University Press

Published in the United States
by Oxford University Press Inc., New York

A catalogue record for this book is available from the British Library

Library of Congress Cataloging in Publication Data
Zann, Richard A.
The zebra finch: a synthesis of field and laboratory studies/
Richard A. Zann.
(Oxford ornithology series: 5)
Includes bibliographical references (p.   ) and indexes.
1. Zebra finch.   I. Title.   II. Series.
QL696.P244Z35   1996   598.8—dc20   95–41163

ISBN 0 19 854079 5

Typeset by Footnote Graphics, Warminster, Wiltshire
Printed in Great Britain by
Biddles Ltd, Guildford & Kings Lynn

Preface

The Australian Zebra Finch is a popular cage bird in many countries, not simply among hobbyists, but also among scientists where it is the preferred subject for laboratory research in a range of biological disciplines. The primary aim of this book is to integrate these diverse laboratory studies and place them in the context of the biology of the animals in the wild so that a more complete picture of the adaptations and life history of the species will emerge. From this I hope new understandings arise that can act as catalysts for better research and lead, on the one hand, to more biologically relevant questions by laboratory workers, and on the other, to new insights into the range of adaptations field workers can investigate. This synergistic interplay between field and laboratory studies is already producing results, for example, in the understanding of paternity and mate choice (Chapters 9 and 11) and there is ample scope for similar advances elsewhere.

Given the vast number of publications on domesticated laboratory Zebra Finches, it is impossible to review all of them, or to keep up with the continuous flow, so I have selected those that complement the field studies and aid understanding of the species. I should say 'subspecies' because almost all research has been conducted on the Australian subspecies of Zebra Finch. Nothing is known about the field biology of the Lesser Sundas subspecies, and only limited aspects of its vocal and reproductive behaviour are known from captive birds. There is some fascinating work to be done on this subspecies; it will not only fill a void, but will provide interesting reflections on the biology of the Australian subspecies.

For most of the last 30 years much of our knowledge of the behaviour of the Zebra Finch in the wild was based on Klaus Immelmann's work. Unfortunately, only a small proportion was published in English, in particular in his wonderful book, *Australian Finches in Bush and Aviary* (1965). Readers of German had access to his massive papers from which the English work was extracted, but if the citations in the literature are any guide, these papers have been largely ignored. Therefore, another of my aims is to make accessible to readers of English additional details of Immelmann's fine fieldwork. Of course, things have not stood still in Australia since his visit and a number of ecological and behavioural studies have been conducted on populations in different parts of the country and I have tried to set these in the context of Immelmann's contribution.

Jiro Kikkawa was responsible for my scientific introduction to Zebra Finches when he brought some pairs into the teaching laboratory at the University of New England where I was enrolled in Zoology in 1964. Minuteness, neatness, wonderful colours and exquisite patterns were the impressions I remember. That same year I helped Jiro mistnet Zebra Finches at his nearby field site, and two years later I began my Ph.D. under his supervision at the University of Queensland on behaviour of relatives of the Zebra Finch, namely, the Long-tailed, Black-throated and Masked Finches. These are possibly more beautiful and charming than 'Zebbies' but have less character. I gained insight into certain aspects of the character of Zebra Finches through a relationship with a male Zebbie I kept during the Ph.D. years. 'Fred' hatched in 1966 and was hand-raised, consequently he became sexually imprinted on me, mainly my fingers, but also my face. In the aviary he would sometimes land on my shoulders and back, but after a few dramatic episodes with the opposite sex he was confined to a small cage on my filing cabinet for safety. Without fail he courted my finger when I placed it inside the cage. His song and dance routine were species-typical, but punctuated with aggressive pecks. He was renowned among my fellow postgraduates for his amusing antics and his clockwork reliability: he always performed for visitors, but only my fingers elicited it. Often on the first courtship of the day he would mount my finger and copulate, occasionally leaving a droplet of semen, but mostly he ended the routine with a vigorous peck instead.

Fred took a wife but they never reproduced despite building a nest. They usually got on fine unless he happened to see me whereupon his sexual preference would take over and he would court my face or fingers then chase and attack his partner until I turned my back at which their bond would be re-established and harmony restored.

Once I was unwittingly involved in his conditioning, or he in mine. He had learned by trial and error to flick a droplet of water from his gravity drinker onto the back of my neck when I was at my desk, a metre away. When I turned around he would begin his courtship routine; presumably my face or fingers or the courtship were reinforcing for him, and naturally, his courtship was reinforcing for me. Poor Fred disappeared in transit between Brisbane and Melbourne in 1972; he was in his sixth year.

Zebra Finches are often taken for granted by Australians because they are numerous, noisy and persistent, but many, including myself, admire them because they typify the 'little Aussie battler'—the small and insignificant, that somehow succeeds by simply hanging on and enduring the vicissitudes of the vast, harsh country of inland Australia.

Finally, for those readers interested in statistical evidence for statements in the text I have only given details where the work has not been published, or is not available in a thesis lodged in a university library.

Acknowledgments

Jiro Kikkawa, my teacher and supervisor, stimulated my love for birds and introduced me to Zebra Finches and their scientific possibilities. The late Klaus Immelmann, my host at Bielefeld in 1980, surrounded me with Zebra Finches and their scientific aficionados. My fieldwork on Zebra Finches would not have been possible without the hospitality of the land owners, in particular, the Danaher, Johnstone, Padgett and Powney families in northern Victoria, and the CSIRO Division of Wildlife and Ecology at Alice Springs.

I am grateful to my collaborators in the field, especially Tim Birkhead, Nancy Burley, Nicky Clayton and Steve Morton. My work on Zebra Finches would have been limited without my students and assistants over the years: Robert Carr, Andrew Dunn, Bruce Male, Andrew McIntosh, Toby Nelson, Martin O'Brien, Bruce Quin, David Runciman, Maurizio Rossetto, Annabelle Roper, Bruce Straw and Elizabeth Tanger.

Naturally I take responsibility for any errors and omissions in the book, but I am especially grateful to the following who read parts of an earlier draft and patiently corrected my mistakes: Tim Birkhead (all chapters), Les Christidis (Chapters 1 and 12), Nicky Clayton (Chapters 2, 3, 10 and 11), Stephen Davies (Chapters 2, 7, 8 and 12), and Peter Slater (Chapter 10).

The Royal Australasian Ornithologists Union kindly provided access to unpublished data from the Field Atlas of Australian Birds, the Australian Bird Count and the Nest Records Scheme. They also granted permission to reproduce the distribution maps which were kindly re-drawn by Jenny Browning. I am grateful to the Queensland Zebra Finch Society who advised on mutations of the Zebra Finch.

Lastly, I want to thank Michael Bamford for his fine pen and ink drawings of Zebra Finch behaviour and postures.

Bundoora, Vic
March 1995   R.A.Z.

Contents

x   *Contents*

Historical note

'After travelling all day and singing about many adventures of the
honeyeaters and other bird ancestors, the Zebra Finch women called softly
to their children to rest; the children however kept their little song about the
Mallee-fowl.
  "My little children with the red ochred noses,
  My little children with the red ochred noses,
  Come and sleep."
The children sang softly until they fell asleep.'

From *The Zebra Finch Journeys*, a dreamtime account of the creation
of the world told by aborigines of the Iwantja Community, South Australia
(Isaacs 1980).

The Zebra Finch first became known to science at the start of the 19th
century when it was collected on one of the earliest and most lavishly
funded voyages of scientific exploration yet to leave Europe. Nicolas
Baudin was commissioned by Napoleon in 1800 to carry out scientific
and geographical surveys of the coasts of Australia in the ships, *'Géo-*
graphe' and *'Naturaliste'*. After mapping the west coast of Australia the
ships visited Kupang, then a village in West Timor, from 18 August to
13 November 1801 and again from 30 April to 3 June 1803 during
which the Zebra Finch was collected by the expedition's naturalists, the
most well known of whom was the headstrong Francois Péron. Zebra
Finches formed part of an enormous collection of 100,000 zoological
specimens that made the hazardous voyage back to France and which
eventually formed the basis of Péron's vast work *'Voyage de découvertes*
aux Terres Australes' published in 1816. Possibly, some live Zebra
Finches survived the journey back to Paris because Louis Jean Pierre
Vieillot, the great taxonomist, made obscure references to behaviour in
his famous book *Les Oiseaux Chanteurs* (Volume 2: 1805 and 1809).
However, Vieillot did not publish the taxonomic name (*Fringilla guttata*)
and the scientific description of the new species until 1817 (*Nouveau*
Dictionnaire d'Histoire Naturelle vol xxi, p. 223). Alfred Russell Wallace
(1864) also collected the Zebra Finch from Timor during his famous
zoological investigations of the region; he named it *'Amadina insularis'*.
  The Australian Zebra Finch was not described until 1837, when the
celebrated ornithologist, John Gould, received the first specimens
collected from the plains of central New South Wales. He named it

'*Amadina castanotis*', which means Chestnut-eared Finch (*Synopsis. Birds of Australia* pt. 1 pl. 10, 1 Jan. 1837). In his *Handbook to the Birds of Australia* (1865) Gould wrote, 'The Chestnut-eared Finch is one of the smallest of the genus (family) yet discovered in Australia; it is also one of the most beautiful, and in the chasteness of its colouring can scarcely be excelled.'

References to Zebra Finches later appear in the accounts of the early explorers and the pioneering settlers of the vast inland of Australia. The strong attachment between Zebra Finches and surface water in the deserts frequently indicated obscure sources of life-saving drinking water to desperate explorers (Davidson 1905) and bushmen (Carter 1903; Lindsay 1963). The pioneering ornith ologists, almost all of them amateurs with amazing qualities of endurance and persistence, found Zebra Finches a conspicuous component of the avifauna of the arid and semi-arid regions of the continent, not least, for the vast numbers that concentrated around the more persistent waterholes during the endemic droughts that are so much a feature of the Australian inland. Early names for the Zebra Finch, besides the Chestnut-eared Finch, include 'Waxbills' and 'Diamond Sparrows'.

Zebra Finches, of course, were significant to the indigenous inhabitants of Australia, the aborigines; in particular, those tribes that inhabited the more arid areas of central and western Australia. Occasionally, Zebra Finches were a source of food, usually being hunted by children, and naturally, during drought they were an important indicator of the presence of surface water hidden away in small rock holes for any wandering bands of nomads in strange country. Zebra Finches were among the few species that did not foul the minute water supplies upon which aborigines often depended for their survival, and were purposely allowed access via small gaps in the rocks that covered some holes. Among some tribes of central Australia, droppings of Zebra Finches, when mixed with herbs, were used medicinally for headaches and other pains (Winfield 1982).

Given its prominence in the landscape and its usefulness, many tribes coined unique names for the species, often based on onomatopoeic renderings of the nasal distance calls. Serventy and Whittell (1976) list seven aboriginal names for Zebra Finches from among the various language groups in Western Australia: 'chiaga', 'newmerri' or 'nee-murri' (= 'red-nosed fellow'), 'yim-eye', 'neamoora', 'nyinnyinka', 'nyi-nyi' and 'nye-nye'. In central Australia the Pitjanjatjara people called Zebra Finches the 'nyi-nyi' (Isaacs 1980; Winfield 1982) or 'njinji' (Cleland 1931) after the grass whose seed they often eat; the Aranda people call them 'nyinka' or 'nienji' (Mountford 1976) and the Walpiri people, 'ithi' (Condon 1955) or 'jindjinmari' (Meggitt 1971).

The Zebra Finch also figured in the mythical traditions and everyday expressions of some groups of aboriginal people. Thus, expressions for abundance would draw similes with a flock of Zebra Finches (Finlayson 1935). In the mountain devil ceremonies of the Aranda tribe, body decorations mimicking the throat stripes of the Zebra Finch male are made on the backs of men of the Zebra Finch totem (Mountford 1976).

Zebra Finches figured prominently in the ceremonial life of the Pitjanjatjara people of the northwest corner of South Australia. In one of their 'dreamtime' stories, which describes the creation of the world, there is a children's song which tells of a great journey undertaken by the numerous Zebra Finch people (the 'Nyi-Nyi') who encounter the human ancestors of other species of birds as they traverse vast parts of the southern deserts visiting important sacred sites until finally returning to their home country bringing a valuable discovery, namely fire for cooking (Isaacs 1980).

Most ornithologists who specialise on a particular species do not remain emotionally detached from their subjects, and this is often the impetus for their research. For the 12-year-old Klaus Immelmann the gift of a pair of Zebra Finches sealed his career for life. Henceforth, the Zebra Finch was his 'bird of fate' ('Schicksalsvogel') (Immelmann 1970) and his love for the species made it the focus of his research for more than 30 years until his premature death at 52. His first major publication on Zebra Finches was his doctoral thesis on the function of species-specific markings which he undertook at the University of Zurich (Immelmann 1959). This massive, and often ignored work, demonstrated a first rate experimental approach, especially when one takes into account the scientific genre of the time. Soon afterwards he spent almost a year in Australia, all of it in the field, studying the behaviour and ecology of grassfinches, the Zebra Finch in particular. He was a pioneer in many respects and was one of the first ornithologists to spend a wet season in northwest Australia. Somehow he seemed to find time to make observations on many other species of animals as well, including the human inhabitants, and published extensively on all these subjects in the years following his return to Germany. His observations on the breeding response of Zebra Finches to drought-breaking rains attracted worldwide attention from ornithologists, although the opportunistic nature of breeding in Zebra Finches had long been remarked upon by Australian observers. Back in Germany he resumed his experimental investigations into domesticated Zebra Finches focusing on the development of behaviour. At Braunschweig and Bielefeld he made pioneering and seminal contributions to two fields of research, sexual imprinting and song learning, both of which are still being vigorously pursued at a number of institutions around the world.

Although Australian Zebra Finches had been held as cage birds in Europe before the 1850s (Neunzig 1965), the first scientific study of the Zebra Finch was not until a century later when Desmond Morris began his behaviour studies of aviary birds under the guidance of Niko

Tinbergen at Oxford University. Morris's much cited paper on reproductive behaviour of captive Zebra Finches, published in *Behaviour* in 1954, is also memorable for its artistic illustrations, entertaining text, and orthodox ethological interpretations. Klaus Immelmann and Desmond Morris together were instrumental in making the domesticated Zebra Finch the avian model of choice for many laboratory investigators in an array of scientific disciplines around the world.

1 Systematics and phylogeny

Zebra Finches belong to the estrildine group of finches of which there are approximately 124 species found mainly in the tropical and subtropical parts of Africa, India, South-East Asia, Australia and Melanesia. Estrildines are the only granivorous group of passerines endemic to Australia and authorities recognise 18 species, of which the Zebra Finch is the most widespread and numerous.

Systematics

If their turbulent taxonomic history is any indication, the grouping of estrildine species into genera and the relationships among genera provided considerable challenge to students of avian systematics. Christidis (1987a,b) provides the most recent phylogeny of the estrildines and I have followed his nomenclature, classification and phylogeny in this book. The virtue of his arrangement is that it is the most extensive and objective to date and it provides convincing solutions to a number of problems ranging from those at the generic level up to those at the family level. Moreover, it corresponds to the main elements of the traditional phylogenies of Delacour (1943) and Mayr (1968) in recognising three lineages within the group. Christidis assigned species and genera according to a combination of morphological, allozyme (protein electrophoresis) and chromosomal characters using computerised cladistic methods to construct phylogenetic trees. Christidis' arrangement in Figure 1.1 depicts the three natural lineages of estrildines that constitute the monophyletic tribes of the subfamily Estrildinae (Blyth 1889) of the family Passeridae (Vigors 1825)—Poephilini (grassfinches), Lonchurini (mannikins) and Estrildini (waxbills). At a higher taxonomic level, the subfamily Estrildinae is shown to be more closely related to the weaver finches (Ploceinini) and sparrows (Passerini) than to old-world finches (Fringillinae) and buntings (Emberizinae).

Despite the comprehensive nature of Christidis' analysis, few specimens of waxbills were available for investigation; consequently his phylogeny and revision were focused principally on species in the Poephilini and Lonchurini. A complete picture of the phylogeny and systematics of the subfamily as a whole must wait until a comparable analysis has been made of the many species that constitute the tribe Estrildini.

Gould (1865) appears to be the first to have coined the term 'Grass-Finches' in reference to the Australian finches. To emphasise differences

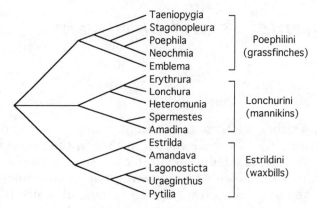

Fig. 1.1 Systematic relationship of estrildine finches to genera and tribe. (After Baverstock *et al.* 1991.)

between ploceine weavers and estrildines, Immelmann (1965a) proposed the word 'grass-finches' or 'grassfinches' to include all species of estrildines, thus, it was the English equivalent of the German word 'Prachtfinken'. However, 'grassfinches' was used by Delacour (1943) in a more exclusive sense to refer to the tribe he called Erythrurae which included all 14 species of Christidis' tribe, the Poephilini, as well as 12 additional species that constitute the genus Erythrura. In this book I will use the term 'estrildines' for all species in the family and reserve the word 'grassfinches' for the 14 species of the Poephilini that inhabit Australia and New Guinea.

The nomenclature and taxonomic affinities of a number of estrildines, the Red-browed Finch, the Plum-headed Finch and the Gouldian Finch above all, have been a source of contention among taxonomists since Delacour's (1943) revision of the subfamily. Earlier opinions on these species and other problems in the subfamily may be found in Steiner (1955, 1960), Wolters (1957) Keast (1958), Morris (1958), Immelmann (1962a), Harrison (1967), Mayr (1968), Schodde (1975), Schodde and McKean (1976) and Ziswiller *et al.* (1972). The phylogeny and systematics of the Australian estrildines, the Poephilini and five of the Lonchurini, will now be considered in detail. The reader is referred to Goodwin's (1982) monograph on estrildines for treatment of non-Australian species.

Waxbills, of which there are approximately 61 species, have radiated extensively in Africa where they occupy a great variety of niches (Goodwin 1982). In contrast, there are only 14 species of Poephilini, all but one Australian, and they occupy a more limited range of niches. The Lonchurini, which appear to be the most recently evolved tribe, has 49 species found mainly in Melanesia and southeast Asia with a few in India and Africa; their range of niches are even more limited than those of the Poephilini.

The Poephilini tribe

The relationships suggested by Christidis between the five genera that constitute the Poephilini are shown in Figure 1.1. The most decisive split in the tribe, using biochemical characters, is between the genus *Taeniopygia* and the rest (Christidis 1987a). Evidence from protein electrophoresis (Christidis 1987a) and chromosome morphology (Christidis 1986a) indicated that the supergenus *Poephila* (Gould 1842) was polyphyletic. The Zebra Finch and the Double-barred Finch have been removed from the supergenus where they had been grouped with the Long-tailed Finch (*Poephila acuticauda*), the Black-throated Finch (*P. cincta*) and the Masked Finch (*P. personata*) by earlier taxonomists (Delacour 1943; Keast 1958; Morris 1958; Schodde 1975). The Zebra Finch and the Double-barred Finch, which share many characters, are now placed in their own genus *Taeniopygia* (Reichenbach 1862), of which the Zebra Finch is the type species, *Taeniopygia guttata* (Vieillot, 1817), and the Double-barred Finch becomes *Taeniopygia bichenovii* (Vigors and Horsfield 1827). The scientific names of the other three *Poephila* species remain unchanged. Christidis' re-arrangement of the *Taeniopygia* and *Poephila* genera is consistent with the subgeneric divisions proposed by Mayr (1968) based on morphological and behavioural attributes, although Mayr (1968) kept *Taeniopygia* as a subgroup of *Poephila*.

The preferred scientific name for the Zebra Finch is now *Taeniopygia guttata*, the one originally advocated by Immelmann (1962a, 1965a), and now recommended for universal scientific use by Clayton and Birkhead (1989). The full scientific name of the Australian Zebra Finch is *Taeniopygia guttata castanotis* (Gould, 1837) and that of the Lesser Sundas Zebra Finch, *Taeniopygia guttata guttata* (Vieillot, 1817). The names come from the Greek: '*tainia*'-band; '*pyge*'-rump, tail; '*guttata*'-spotted; '*castan*'-chestnut, '*otis*'-ear (Cayley 1959).

The branching sequence in the phylogeny shows that the *Poephila* group (*sensu stricto*) have shared a more recent ancestor with the *Neochmia* and *Stagonopleura* groups than the *Taeniopygia* group. The affinities of species of the *Neochmia* group, namely the Crimson Finch *Neochmia phaeton*, the Star Finch *N. ruficauda*, the Red-browed Finch *N. temporalis* and the Plum-headed Finch *N. modesta*, have long been a source of contention among taxonomists, but Christidis argues that they are very similar in structure of proteins and chromosomes and form a tight natural group. Moreover, Goodwin (1982) reached a similar conclusion in his assessment of their behavioural and morphological characters. The *Stagonopleura* group consists of the homogeneous firetails: the Diamond Firetail *Stagonopleura guttata*, the Beautiful Firetail *S. bella* and the Red-eared Firetail *S. oculata*; this arrangement concurs with that of most previous workers. The last species in the tribe, the Painted Finch

Emblema pictum is not closely related to the other genera and is placed in its own monotypic genus.

The Lonchurini tribe

Three of the five genera in the Lonchurini have indigenous species in Australia. *Erythrura*, a supergenus, contains the Gouldian Finch *Chloebia gouldiae* and the Blue-faced Parrot-Finch *Erythrura trichroa*; a further 10 species of *Erythrura* live in Malesia and Melanesia. The genus *Heteromunia* has only one species, the Pictorella Mannikin *Heteromunia pectoralis*, and the *Lonchura* genus has two species, the Chestnut-breasted Mannikin *Lonchura castaneothorax* and the Yellow-rumped Mannikin *Lonchura flaviprymna*.

Evolutionary and Biogeographical History

Christidis (1987b) concluded that the Poephilini have the most primitive characters in the subfamily and that the *Taeniopygia* group are the most primitive of these in possessing a pattern of chromosome morphology that is present in species from all three tribal lineages and may best approximate the ancient condition in the subfamily (Christidis 1986a). Therefore, in some respects it could be argued that, of all the species of estrildines living today, the Zebra Finch resembles most closely the archetype from which other members of the subfamily evolved. In a recent analysis of immunological evolution of albumin in a selection of estrildine finches, Baverstock *et al.* (1991) found that the albumin molecule of the Zebra Finch was exceptionally different from that of all other species due to an extremely rapid change in its immunological and electrophoretic properties. Nonetheless, the significance of this discovery with respect to systematics has yet to be determined.

African origins

Most authorities (Morris 1958; Immelmann 1962a; Harrison 1967; Goodwin 1982) agree that estrildines originated in Africa and spread to southern Asia and Australia; however, the number of waves of colonists and their timing are in contention. The reasons for an African origin are as follows:

(1) the affinities of estrildines with the ploceine weaver finches, which are predominantly African, imply a common African ancestor for both subfamilies,

(2) estrildines have undergone greater speciation and adaptive radiation in Africa relative to that of southern Asia and Australia (Goodwin 1982), and

(3) inter-specific brood parasitism of estrildines only occurs in Africa (by parasitic whydahs—Viduinae) and has not yet evolved elsewhere (Kunkel 1969).

Table 1.1 Taxonomy of indigenous Australian estrildines including their extra continental subspecies (modified from Christidis 1987b)

Tribe[a]	Genus[a]	Subgenus[a]	Species[a]	Subspecies[b]	English name[a]
Poephilini	*Taeniopygia*		*guttata*	*guttata*	Lesser Sundas Zebra Finch
				castanotis	Australian Zebra Finch
			bichenovii	*bichenovii*	Double-barred Finch
				annulosa	
	Poephila		*acuticauda*	*acuticauda*	Long-tailed Finch
				hecki	
			cincta	*cincta*	Black-throated Finch
				atropygialis	Diggles Finch
				nigrotecta	
			personata	*personata*	Masked Finch
				leucotis	White-eared Masked Finch
	Neochmia		*phaeton*	*phaeton*	Crimson Finch
				evangelinae[c]	Pale-breasted Crimson Finch
			ruficauda	*ruficauda*	Star Finch
				clarescens	
			modesta		Plum-headed Finch
			temporalis	*temporalis*	Red-browed Finch
				minor	
				loftyi	
	Stagonopleura		*guttata*		Diamond Firetail
			bella		Beautiful Firetail
			oculata		Red-eared Firetail
	Emblema		*pictum*		Painted Finch
	Oreothruthus		*fuliginosus*	*fuliginous*[d]	Red-bellied Mountain Finch
				hagenensis[d]	
				pallidus[d]	
Lonchurini	*Lonchura*		*flaviprymna*		Yellow-rumped Mannikin
			castaneothorax	*castaneothorax*	Chestnut-breasted Mannikin
				uropygialis[d]	
				sharpeii[d]	
				nigiriceps[d]	
	Heteromunia		*pectoralis*		Pictorella Mannikin
	Erythrura		*trichroa*	*sigillifera*[c]	Blue-faced Parrot-Finch
		Chloebia	*gouldiae*		Gouldian Finch

[a] After Christidis and Boles (1994).
[b] After Boles 1988.
[c] Found in northern Australia and New Guinea. Ziswiller *et al.* (1972) list a further nine subspecies of *sigillifera* in Melanesia and Micronesia.
[d] New Guinea.

A single primary wave of ancient estrildines spread from Africa to southern Asia and Australia giving rise in the latter to the present day Poephilini; secondary, and much more recent invasions, populated northern Australia with representatives of the Lonchurini. The ancestors of the Lonchurini also came out of Africa but first radiated in South-East Asia from where they subsequently invaded Australia, Melanesia and Africa. Consequently, modern Estrildini and Poephilini share more primitive characters with one another than they do with the modern Lonchurini (Christidis 1987a).

The principal patterns and processes of bird speciation in Australia were first elucidated by Keast (1961) in a pioneering monograph on the subject. However, speciation within the Poephilini is known only in broad outline although more detail can be deduced for more recent evolutionary changes evident within the species (Keast 1958; Cracraft 1986). Similarly, speciation of the recently arrived mannikins can also be reconstructed in more detail and will be described first.

Differentiation of the Lonchurini (mannikins)

Five separate, relatively recent invasions from the north established the Australian Lonchurini; none have formed subspecies (Boles 1988). All the Australian Lonchurini have main distributions in the north of the continent. The two earliest invaders are presumably those that are the most differentiated of the extant members of the Lonchurini, namely the Pictorella Mannikin, an inhabitant of open savanna and the most arid-adapted and terrestrial of all the mannikins and the Gouldian Finch, an inhabitant of tropical savanna. The Yellow-rumped Mannikin, a derivative of the Pale-headed Mannikin *Lonchura pallida* of the Lesser Sundas, is probably a relatively recent colonist to northwest Australia (Immelmann 1965a). Finally, two recent invaders come from New Guinea—the Chestnut-breasted Mannikin and the Blue-faced Parrot-Finch. The former species arrived before the latter and it is widely distributed across the north and eastern parts of the continent. The Blue-faced Parrot-Finch is a subspecies found widely in New Guinea, but has a very restricted distribution in the rainforests of north Queensland.

Differentiation of the Poephilini (grassfinches)

Immelmann (1962a) contends that the ancestor of the Poephilini probably resembled species of the *Stagonopleura* group which he believed were the most primitive of the tribe. His conclusion was based on morphology and the fact that the Red-eared Firetail and the Beautiful Firetail were the most territorial and least social of the Poephilini. In the light of Christidis' (1986a,b, 1987a,b) more comprehensive analysis, one must conclude that both the morphology and behaviour of these species are probably derived rather than primitive conditions. Thus, it is more likely that a Zebra-Finch-like progenitor formed the stock from which the

Poephilini evolved, although it is impossible, of course, to determine the sequence of their speciation.

After early establishment in Australia, the proto-grassfinch radiated into at least 14 species (Table 1.1), which now occupy most habitats and climates and range throughout the continent, in one instance as far as Tasmania. One species, the Red-bellied Mountain Finch *Oreostruthus fuliginosus*, is endemic to New Guinea but is believed to have originated from Australian stock, possibly from a form of *Stagonopleura* (Schodde 1982; Christidis 1987b). Of 18 subspecies recognised by Boles (1988) in the Australian Poephilini, two have colonised islands beyond continental Australia, namely, *Taeniopygia guttata guttata* (Vieillot, 1817) in the Lesser Sundas and *Neochmia phaeton evanganelinae* (D'Albertis and Salvadori, 1879) in New Guinea. The latter, a white-breasted form of the Crimson Finch, has reinvaded the savannas of northwestern Cape York and thus constitutes a case of 'reverse colonisation'.

The origins of the five genera of Australian grassfinches (*sensu* Christidis 1987b) from the ancestral stock are unknown, but it is possible to identify likely geographic centres of differentiation for the genera and to reconstruct patterns of speciation within genera and species if their present day distributions are taken into account. *Poephila* are clearly of northern origin whereas the *Stagonopleura* are from southern Australia. However, *Neochmia* has two species in the north (*N. ruficauda* and *N. phaeton*) and two in the southeast (*N. modesta* and *N. temporalis*). *Taeniopygia g. castanotis* has an extensive and continuous distribution away from the periphery of the continent and provides no clues to the geographic origin of the genus although the distribution of its congener, *T. bichenovii*, would suggest that either the north or southeast is the region of origin for the genus. Considering the primitive attributes of *Taeniopygia* described earlier and the likely point of entry to Australia of the ancestral grassfinch it is reasonable to postulate that *Taeniopygia* originated in northwest Australia. Schodde (1982) reached the same conclusion. A hypothetical re-construction of the point of entry to, and dispersal routes throughout Australia are shown in Figure 1.2.

Grassfinches, which have fairly simple habitat requirements, namely areas of grassland for feeding and accessible sites for drinking, nevertheless vary in their preference for, or tolerance of, shrub and tree cover which they use for shelter and breeding. They occupy four main vegetation formations in Australia that grade in density of bush and tree cover according to diminishing rainfall from the continent's periphery to its interior: sclerophyll forest, savanna woodland, savanna grassland and grassland steppe. In all but the last vegetation type, speciation of grassfinches has probably occurred through the geographic isolation of large tracts of land on the continent's periphery by geomorphological and climatic barriers during episodes of extreme aridity in the late Pliocene and early Pleistocene (Cracraft 1986); some of these still persist in diminished

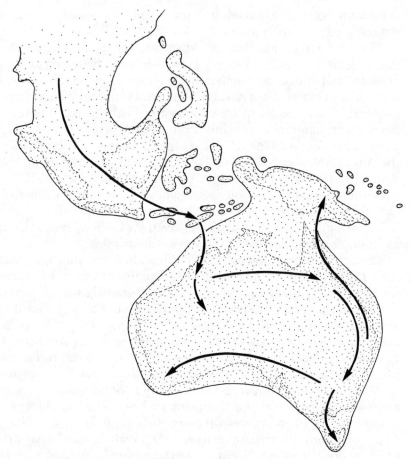

Fig. 1.2 Hypothetical reconstruction of the likely route of invasion of ancestral estrildines to Australia and their routes of dispersal throughout the continent. Approximate shorelines of Australia, New Guinea and South-East Asia during Pleistocene glaciations are shown; the dotted lines show the present coasts which approximate to those during the Pliocene.

form to the present day. In the arid interior, dominated by desert and grassland steppe, no fragmentation and subsequent speciation has occurred to any extent due to the absence of geographic barriers (Keast 1961; Schodde 1982) although Cracraft (1986) postulates the existence of inland barriers in pre-Pliocene epochs. High levels of nomadism in arid species ensures a continuous distribution and thus militates against isolation and speciation (Keast 1961).

Today most species of grassfinches live in savanna woodlands and these have distinct northern ('Torresian' biogeographic region, Schodde 1982) and southeast ('Bassian') components that roughly correspond to tropical and temperate woodlands. The tropical northern eucalypt woodlands with their understorey of tall grasses form a broad belt across the north of the continent. They are subdivided into northwestern (Kimberley and Arnhemland) and northeastern (Cape York and northeast Australia) components by a tongue of arid country at the head of the Gulf of Carpentaria, which acted as a barrier to movement. According to Keast (1958), subsequent geographic isolation has led to speciation in the case of *Poephila acuticauda* and *P. cincta* and subspeciation in *Poephila personata* (*personata and leucotis*), *Taeniopygia bichenovii* (*bichenovii and annulosa*) and *Neochmia phaeton* (*phaeton and iredalei*), but Boles (1988) has recently merged *iredalei* into *phaeton*.

Five grassfinch species live in the eucalypt and acacia woodlands of the eastern periphery. *Poephila cincta* and *N. ruficauda* originated most likely from northern forms, whereas *N. modesta* and *N. temporalis* probably have southern or eastern origins. *Neochmia modesta* prefers drier less dense woodlands, whereas *N. temporalis* tolerates a range of habitats from sparsely vegetated woodland to moist, closed forests and thickets (Immelmann 1965a). Keast (1958) concluded that an arid barrier at the base of Cape York isolated northern and southern populations, which have differentiated to the level of subspecies in *P. cincta* (*atropygialis and cincta*), *N. ruficauda* (*clarescens and ruficauda*) and *N. temporalis* (*minor and temporalis*). There is a third subspecies of *N. temporalis* in South Australia where the Coorong arid tract isolated *loftyi* from *temporalis* (Keast 1958).

The Diamond Firetail *Stagonopleura guttata* is the fifth species of grassfinch of the southeastern woodlands. It belongs to a genus with southern origins adapted to moist closed forests, but has made an adaptive shift to drier but still fairly densely wooded habitats often bordering watercourses and has an extended distribution over much of the woodlands of the south and east. However, no subspecies have differentiated due to the absence of arid barriers across its range. The Beautiful Firetail *S. bella* from the southeast coast of the mainland and Tasmania and its counterpart the Red-eared Firetail *S. oculata* from southwest corner of Western Australia are the only species of grassfinches to specialise in damp, thick vegetation, in and adjacent to eucalypt forest. Keast (1961) believes that they differentiated into eastern and western species when isolated by the vast arid Nullabor zone during post-Pleistocene climatic changes.

Origin of the Lesser Sundas Zebra Finch

The Australian Zebra Finch is thought to have crossed to Timor from the Kimberley region sometime during a Pleistocene glaciation when the sea level dropped some 100–150 m thus extending the coastline to within some 72 km of Timor (Mayr 1944 a, b). The high mountains of west Timor may have been visible to birds blown out to sea by cyclones from the northern coast of Australia. This might explain why Zebra Finches dispersed to Timor and why Timor estrildines (*Amandava amandava, Lonchura quinticolor, L. punctulata, L. molucca* and *L. fuscata*) that are well adapted to grassland and open country, failed to colonise Australia: they could not see landfalls from coastal Timor.

The Zebra Finch is clearly the most superior disperser among the Australian estrildines and this probably holds for most of the Indonesian estrildines as well. Mayr (1944a,b) made the definitive zoogeographical study of the transitional avifauna of Wallacea, the biogeographical zone that separates the Oriental and Australian avifaunas and new data have been reviewed by White and Bruce (1986). On the one hand, Australia received 20–22 species of birds of oriental origins via Timor, yet no estrildines presently found on Timor are among the 10 species of passerines that colonised Australia. On the other hand, Timor received 17 Australian species of which seven are passerines including the Zebra Finch. Of the 12 species of estrildines listed by White and Bruce (1986) for the Lesser Sundas archipelago, the Zebra Finch has the most extensive distribution by far (Chapter 2). Nonetheless, it has not crossed Wallace's line and made it to Bali, with only three species of passerines of Australian origin having done so.

Colonisation of Australia by ancient estrildines

According to Keast (1981), the ancestral Estrildini that colonised Australia and radiated into the species that constitute the Poephilini reached the northern part of the continent in the Pliocene epoch, about 5-1.5 million years ago. While this is a reasonable guess, there is no fossil evidence to support this or any other time of arrival; it may have been earlier or later. Nevertheless, the chain of islands that constitute the Lesser Sundas archipelago probably acted as stepping-stones from South-East Asia to northern Australia and New Guinea for the ancestral estrildines and other northern invaders. This route no doubt led to an exchange of the Asian and Australian faunas which is believed to have begun around the mid-Miocene when the northward-drifting leading edge of the Sahul Shelf bearing Australia and New Guinea collided with the Sunda plate of South-East Asia thus narrowing the water gap over which the avian colonists must have crossed from 4,000 km to less than 800 km (Powell *et al.* 1981). Ancestral Poephilini probably did not reach Australia until after the first interchange of the Asian and Australian avifaunas, but well before most of the other recently arrived species that used Timor as the final departure point for Australia (Mayr 1944a).

Origin of Australian avifauna

On arrival in Australia, the ancestral Poephilini would have already found a highly evolved and differentiated group of 'old endemic' birds, both passerine and non-passerine, which arose from massive speciation of ancient stock back in the Eocene epoch, more that 55 million years ago, when the Sahul Shelf was isolated in its northward drift after the breakup of Gondwana.

Of the indigenous avifauna now breeding in Australia, approximately 58% (135 species) of the non-passerines are of Gondwanan origin with the remainder having immigrated to Australia via Asia (Schodde and Calaby 1972). By means of DNA–DNA hybridisation techniques, Sibley and Ahlquist (1985) determined that 79% of Australian passerine species belong to the 'old endemic passerines', collectively called Parvorder Corvi. The remaining endemic passerines, some 43 species belonging to 10 miscellaneous families with strong African connections, were very distantly related to the Parvorder Corvi and represent the 'new endemic passerines' which have invaded Australia from the north and collectively form the Parvorder Muscicapidae. Eighteen of these 43 species are estrildine finches while the remaining 25 species belong to the nine miscellaneous families. Therefore, of the new endemics, the estrildines are predominant in numbers of species (and subspecies), which suggests, but in no way proves, that they have had longer to differentiate and presumably were in the vanguard of immigrants to have reached Australia from the north.

As one might expect of small, fine-boned birds, fossil remains of estrildines tell us little of their evolutionary history in Australia. Estrildine remains have been identified from three cave deposits from the Late Pleistocene (approximately 20,000 years ago; R. F. Baird, pers. comm.): two in the State of Victoria and one in the State of South Australia (Baird 1991). Van Tets (1974) concluded that post-cranial remains recovered from Weeke's Cave on the coast of the Nullarbor Plain were those of Zebra Finch whereas Baird (1991) contends that none of the material allows a determination below the subfamily level.

Granivorous competitors

On arrival in Australia the ancestors of the grassfinches would have found no specialist granivorous competitors among the old endemic passerines, although a few insectivorous species (e.g. Grasswrens *Amytornis* spp., Whitefaces *Aphelocephala* spp., Quail-thrushes *Cinclosoma* spp.) had become secondary seed-eaters in that they eat roughly equal proportions of seed and other items such as insects (Joseph 1986). These species have evolved typical granivore adaptations: stout finch-like bills and strongly muscled gizzards (Keast 1961; Schodde 1982). Many species of ants, pigeons and parrots specialise in eating seeds of grasses, herbs and shrubs, although only those that specialised on grasses would be in direct competition with grassfinches, which are mainly graminivorous (Immelmann 1962a; Chapter 4). Morton (1985) found that ants are the dominant granivore in arid regions. Thus, the main granivorous competitors for the ancestral estrildines would have been ants, with parrots and pigeons secondary competitors.

Evolution of the Australian climate

To understand the adaptations of the ancestral grassfinches and the processes that led to their differentiation, it is necessary to appreciate the changes to the Australian climate and environment in the late Tertiary and early Quarternary periods (see Kemp 1981; Bowler 1982; Nix 1982; Frakes *et al.* 1987; among others).

At the end of the Miocene epoch (10–5 million years ago) global cooling formed high pressure mid-latitude systems which blocked moist maritime air masses from entering the interior of Australia thereby causing extensive desiccation (Bowler 1982; Frakes *et al.* 1987). There was a pronounced, but gradual transition in vegetation from widespread rainforest cover, to open forests and woodlands, to dry scrublands, and, finally, to grasslands. Aridification continued into the mid- to late Pliocene (Galloway and Kemp 1981) and reached its present state in the Late Pliocene to Early Pleistocene, about 2 million years ago (White 1986).

Increased seasonality in moisture and temperature arose due to the formation of a monsoonal climatic regime of summer rainfall and dry winters. In southern Australia this later shifted to the opposite regime of wet winters and dry summers. Nix (1982) believes that this seasonal deficiency in water is the primary factor in the evolution of the Australian biota found today and that these forces were greatest in the Tertiary and relatively minor in the Quarternary. In contrast, Frakes *et al.* (1987), postulate that the dramatic oscillations in humidity that began in the Pliocene and intensified in the Pleistocene glaciations, especially in the last 400,000 years when the frequency and amplitudes of the oscillations were at their highest, have had the greatest ecological impact on the flora and fauna of Australia. During the Quarternary, the Australian deserts expanded on the northern and southern margins during glacial periods but became so humid during the interglacial intervals that the arid belt may have disappeared. At present the climate is about mid-cycle heading from a cool dry period for a more humid phase (Bowler 1982). It is reasonable to conclude that the climatic extremes, especially that of aridity, either of a seasonal or more permanent nature, caused severe ecological stresses which have sifted the flora and fauna by means of differential adaptation and extinction of species, thereby leaving but a fraction of the Miocene biota to be encountered by humans on their arrival on the continent.

With this scenario in mind one can speculate on the type of climate a proto-grassfinch arriving in northern Australia during the Plio-Pleistocene epochs might have experienced. While the climatic data are more sketchy than those for southern Australia, it is plausible to assume that grassland and shrubland habitats were expanding across northern Australia as the continent became increasingly arid and seasonal. Apparently, the present climate and vegetation resemble those of the late Tertiary (Nix 1982). Thus, granivorous niches may have been largely unoccupied and increasing in availability as new ones were created around the time of arrival of the proto-grassfinch. The colonisers may have even been pre-adapted to savanna conditions as Morley and Flenley's (1987) reconstructions indicate that Late Tertiary paleoclimates of the Malay Peninsula had savanna corridors along which savanna species could have travelled to reach the stepping-stones of the Lesser Sundas and the departure points for north-west Australia.

The first grassfinches to arrive in north-west Australia may not have needed to make dramatic adaptive changes in order to survive and disperse. From there they would have spread across the savanna woodlands of northern Australia and down the east coast. This widespread ancestral species was subsequently subdivided by geomorphological and ecological–climatic barriers into isolated areas where species differences arose as forms adapted to more mesic conditions in some regions. Speciation could have occurred repeatedly during the early glacial periods of the Pleistocene when extreme aridification isolated patches of humid forest in the south and east on the one hand, and savanna woodland in the north on the other. Cracraft (1986) has demonstrated a coincidence of patterns of phylogenetic differentiation among the *Poephila* grassfinches and the isolation of areas of endemism by barriers that formed in the late Pliocene. During the Pleistocene many species must have been driven to extinction. Species, such as the Zebra Finch and the Painted Finch, which developed adaptations to the arid environments, may have moved to the interior directly from the savanna woodlands of the northwest. During periods of extreme aridity, Schodde (1982) believes these two species took refuge in the mountain ranges of Central Australia (MacDonnell, Tomkinson, Petermann, Musgrave and Everard Ranges), Western Australia (Hamersley and Pilbara), South Australia (Flinders Ranges) and Queensland (Selwyn Ranges) although the evidence for this is unstated.

Summary

Two subspecies of Zebra Finches exist, the Lesser Sundas Zebra Finch *Taeniopygia guttata guttata* (Vieillot, 1817) from eastern Indonesia, and the Australian Zebra Finch *Taeniopygia guttata castanotis* (Gould, 1837) from continental Australia. They belong to the Poephilini tribe of the subfamily Estrildinae, family Passeridae. There are thirteen additional species of Australian grassfinches in the Poephilini, of which the Double-barred Finch *Taeniopygia bichenovii* is the only congener. The Zebra Finch has a number of primitive phylogenetic characters, and may resemble the ancestral estrildine that colonised Australia from the Lesser Sundas archipelago sometime during an early Pleistocene glaciation. The ancient estrildines were in the vanguard of the 'northern elements', mostly of African origin, that invaded Australia when it drifted north-ward and contacted the Asian plate. The ancient estrildines radiated in Australia and have occupied most habitats and climates. This was aided by climatic changes in Australia where increasing aridity extended grass-lands and shrublands, the presumed habitat of the proto-grassfinches. Among the old Australian endemics there are no passerine granivores that might compete with proto-grassfinches, although pigeons, parrots and ants would have provided some competition for seed. The Lesser Sundas Zebra Finch is believed to have differentiated when Timor was colonised by Zebra Finches from northwest Australia during a late Pleis-tocene glaciation when the water gap narrowed. When sea levels rose the population was isolated from the mainland but was able to disperse westward from Timor, reaching as far as Lombok. A detailed biochemi-cal analysis of geographic variation in both subspecies of Zebra Finches is needed.

2 Distribution and habitat

'This bird appears to be almost peculiar to the interior of Australia;'
                                                                    J. Gould 1865.

To understand the distribution of the Zebra Finch we not only need to appreciate the environment of its recent evolutionary past (Chapter 1), but more importantly we need to understand its present day patterns of distribution under the present day environment, especially the plant environment. Zebra Finches have a very wide distribution. They are distributed throughout continental Australia and most of the islands that form the Lesser Sundas archipelago of eastern Indonesia (Figure 2.1).

Lesser Sundas Zebra Finch

The Lesser Sundas Zebra Finch *T. g. guttata* is found on 18 of the 21 large islands of the Lesser Sundas archipelago (White and Bruce 1986). The islands extend from Sermata (8°13'S, 128°55'E) in the east to Lombok (8°45'S, 116°30'E) in the west (Table 2.1). While it has an almost continuous distribution between these limits, it has not been recorded from three small islands between Flores and Alor: Lomblen (8°25'S, 123°30'E), Pantar (8°25'S, 124°07'E) and Solor (8°27'S, 123°05'E) (Figure 2.2). Its recorded absence here may simply be a function of inadequate collecting. Zebra Finches have not colonised two archipelagos in the eastern extreme of the Lesser Sundas, Kepuluan Tanimar and Kepuluan Babar, nor have they reached all islands on the Barat Daya archipelago northeast of Timor, getting no farther east than the large island of Wetar. It is possible that new surveys may extend the range in this region. Ten other species of estrildines are also found in the Lesser Sundas (White and Bruce 1986), but the Zebra Finch is by far the most widespread (Table 2.1).

Presumably, dispersal of the Zebra Finch from Timor, its point of entry from the source population in northern Australia, was a function of the width of the water gaps between adjacent islands en route. During Pleistocene glaciations these gaps were reduced considerably with distances smaller on the western route from Timor to Lombok than on the eastern route from Timor to Sermata. This may account for the more extensive movements to the west. Although the Bali–Lombok strait is only 32 km wide this gap has been an effective limit to the western dispersal of the Zebra Finch. It is also a barrier to dispersal for many other

Fig. 2.1 Distribution (stippling) of the Zebra Finch in Australia and in the Lesser Sundas archipelago of eastern Indonesia.

species of animals and it forms the southern limit of Wallace's line, the boundary of major discontinuity in the distribution of Australasian and Oriental avifaunas.

Climate and habitats of the Lesser Sundas

The Lesser Sundas lie in the rain shadow of northern Australia and have the driest, most seasonal climate in Indonesia. They resemble the wet–dry tropics of northern Australia in that they receive heavy monsoon rains from December to March but are completely dry for the rest of the year (Nix 1976; White and Bruce 1986). Rainfall throughout the archipelago becomes progressively less from the northwest to the southeast so that eastern islands are more arid than those of the west.

Table 2.1 Distribution, habitat and altitudinal range of the Lesser Sundas Zebra Finch and other sympatric species of estrildines in the Lesser Sundas archipelago ranging over 21 islands from Lombok in the west to Sermata in the east (see Figure 2.2)

Species	Habitat[a]	Altitude (m)[a]	Islands[d]
Taeniopygia guttata guttata	Grassland, cultivation and woodland[b]	0–2300[c]	1–11, 14–21
Amandava amandava flavidiventris	Grassland, cultivation	0–2400	3, 8, 14, 19, 20
Erythrura hyperythrura intermedia	Woodlands and forests	300–1000	3, 8, 19
Erythrura tricolor	Thickets	0–1200	20, 21
Lonchura leucogastroides	Cultivation	0–1000	8
Lonchura molucca	Cultivation, grasslands	0–1000	3, 12, 13
Lonchura punctulata blasii	Cultivation, grassland,	0–2000	3, 4, 6–8, 15, 18–20
Lonchura quinticolor	Grassland, cultivation shrubbery	0–1200	1, 3, 8, 14, 17, 18–20
Lonchura pallida	Cultivation, grassland, shrubbery	0–1000	1, 2–4, 8, 14, 15, 17, 19
Padda oryzivora	Cultivation?	0–400	8, 20
Padda fuscata	Cultivation?	0–160[c]	14, 15, 20

[a]White and Bruce (1986). [b]Clayton, *et al.* (1991).
[c]Mayr (1944b).
[d]1—Alor, 2—Dao, 3—Flores, 4—Kisar, 5—Komodo, 6—Letti, 7—Lomblen, 8—Lomok, 9—Luag, 10—Moa, 11—Padar, 12—Paloe, 13—Pantar, 14—Roti, 15—Sawu, 16—Semau, 17—Sermata, 18—Sumba, 19—Sumbawa, 20—Timor, 21—Wetar.

Originally covered with monsoon forests of various kinds, much of the Lesser Sundas now consist of grassy hills because of massive forest clearance for cultivation by the large human populations. In East Timor, clearing and cultivation had already extended to the peaks of the tallest mountains by the 1930s (Mayr 1944b). Grasslands consist of tall stands of coarse grass called alang alang (*Imperata* sp.) with a sparse scattering of stunted, thorny bushes (*Zizyphus* sp.), and palms (*Borassus* sp.) (White and Bruce 1986). This is prime habitat for estrildines, the Zebra Finch in particular. According to White and Bruce (1986) the Zebra Finch is confined to altitudes below 500 m. However, in 1931–1932 Mayr's (1944b) collector, Georg Stein, took specimens from sites ranging from sea level to medium altitudes (1200 m) in West Timor; in East Timor they were collected from the 2000–2300 m level on Mt Ramelan. Clayton *et al.* (1991) observed and trapped Zebra Finches on the dry coastal grasslands and cultivated areas of Lombok, Sumbawa, Flores and Timor. Birds were often found in paddy fields and near streams and waterfalls. From these observations it appears that the Lesser Sundas Zebra Finch has its stronghold on the low coastal margins of the islands but will move to high altitudes in order to exploit expanding cultivation and the grasslands that follow.

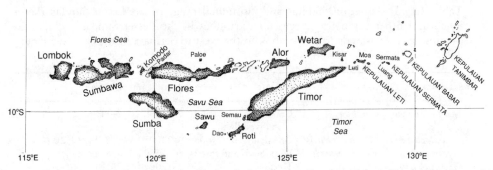

Fig. 2.2 Distribution (stippling) of the Lesser Sundas Zebra Finch *Taeniopygia gutata guttata* in eastern Indonesia.

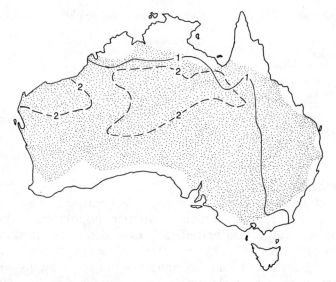

Fig. 2.3 Main distribution of the Australian Zebra Finch *Taeniopygia guttata castanotis*. The edge of the stippled area encompasses 507 one-degree blocks in which Zebra Finches were found breeding or were observed in more than 10% of reports during the 1977–1981 Field Atlas of Australian Birds (Blakers *et al.* 1984). This area constitutes the main distribution of the species although not all one-degree blocks in this area reported the presence of Zebra Finches. There were non-breeding records in an additional 48 one-degree blocks peripheral to the main distribution. Density of birds was relatively low (report rate <10%) in the peripheral distribution that mainly extended along the eastern coast of the continent and north into the Top End of the Northern Territory. Isoclines of two species of estrildines that have extensive overlapping distributions with the Zebra Finch are shown. Isocline 1 shows the western and southern limits of the distribution of the Double-barred Finch, which lives in the north and east of the continent; and isocline 2 shows two zones of distribution of the arid-adapted Painted Finch which lives in the centre and west of the continent. (Modified from Blakers *et al.* 1984.)

Australian Zebra Finch

The most recent and accurate map of the distribution of the Australian Zebra Finch (Figure 2.3) was compiled for the Field Atlas of Australian Birds (Royal Australasian Ornithologists Union; Blakers *et al.* 1984). Zebra Finches are absent from Tasmania, but were observed over 75% of the area of the mainland. Of all Australian estrildines the Zebra Finch has the greatest distribution by far, with its congener, the Double-barred Finch, having the next most extensive distribution (32%), followed by the Red-browed Finch (17%), Diamond Firetail (15%), Chestnut-breasted Mannikin (13%) and the Plum-headed Finch (11%). The 12 remaining species have more restricted distributions ranging from 2-10% of the mainland. Zebra Finches were found breeding over 45% of the mainland and had the fourth largest breeding distribution among 656 species censused. This not only reflects a tolerance of a wide range of ecological conditions for breeding but it also indicates that their nests are easily detected by humans. Nevertheless, it is likely that breeding is more widespread than this figure suggests.

Australian Zebra Finches have been introduced successfully to Nauru (Pacific Ocean) and unsuccessfully to Kangaroo Island (South Australia), New Zealand and Tahiti (Long 1981).

Limits of distribution

When the peripheral distribution is taken into account, the Australian Zebra Finch ranges from the most eastern point of Australia (Cape Byron, 28°45′S 153°30′E) to its most western point (Dirk Hartog Island 25°50′S, 113°05′E). Darwin (12°25′S, 130°51′E) is the most northern limit and the You Yangs (50 km southwest of Melbourne 38°30′S, 144°22′E), the most southern (Figure 2.4). Despite this extreme coast to coast coverage, the main distribution is found away from the coastal margins, except for the arid western edge of the continent. It has one almost continuous range throughout the interior. On all but the western edge of their distribution, density of birds in the periphery of the range, as reflected in their reporting rate during the atlas project, diminished to less than 11% of the maximum (79% of reports in 18 1° blocks) (Blakers *et al.* 1984). Low densities in the coastal periphery are probably due to two factors—a recent extension of range and sub-optimal habitat.

Zebra Finches have reached a number of islands off the coast of Australia: Groote Eylandt (14°0′S, 136°40′E) in the Gulf of Carpentaria, North Keppel island (23°3′S, 150°52′E) off the east coast of Queensland and Goose island (34°27′S, 137°21′E) in the Spencer Gulf, South Australia. They have been recorded from the following islands off the western coast of Western Australia: Dirk Hartog (25°50′S, 113°05′E), Barrow (20°45′S, 115°20′E), Thevenard (21°25′S, 115°05′E) and at least four islands of the Dampier archipelago (20°35′S, 116°35′E) namely,

East Enderby, East Lewis, West Lewis and Dolphin. The widest water gap is approximately 60 km from the mainland to Barrow island. It is believed that the island populations are only transitory having being blown there from nearby mainland sites by high winds during cyclones and heavy storms (S. J. J. F. Davies and S. Ambrose, pers. comm.)

Zebra Finches are conspicuously absent from Cape York Peninsula of Queensland and are also absent from a major part of the Kimberley region of northern Western Australia (Figure 2.3). One can only speculate why they have not invaded these regions. They are found in areas that are harsher and drier than the Kimberley and in fact are abundant at Wyndham (15°33'S, 128°03'E) and Derby (17°18'S, 123°28'E) on the eastern and southern edges of the Kimberley respectively. Evans and Brougher (1987) found Zebra Finches at waterholes scattered along a direct line between Wyndham and Derby, which is further north than the main distribution shown in Figure 2.3; this extension may be part of the non-permanent peripheral distribution of the species in this region. The Kimberley is known as the 'Land of the Finches' for its abundance and variety of species throughout, yet oddly, the most abundant and

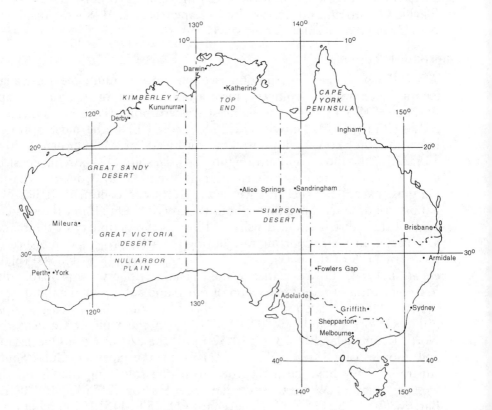

Fig. 2.4 Main site localities mention in the text.

widespread species of all, the Zebra Finch, has a limited range in this region. Presumably, the northern Kimberley was the departure point for the colonisation of Timor and the Lesser Sundas.

The Environment of the Australian Zebra Finch

Climate

Australian Zebra Finches are distributed over some sixteen degrees of latitude, from 14°S to 38°S (Figure 2.4) and encompass all climatic zones that prevail on the continent. Furthermore, they are found in 14 of the 16 avifaunal regions identified during the mapping of the distribution of Australian birds (Blakers *et al.* 1984).

Australia has four of the world's climatic types identified and described by Walter *et al.* (1975):

(1) *Type II Tropical*—some seasonality in the mean daily temperature with rainfall concentrated in the summer months (e.g. Wyndham, Broome, Darwin);

(2) *Type III Subtropical*—very low rainfall, high daytime temperature in summer with low winter minima (e.g. Alice Springs, Mileura);

(3) *Type IV Transitional zone with winter rain*—very little summer rainfall, but cyclonic rains in winter; typically no cold season, but permanent summer drought (e.g. Perth, Adelaide);

(4) *Type V Warm Temperate*—no noticeable winter, with year-round rainfall (e.g. Sydney, Melbourne, Canberra, Griffith and Shepparton).

A transition zone between types II and V occurs from northern New South Wales to southeastern Queensland (e.g. Armidale, Brisbane).

The Australian climate is distinguished by its dryness. Rainfall in tropical Australia comes in the summer from monsoons and occasional cyclones while in the south it comes from winter storms moving up from the Antarctic. In addition, the Great Dividing Range that runs along the entire eastern seaboard traps rain from the southeasterly trade winds that move in from the Pacific Ocean. Consequently, there is an asymmetrical concentric zonation of rainfall over the continent so that it rains more on the coast and progressively less towards the centre. Rainfall is higher in the northern and eastern margins than those in the south and west. In terms of mean annual precipitation, only one third of the area of the continent (the periphery) has more than 500 mm while the inner third has less than 250 mm (Nix 1982). Strictly speaking, areas with a mean annual rainfall of less than 250 mm are termed 'arid' or 'desert', and those between 250–500 mm are termed 'semiarid' or 'steppe' (Serventy 1971). Zebra Finches are distributed throughout both arid and semiarid areas (Figure 2.5); however, 15% of the main distribution

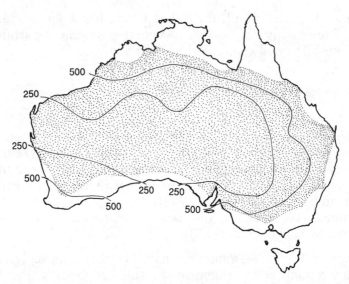

Fig. 2.5 Isohyets of mean annual rainfall in millimetres for arid (≤250 mm) and semiarid (251–500 mm) zones superimposed on the main distribution (stippled) of the Australian Zebra Finch. (Modified from Leeper 1970.)

extends beyond the 500 mm isohyet to areas of high rainfall towards the periphery on the east and north of the continent, but high rates of evaporation reduce plant growth here and conditions are dry.

Vegetation associations

Rainfall, in combination with regimes of light and temperature, and topography are the primary determinants of vegetation patterns in Australia (Nix 1982). The high rainfall fringes on the eastern and south-western part of the continent were originally covered in tall, closed forests, much of which has been cleared for agriculture since European settlement. In dry areas farther inland, woodlands dominate in a broad concentric zone and these gradually give way to low woodlands as it becomes drier towards the centre; shrublands and grassland dominate in the areas of the lowest rainfall that form the vast arid interior. The vegetation is distinguished by the dominance of the following three genera: *Eucalyptus* in the forests and woodlands, *Acacia* in the shrublands and *Triodia* (spinifex) in the grasslands. The height of the vegetation becomes progressively lower in concentric zones from the moist margins to the arid centre of the continent. A detailed description of vegetation in Australia can be found in Bridgewater (1987) and Specht (1981).

Ecology of arid Australia

By comparison with other arid regions of the world the Australian arid zone, the stronghold of the Zebra Finch, is unusual in climate and ecology. An abundance of perennial plants, especially trees and shrubs, is the outstanding feature. Although the climate is only moderately arid by world standards, the unpredictability of climatic extremes over a vast geographic scale is exceptional (Stafford Smith and Morton 1990). This is driven by rainfall, which is highly unpredictable in space and time, so that there is no way to predict the timing and severity of drought. The occasional big fall of rain, which comes at irregular intervals, is also important because it structures the landscape so that a mosaic of areas of varying fertility and moisture is produced. Floodplains and floodouts are the major areas where nutrients and water are concentrated and here plant productivity can be higher than the surrounding areas. Nevertheless, there are still great extremes in soil moisture and these dominate the lives of plants and herbivores. In regions with some topographic relief, such as the central ranges, water is redistributed in a more concentrated and dependable way to so-called 'runon' areas. Here even small falls of rain can be useful. Therefore, what appears to be a vast, superficially uniform landscape, is in reality, a mosaic with 'fertile or reliable sites scattered like islands in a sea of exceptionally infertile and unreliable conditions' (Stafford Smith and Morton 1990). These islands of higher plant production vary in scale from cracks in rocks to whole floodplains and drainage lines, and together form that part of the landscape upon which Zebra Finches, and many other species of birds and mammals, rely, especially during drought. These fertile sites provide the grass seeds and the nesting bushes for the finches.

Davies (1977a, 1986) maintains that plant productivity in arid regions is more regular than it appears because the redistribution of water to fertile sites can make even light falls of rain effective; moisture can be stored until temperatures in spring and summer permit seed germination. Indeed, spring breeding on a fairly regular basis appears to be the pattern in most species of birds in arid southwest Australia (Davies 1979) and might also be the case in the eastern parts of the arid zone (Serventy 1971).

Habitat requirements of the Australian Zebra Finch

Immelmann (1965b) published a comprehensive analysis of the ecological factors that affect the distribution of the Zebra Finch across the continent with a particular emphasis on those that control the timing and extent of breeding in different climatic regions. This important article was published in German and has not been widely cited since publication. However, it is worth detailed examination particularly in the light of additional data accumulated over the last 30 years. Detailed ecological and physiological considerations will be made in Chapters 4 and 5.

Grass and surface water

Daily access to supplies of grass seeds is the essential requirement for the survival of Zebra Finches and, although they are renowned for their physiological adaptations to aridity and indeed some populations may tolerate several days, or longer, without it (Chapter 5), their presence is strongly linked to supplies of drinking water. In arid regions, the concentration of Zebra Finches in the vicinity of surface water of any one of a variety of sources, both artificial and natural, has been invariably remarked upon by observant explorers, pioneers and ornithologists. However, this dependence has not limited the geographic distribution of the species, since it is found in even the most arid parts of the continent, but it does determine its local distribution within the landscape and it may explain aseasonal movements reported in the literature.

Temperature

For a bird with an extensive tropical and subtropical distribution, the Zebra Finch is remarkably tolerant of low temperatures. In the centre of the continent it is exposed to low overnight temperatures, and in the southeastern part of the range in the State of Victoria it can occasionally endure light frosts for the whole day. Zebra Finches breed on the New England Tableland (30°30'S, 151°40'E) in the State of New South Wales, much of which is above 1000 m elevation, and up to 50 days of frost are expected each year (Kikkawa 1980). Zebra Finches are also recorded at altitudes of 1000 m or higher in the Southern Alps along the eastern coasts of the States of New South Wales and Victoria but breeding records from the Field Atlas are few. They occupy most locations with elevations below 1000 m. Immelmann (1965b) states that they leave an area if mean daily temperatures are less than 6°C for any extended period, but provides no evidence for this. Analysis of unpublished maps compiled during the Field Atlas of Australian Birds shows that fewer Zebra Finches were sighted during winter months in habitats above 1000 m than during other months of the year. This suggests that there may be some local movements away from the higher altitudes at the end of summer and provides support for Immelmann's assertion.

According to Immelmann (1965b), the mean minimum daily temperature required for breeding by Zebra Finches is 12°C. Breeding is abandoned if the temperature falls below this. However, in Alice Springs, birds were found breeding in the middle of winter in 1989 when mean minimum temperatures were about 4°C (Chapter 7). Low winter temperatures inhibit breeding in southern Australia (Davies 1979; Kikkawa 1980; Serventy and Marshall 1957; Zann and Straw 1984a; Zann 1994a). Different populations across Australia may have different degrees of sensitivity to low temperatures.

Rain

Zebra Finches feed almost exclusively on grass seeds (Chapter 4) and can survive for many months on dry seeds that have fallen to the ground six or seven months previously. Half-ripe seeds and green leaf material are probably necessary for the raising of young (Immelmann 1962a, 1965b; Chapter 7). The breeding distribution of the species across Australia depends on the germination, growth and seeding of grasses which, in turn, depends on the amount and timing of precipitation. Warm, moist conditions are optimal for the seeding of most species of grasses.

Immelmann (1965b) asserts that Zebra Finches are adverse to heavy sustained rainfall. In comparison with other species of estrildines living in the Wyndham area of the extreme northern coast of Western Australia, he found that Zebra Finches were less active during the continuous downpours of the wet season. They avoided foraging in wet vegetation and avoided contact with damp leaves and grass. Although I could not confirm these observations in central Australia (Alice Springs) or in northern Victoria, I could confirm Immelmann's observation that the thinly roofed nests provide no protection against heavy rain, which can easily penetrate the nesting chamber, and occasionally lead to the death of nestlings or to the desertion of clutches (Immelmann 1962a).

During the northern wet season in 1959, Immelmann (1965b, 1970) found that Zebra Finches disappeared from the Wyndham area with the first heavy falls (November), but returned at the start of April after the heavy rains had diminished. Breeding was squeezed in during the few weeks of light scattered showers that preceded and followed the storms of the main wet season since any nests of Zebra Finches that remained were knocked to the ground by violent storms and all breeding attempts failed. Immelmann concluded that this inability to adapt to heavy precipitation and wind characteristic of the monsoons of northern Australia is responsible for movements away from these regions during the wet season—the heavier the wet season, the further south they move to avoid the rain. He maintains that across a broad belt of northern Australia, from Derby ($17°18'$S, $123°38'$E) in the west, to Katherine ($14°40'$S, $131°42'$E) in the east, Zebra Finches move northward occupying coastal regions during the dry season (May to October) while at the start of the wet season they retreat inland again. Consequently, there is a zone along the north-west and northern coasts of Australia where the Zebra Finch only occurs during the dry season. Immelmann provided no evidence in support of this seasonal movement—no banded birds were followed nor any seasonal changes in distribution reported. During fieldwork for the Field Atlas of Australian Birds (1977–1981), observations of Zebra Finches were much lower during the summer months over the whole distribution, especially in the far north of the continent when few observers are about. Therefore, there is a bias toward more frequent records during the winter months. In spite of this, data from four sites on the northern limits of the range of distribution show that there is a significant increase in the proportion of Zebra Finches observed during the dry winter months and fewer during the wet summer months (Table 2.2). This finding is consistent with Immelmann's hypothesis. These movements are

Table 2.2 Occurrence of Australian Zebra Finches on the northern limits of the main distribution in the wet–dry tropics by month of observation[a]

	Western Australia (Derby)	Western Australia (Wyndham)	Northern Territory (north of 16°S)	Queensland (north of 17°S)	Total
January	3.6	3.3	6.2	0.0	4.1
February	1.8	1.6	3.1	2.7	2.4
March	7.3	4.9	3.1	0.0	4.1
April	9.1	6.6	3.1	2.7	5.2
May	12.7	9.8	7.2	8.5	8.8
June	14.5	50.0	13.4	5.5	15.3
July	12.7	18.0	14.4	16.7	15.7
August	9.1	11.5	24.7	27.8	18.9
September	12.7	6.6	13.4	25.0	13.3
October	3.6	8.2	5.1	8.3	6.0
November	3.6	3.3	2.1	2.7	0.4
December	7.3	1.6	4.1	0.0	3.6
	55	61	97	36	249

[a] Data were extracted from the Field Atlas of Australian Birds, 1977–1981 (RAOU). Numbers show per cent of observations per month for four northern regions. The main wet season extends from December to April. Greater absolute numbers of observations of all species were made in the months of winter (June to August) and spring (September to November) and fewer in autumn (March to May) and summer (December to February). Nevertheless, a significantly greater proportion of Zebra Finches were observed on the northern limits of the distribution during the winter (dry season) months and fewest in the summer (wet season) months (G = 47.51, df = 3, P < 0.0001). This supports the suggestion that Zebra Finches move away from the northern extremes of the distribution during the wet season.

not conspicuous; they were not detected in the formal analysis of the atlas data (Blakers *et al.* 1984), nor were they confirmed by contributors to Australian Bird Count (RAOU, unpublished data), a project specifically aimed at detecting movements by means of changes in seasonal abundance. It is not surprising that movements are difficult to detect in the far northern parts of Australia since observers are sparsely distributed and travel during the wet season is difficult because of the inundation. Final confirmation of these seasonal movements must wait until banded birds are retrapped or recovered; however, they are consistent with Nix's (1976) predictions based on climatic factors.

Immelmann (1965b) also hypothesised that the intensity and timing of the monsoon in the northeastern part of Australia, in the Cape York Peninsula region of Queensland, prevented Zebra Finches from breeding at all, and so constrained the species from advancing north of latitude 17°S. He argued that the heavy downpours that began suddenly at the onset of the wet season made breeding impossible; furthermore, even in the dry season, occasionally rain was too heavy for breeding. However, in north Queensland, Zebra Finches normally breed throughout the year except for the dry winter months (Chapter 7) and they have been recorded breeding at Ingham (18°43'S, 146°10'E) in March—one of the wettest months of the rainy season in one of the wettest regions of Australia (mean annual rainfall of 2,000 mm; White 1946). Clearly, Zebra Finches in Queensland are not prevented from breeding by heavy rain *per se*; nevertheless, some environmental factor(s) prevents them from breeding north of the 17°S latitude.

Habitat selection

Zebra Finches prefer open grassy country with a scattering of trees and bushes. Grasses provide food, and trees and bushes provide nesting sites and shade. Forests, dense woodlands, including mallee, are not penetrated by Zebra Finches and are barriers in some peripheral parts of the continent to areas that might otherwise be suitable (Table 2.3). These formations, which have limited seeding grasses, are also avoided by Zebra Finches when encountered within the distribution. In the savanna of northern and northwestern Australia, Zebra Finches also avoid the dense forests and woodlands that border the banks of rivers and creeks. It is possible that the dense low tree cover in Cape York Peninsula is one factor that has prevented colonising.

Zebra Finches have extended their range in the southeastern (Davies 1977b) and southwestern (Immelmann 1965b) margins of the distribution where forests and woodlands have been cleared or thinned for agriculture and pasture grasses cultivated. Here they are usually found where suitable nesting trees and bushes are growing. Densities in rural land are highest in irrigated regions where agriculture is most intense. Grasses seed in these regions throughout much of the year and this regular supply of food is shielded from the effects of droughts that prevail throughout the range. Nesting sites for breeding and roosting are provided in shelter belts planted around farm houses and outbuildings and in the numerous fruit trees (especially citrus) and vines cultivated in the fruit growing regions. It is not uncommon for wild Zebra Finches to live on the outskirts of towns and in villages, frequently invading parks and residential areas. I have even seen them breeding in dense shrubs planted in traffic islands in small shopping centres where they fearlessly go about their business feeding on cultivated lawns, and raising families oblivious to people and passing vehicles.

Although Zebra Finches can roost like most birds by clinging onto twigs or branches they show a strong preference for roosting in nests, either complete or partially complete. Sites for breeding are needed especially those that can provide a firm anchorage for the flimsy nests. Densely branching, preferably thorny, trees and shrubs provide the best sites but other structures, both man- and animal-made will do (Chapter 6). Extensive areas of saltbush and spinifex that are devoid of shrubs and trees are avoided because the required sites for nesting are not present

Table 2.3 Vegetation associations of the Australian mainland within and beyond the area of the main distribution of Zebra Finches; the quantitative breakdown is based on the dominant vegetation in 1° blocks[a] in which Zebra Finches were recorded present or absent during the Field Atlas of Australian Birds from 1977–1984 (Blakers *et al.* 1984); Zebra Finches are not found in forest, but will invade margins of forests that have been cleared

Vegetation association	Number of 1° blocks in the main distribution in which Zebra Finches were recorded				Number of 1° blocks beyond the main distribution
	present	(%)[b]	absent	(%)[c]	
Rainforest	2	(0.3)	0	(0)	10
Forest	11	(1.8)	0	(0)	18
Woodland	159	(26.2)	33	(17.2)	131
Acacia scrub	217	(35.8)	2	(0.9)	0
Mallee	44	(7.2)	14	(24.1)	5
Saltbush	45	(7.4)	6	(11.8)	0
Spinifex	82	(13.5)	11	(11.8)	0
Tussock grassland	47	(7.7)	0	(0)	0
Total	607	(100.0)	66	(65.8)	164

[a] Measures one degree of latitude by one degree of longitude giving an area of 10,000 km^2.
[b] Proportion of 607 blocks in which Zebra Finches were recorded.
[c] Proportion of blocks of each vegetation formation where Zebra Finches were absent within the area covered by the main distribution.

(Table 2.3). Furthermore, there is some evidence that seeds of spinifex are not eaten by Zebra Finches (Chapter 4). Despite these limitations, Zebra Finches have managed to exploit most areas at one time or another by basing their nesting and roosting activities in the thinly scattered shrubs and trees that line the ephemeral water courses and run-on areas that penetrate most of these regions. Similarly, in the extensive acacia scrublands that cover vast areas of the inland, nesting sites are abundant, but grass may be rare or absent in some locations; consequently, Zebra Finches again base their activities along the dry water courses and run-on areas where grass is more plentiful.

Davies (1986) noticed that Zebra Finches are selective in their use of the landscape in arid regions in Western Australia. First impressions suggest they are everywhere, but careful observation shows that their activities are focused on fertile sites, the small watercourses where the grasses grow and surface water is found. In good seasons they move beyond these areas, following the advancing grass and retreating when seasons are poor. In the Simpson Desert, the driest region in Australia, Zebra Finches are again selective in their use of the landscape. Here they base their activity in the swales between the dunes where most grasses grow and seed is found; hakeas and other bushes also grow here and provide sites for shelter and for breeding and roosting nests (P. B. Taylor, pers. comm.). On a larger scale, Ford and Sedgewick (1967), in their survey of the Nullarbor Plain and the Great Victoria Desert, found Zebra Finches restricted to certain habitats, namely occasional depressions (uvalas) where a few tall bushes and low trees grew, and in the Great Victoria Desert Zebra Finches were only seen in the breakaway country where water existed in gorges and valleys; they were absent from the mallee and dune country.

When present, Zebra Finches are normally one of the more conspicuous components of the avifauna, especially when large flocks mill around isolated surface water or thorny bushes. Feeding and resting flocks are fairly cryptic. The predominantly grey-coloured body blends in well with most types of soils and vegetation so what can initially appear to be empty, desolate country may suddenly echo with the familiar sounds of a flock on the move.

Summary

Zebra Finches have an extensive distribution in Australia and Indonesia. The Lesser Sundas Zebra Finch is found on 18 of the 21 main islands that make up the archipelago. It is found from Sermata in the east to Lombok in the west, from sea level to elevations of 2,000 m. Grasslands, rice fields, and secondary woodland are the preferred habitats. Clearing of forests for timber and cultivation has extended the area of suitable habitat and provided sources of grain. The Lesser Sundas Zebra Finch has the most extensive distribution of 11 species of estrildine found in the Lesser Sundas archipelago.

The Australian Zebra Finch is found over 75% of the mainland; it is absent from Tasmania, but has reached a number of islands within 60 km of the mainland coast. It is the most widely distributed Australian estrildine; its congenor, the Double-barred Finch, is the next most widespread species occupying 32% of the mainland. The Zebra Finch is not found in Cape York Peninsula or the northern Kimberley, but occupies all the arid zone of the interior and most of the semiarid zone towards the periphery of the continent. Dense vegetation and heavy monsoonal rains limit the extent of its distribution in the north and east, but clearing of vegetation has extended the range into semiarid peripheral regions. Low temperatures may restrict its spread to cleared areas in southern parts of the range. Zebra Finch habitats must have accessible surface water for drinking, grass seeds for food, and bushes and shrubs for nesting and roosting. In much of arid Australia these resources are patchily distributed and concentrated mostly on sites of higher soil fertility and moisture that occur in flood plains and along water courses.

3 Morphology, domestication and moult

Morphological variation

Subspecific variation

Australian Zebra Finches are significantly bigger than Lesser Sundas Zebra Finches (Clayton *et al.* 1991). Samples from northern Victoria (Danaher colony) in southeastern Australia were heavier (males 1 g heavier on average, females 1.5 g) than those trapped on Lombok, Sumbawa, Flores and Timor; wings and bills were also longer (5 mm and 1 mm on average, respectively). Bill-depth was significantly smaller only in the Flores and Timor samples (Figure 3.1). Bills of Australian males were redder, darker and more intensely coloured than those from the Lesser Sundas, but there were no significant differences between females. Within each subspecies the bills of males are, on average redder, darker and more intense than those of females which are an orange-red colour. Considerable overlap exists in bill colour of the sexes (Burley and Coopersmith 1987). Detailed comparisons of iris colour of the two sub-species have not been made but both are a deep reddish orange in wild-caught birds, but a dark brown in aviary-bred ones.

Males of the two subspecies, but not females, also differ in plumage on the throat, fore neck and breast. The fine black barring of the throat and fore neck found in Australian males is absent from Lesser Sundas males which simply have a pale grey ground colour. The size of the male black breast band varies considerably among Zebra Finches from the different islands of the Lesser Sundas, but in every case the absolute and relative size of the band is significantly smaller than that found in the Australian subspecies which is also highly variable. Finally, males of the Lesser Sundas subspecies have an abdomen off-white in colour, thus resembling that of females with their honey-coloured wash, whereas Australian males have a pure white abdomen, when free of dirt. In museum skins, ageing causes the white abdomen of the Australian males to brown slightly and this tends to make the abdomen resemble that of the females. Sexual dichromatism is thus more extreme in the Australian subspecies.

Cross-fostering between subspecies in aviaries does not affect size, plumage and bill colour differences. Hybrids of both combinations (male *guttata* × female *castanotis* and male *castanotis* × female *guttata*) were indistinguishable in size, with bill and plumage intermediate between the two parental subspecies (Clayton 1990a). Interestingly, it was easier to

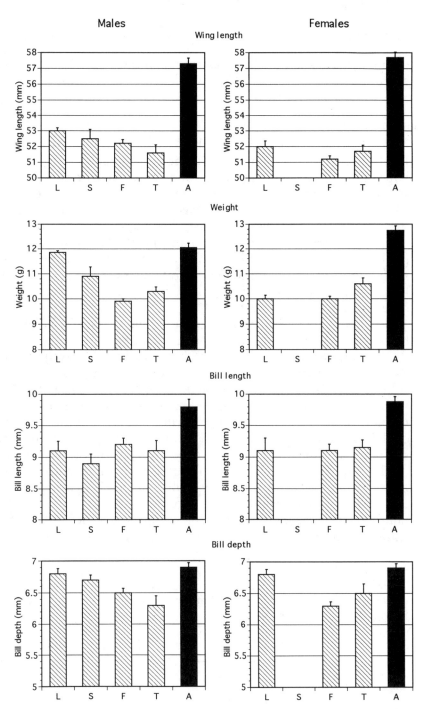

Fig. 3.1 Size comparison of (A) Australian *n* = 55 males and 45 females) and (L,S,F,T) Lesser Sundas Zebra Finches (hatched bars) based on four measurements of free-living birds. Bars show means plus one standard deviation. L = Lombok (11 males, five females), S = Sumbawa (four males), F = Flores (12 males, 14 females), T = Timor (eight males and five females). (Modified from Clayton *et al.* 1991.)

obtain hybrids from the latter pairing than the former (N. Clayton, pers. comm.).

Variation in the Australian Zebra Finch

Although Mathews (1913) described six subspecies of Australian Zebra Finches, Keast (1958) found no significant geographic variation among a large number of museum skins, and accordingly recognised no subspecies; this conclusion has been followed by most modern authors. Museum skins suffer from shrinkage and colour changes, especially in the softer parts, and there is a need for a systematic re-examination of geographic variation in the Australian Zebra Finch using standard morphological methods as well as modern molecular techniques. During my studies in northern Victoria and central Australia I made some morphological measurements on living birds that suggest that some geographic differences in size may exist.

Weights of free-living adults ranged between 10–17.5 g. The heaviest birds were females with full crops and an egg ready to lay. Birds from the Danaher colony in northern Victoria were significantly heavier than those at Alice Springs (Research Centre colony) (Table 3.1). Birds at each colony were baited with seed *ad libitum*, so differences in immediate food supply were not responsible for differences in condition. Levels of mobility were higher at the Alice Springs colony so birds may not have exploited the feeder for as long as those at Danaher, which were more sedentary (Chapter 8), and consequently, may have been under more food stress. Within each colony females were significantly heavier than males (Danaher $F_{1,478} = 21.5$, $P<0.0001$; Research Centre $F_{1,1079} = 20.6$, $P<0.0001$). When weight of the sexes was compared in breeding and non-breeding seasons there was no significant season effect for either colony (Danaher $F_{1,547} = 1.2$, *n.s.*; Alice Springs $F_{1,1077} = 0.01$, *n.s.*); how-

Table 3.1 Weight (g) of free-living adult Australian Zebra Finches from two distant regions—northern Victoria (Danaher colony) and central Australia (Alice Springs Research Centre colony)[a]

	Danaher colony [b]		Alice Springs colony [b]	
	Males	Females	Males	Females
Mean	12.4	12.7	11.9	12.2
s.d.	0.76	0.98	0.81	0.94
Range	10.5–15.1	10.0–16.2	10.0–15.2	9.4–15.7
n	295	256	560	521

[a] Differences between colonies $F_{1,1627} = 88.2$, $P < 0.0001$.
[b] Differences between sexes $F_{1,1627} = 40.1$, $P < 0.0001$. Sex × colonies interaction $F_{1,1627} = 1.4$, *n.s.*

ever, the sex by season interaction was significant for Danaher ($F_{1,\ 547}$ = 5.2, P = 0.02) where sex differences were greater in the breeding season. Thus, the sex differences in weight are not entirely due to differential investments in breeding effort although Skagen (1988) found that domesticated Zebra Finch females lost a greater percentage of their weight over a breeding cycle than did males (11.4% *vs.* 7% respectively) and there was a greater difference when food was limiting.

Despite being lighter, males were bigger than females in wing-length, bill-depth, head–bill length, but not in length of tarsus in the sample from northern Victoria (Table 3.2a); only bill-depth was significant in the smaller Alice Springs sample, but head–bill approached significance (Table 3.2b). Boag (1987) made a similar finding in his morphometric analysis of domesticated Zebra Finches; males had smaller bodies than females but larger bills. Alice Springs birds were significantly bigger in all dimensions than Danaher birds except for wing-length where there was no significant difference; there was no significant location by sex interaction for any of the four characters (Figure 3.2).

Table 3.2a   Means ± *s.d.* and (range) of four measurements of adult Zebra Finches from Danaher, northern Victoria

	Males		Females	
Head–bill[a]	23.1 ± 0.50	(21.8–24.4)	22.8 ± 0.46	(21.5–24.0)
Wing[b]	55.2 ± 1.34	(49.5–59.0)	54.7 ± 1.43	(49.0–58.0)
Bill-depth[c]	7.9 ± 0.50	(6.0–8.3)	6.9 ± 0.50	(5.9–8.7)
Tarsus[d]	14.3 ± 0.55	(12.4–15.9)	14.3 ± 0.56	(12.1–15.8)

[a] Differences between the sexes: n = 242 males, 237 females $F_{1,477}$ = 9.6, $P<0.0001$.
[b] n = 251 males, 241 females $F_{1,490}$ = 12.0, P = 0.0006.
[c] n = 205 males, 205 females $F_{1,408}$ = 9.6, $P < 0.002$.
[d] n = 209 males, 287 females $F_{1,421}$ = 0.05, P = 0.8.

Table 3.2b   Means ± *s.d.* and (range) of four measurements of adult Zebra Finches from Alice Springs

	Males (n = 48)		Females (n = 48)	
Head-bill[a]	23.6 ± 0.61	(22.4–25.0)	23.4 ± 0.54	(22.3–24.6)
Wing[b]	55.3 ± 1.30	(53.0–58.5)	55.0 ± 1.71	(51.8–58.0)
Bill-depth[c]	8.9 ± 0.39	(8.2–10.0)	8.6 ± 0.47	(7.8–9.5)
Tarsus[d]	15.0 ± 0.55	(13.7–16.0)	15.0 ± 0.47	(14–16)

[a] Differences between the sexes $F_{1,98}$ = 3.3, P = 0.07.
[b] $F_{1,96}$ = 1.0, P = 0.31.
[c] $F_{1,94}$ = 19.0, $P < 0.0001$.
[d] $F_{1,96}$ = 0.2, P = 0.67.

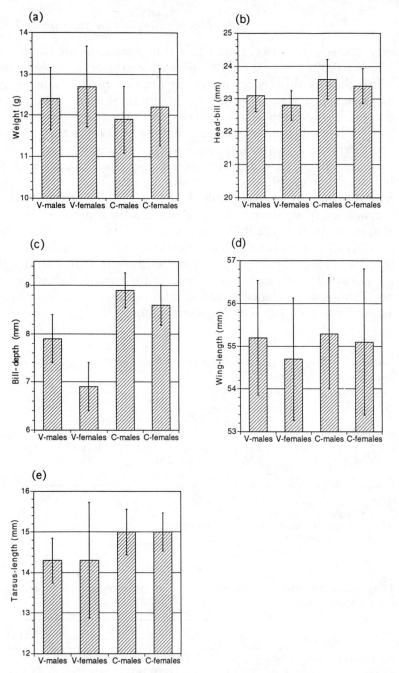

Fig. 3.2 Size comparisons of four morphometric characters for male and female Zebra Finches from two widely separated localities, northern Victoria (V) and central Australia (C). Means ($\pm$ *s.d.*) are given. Central Australian birds were significantly lighter than birds from northern Victoria ($F_{1,1627} = 40.1$, $P<0.0001$) but larger in length of head–bill ($F_{1, 576} = 89.3$, $P<0.0001$), tarsus ($F_{1, 518} = 134.0$, $P<0.0001$) and bill-depth ($F_{1, 507} = 1030.0$, $P<0.0001$); there was no significant difference in wing-length.

Domestication

Because it is hardy and easy to keep in captivity the Australian Zebra Finch has long been one of the most popular cage birds in the world. There is some suggestion that Louis Jean Piere Vieillot held Timor birds in captivity (Immelmann 1962b, 1965a), but this cannot be confirmed (Rogers 1986). Captive breeding of the Australian subspecies became well established in Europe by the 1870s as indicated by the low market price of the times (Sossinka 1970). Thus, imports of wild stock from Australia became superfluous towards the end of the nineteenth century and gradually diminished, with few reaching Europe after the First World War (Immelmann 1962c). Export bans were implemented by the Australian government on all native wildlife in 1960 and no wild Zebra Finch stock has been legally imported by overseas bird keepers since that time. By 1962, Immelmann calculated that some 50–80 generations of Zebra Finches had been bred in Europe without any significant input of wild blood and a fully domesticated strain had arisen.

The first colour morph identified was the 'fawn', which was discovered in wild birds from South Australia in 1927. This formed the breeding stock from which many other mutant strains were developed, and it is still one of the most popular shown in competitions today. About 30 main colour morphs, or plumage mutations, of domesticated Zebra Finches currently exist and there are about 44–50 combinations of these (Immelmann 1965a; Martin 1985; Rogers 1986; Corbett 1987; Appendix 1). Homozygous strains are rare. Some mutations change the colour of all the plumage in both sexes (e.g. white, silver, cream), while others change particular parts, principally the dorsal surface (e.g. fawn, penguin), back (light- and saddle-back), breast and throat (black- and orange-breasted), cheeks (silver-, grey-, fawn- and black-breasted) and eye/tear stripe (e.g. black-faced). Relatively few mutations have altered the colours of the flanks, tail, upper-tail coverts and rump. Some mutations are sex-linked (e.g. fawn, chestnut-flanked and saddle-back), and while most are recessive some autosomal mutations are dominant (e.g. crested, silver and cream). Those strains that change the ground colour of the plumage (e.g. fawn, silver) produce different combinations with other strains that change the colour of particular markings. Some mutations are based on the absence of pigment, totally (white) or in part (pied) and the latter can be combined with a number of other mutations. One autosomal recessive mutation makes the bill colour yellow. Finally, there is one morph that is not a colour variant, namely 'crested', which causes a flat swirl of feathers to grow on the forehead. Breeding of these morphs and the development of new ones are the principal pleasures of many bird keepers, and Zebra Finch societies have been formed in Australia, the United Kingdom, Germany, Denmark and elsewhere to foster this hobby. These societies publish the standards for the various morphs and the crossings necessary to obtain them and conduct show competitions.

Immelmann (1962c) and Sossinka (1970, 1972a, 1974, 1980) investigated morphological, physiological and behavioural changes in Zebra Finches that have arisen as a result of domestication in Zebra Finch stocks held in Europe. Offspring of domesticated Zebra Finches were found to be significantly heavier than aviary-reared offspring of wild-caught birds. Domesticated females had longer wings, and longer and deeper bills than their wild counterparts and took longer to reach adult size, but the males did not differ significantly in these measurements. These differences arose through isolation of the captive populations, through chance fixation of genes, and through selection in the captive environment. Immelmann assumed that most of the ancestral wild-stock that formed the basis for the domesticated line came from northwest Australia, but unfortunately no morphological measurements have been made on birds from this region. Selection by bird keepers is considered the strongest force in the domestication of the Zebra Finch (Sossinka 1982).

In contrast to their European counterparts Australian aviculturalists have not selected for larger body size. Whites, fawns and mixed morphs were all smaller than wild-caught Zebra Finches and their first generation aviary-reared offspring (Carr and Zann 1986). They were lighter, had shorter wings, head–bills, tarsi and bill-depths. There is no plausible explanation for these contrasting directions in the domestication process. Conceivably, differences may have arisen because of difference in sources of wild stock, in climatic conditions, and in show standards for size.

Morphological differences can arise within one generation of captive rearing. Colour of the irides is the most noticeable difference: first-generation aviary-bred birds do not develop the red iris of free-living adults; a few develop reddish-brown shades, but most retain the dark grey-brown colour of juveniles (Sossinka 1970). First-generation aviary-bred birds also have smaller wing-lengths and shorter tarsi than wild-caught birds (Carr and Zann 1986). No differences with wild birds are found in bill colour (Burley *et al.* 1992), but the grey dorsal surface of the back of domesticated wild-type males has less of the brownish wash found in the free-living ones; a consequence of having been selected against by bird keepers (Sossinka 1970).

Moult

Like all estrildines Zebra Finches have nine primaries and nine secondaries. The tenth primary is much shorter, being only about one fifth as long as the rest. The more proximal primaries, numbers 1, 2 and 3 (41–42 mm), are shorter than more distal ones (44–47 mm), of which numbers 6 and 7 are the longest. All twelve rectrices are the same length (35–38 mm). The feather tracks (pterylae) for the contour feathers follow the conventional passerine pattern (Ginn and Melville 1983). Only about 20 down feathers grow from the large ventral apterium. Details of the pterylography of the Zebra Finch can be found in Parsons (1968).

In some males the juvenile body plumage may begin to be replaced by the secondary sexual plumage as early as 20 days after hatching, but in most this occurs between days 35 and 40. The first indication that a juvenile is a male will be a few flecks of colour among the grey plumage, for example, a black spot on the breast band, an orange spot on the ear coverts, or odd striped feathers on the throat. Around 40 days, patches of male-specific feathers begin to appear on the throat and flanks and the characteristic adult plumage is complete by days 55 to 60 but the colours, which are somewhat faded at first, intensify in the next few weeks. For unknown reasons the development of the male plumage is often delayed for several months in wild birds hatched in late autumn in south-east Australia.

Acquisition of male plumage marks in the Lesser Sundas Zebra Finch have not been described.

Wing moult

Wing moult of the Australian Zebra Finch was investigated at the Padgett colony (Figure 3.3) from 1976 to 1983 (Zann 1985a). Multiple captures of individuals during the same moult cycle permitted accurate estimates, through extrapolation, of the starting and completion dates of primary moult in adults and young. The major finding was that Zebra Finches moult their primaries extraordinarily slowly in a continuous step-wise moult. Feathers are lost in a rigid, non-overlapping sequence in ascending order from number 1 (the innermost) through to number 9 (the outermost), that is, from proximal to distal. Primaries are not lost until the preceding one is fully grown, whereas most species of birds will loose and regrow two or three primaries simultaneously, and so complete the cycle more rapidly. On average, the smaller inner primaries of Zebra Finches take about 21 days from loss of the old feather to the full growth of its replacement, and the larger outer ones take about 26 days. The interval between the full growth of one primary and the loss of its more distal neighbour was highly variable, ranging from a few days to several weeks. It was extremely rare for moult to stop during mid-growth of a primary. Secondaries were also replaced very slowly, but did not begin before the primary moult commenced: their sequence of replacement of individual feathers was not so rigid as that of the primaries, but the approximate order was the conventional one found in most passerines, namely the tertials first, usually the middle one (number 8), then the inner (number 9) and then the outer (number 7);

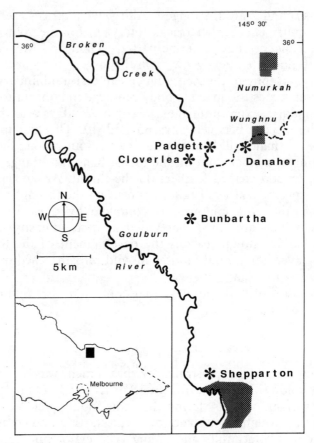

Fig. 3.3 Locations (∗) of five Zebra Finch colonies studied in northern central Victoria.

the secondaries then moult in order from the most distal to the most proximal (number 1 to number 6).

This slow moulting occurs continuously so that there is only a short period between successive cycles. Moreover, in many individuals the new cycle will begin before the previous cycle is complete, that is primary number 1 will be lost before number 9 has been lost or finished growing, consequently there will be two actively moulting centres within the primaries of each wing. This type of overlapping moult sequence is termed 'continuous step-wise' moult, 'Staffelmauser' or 'serially descendent' moult (Ginn and Melville 1983). During any month of the year a majority of Zebra Finches at Padgett were found with some primaries in pin or regrowth including those months when breeding was at its peak.

The loss of primary number 1 of the juvenal moult occurred at a mean age of 80 days in both males and females, but the latter took significantly longer to complete the cycle (males 204 ± 54 days *vs.* females 224 ± 56 days). The date of hatching had a significant effect on the age when juvenal moult began. Young hatched in the spring half of the breeding season commenced their moult before 80 days of age, some as early as 30 days. Those that hatched in the second half of the season waited until they were 80–100 days of age, and those that hatched in April, the last month of the breeding season, waited until winter was over before they lost their first primary. In many young the next cycle of primary moult began before primary number 9 of the juvenal moult was complete so that the two cycles overlapped.

The duration of primary moult in adults at Padgett was slightly longer than that of the juvenal moult, but there were no significant differences between the sexes (males 229 ± 62 days *vs.* females 239 ± 79). The interval between successive moults was longer in females (mean 6 days) than males (mean -16 days), that is, their successive moult cycles overlapped less; consequently, a greater proportion of males than females captured in the course of the study had primaries in moult (males in moult 68% *vs.* females 58%, $n = 3,080$). There were no significant differences in duration between successive cycles, either from juvenal to first, or from first to second. Primary moult in adults could be initiated in any month of the year, but rate of feather growth and replacement was slowest in winter and early spring, presumably because of low temperatures and food shortages. The amount of nesting activity each month was inversely correlated with the rate of primary moult, which suggests that the two processes are competing for the same sources of energy (Zann 1985a). This appears to be more so in females than males since significantly more males were found in moult than females during the three peaks of the breeding season at this colony (Chapter 4). Nevertheless, individuals could breed and moult simultaneously, as I found when I raided breeding pairs in their nests. Of 12 parents caught one night in March 1981, seven (four males and three females) were in active stages of feather re-growth and one female was growing two primaries (numbers 9 and 1) simultaneously.

Zebra Finches at other colonies in northern Victoria (Shepparton, Cloverlea and Danaher; Figure 3.3) had a similar sequence and duration of primary moult to that found at the Padgett colony. Furthermore, Zebra Finches at Alice Springs demonstrated a basically similar pattern of moult to that found in northern Victoria (R. Zann and N. Burley, unpublished observations); however, a smaller proportion of trapped adults were found in active moult at any one time (28% at Alice Springs *vs.* 63% in northern Victoria) and more had moult stopped mid-cycle so there were fully grown primaries of two or more distinct ages (based on colour and degree of wear). Complications often arose when several simultaneous cycles were stopped and re-started several times on the one wing. Moult usually restarted at the next distal primary in both cycles ('suspended moult'), but in some instances it also initiated a new cycle with the loss of primary number 1 again ('arrested moult'). Based on growth rates of adjacent primaries, the mean duration to moult all nine primaries was estimated at 287 ± 168 days ($n = 126$). Actual values were probably much higher due to frequent halts and delays in the moulting cycle between full growth of a primary and loss of the next.

Seasonal conditions strongly affected the incidence of wing moult in Zebra Finches in central Australia. Where seasons were poor and birds could not maintain reasonable condition (low weight, ragged plumage) new cycles of moult were not initiated and ongoing cycles halted. In June 1986, we found birds significantly heavier at the Research Centre in Alice Springs than at Undoolya Bore, 31 km east (Research Centre: 12.3 g $\pm$ 0.90 $vs.$ Undoolya: 11.5 $\pm$ 0.79, t_{122}=5.01, $P<0.001$) and their plumage was less faded and tattered; correspondingly, a significantly greater proportion of the Alice Springs birds were regrowing primaries than were Undoolya birds (G_2=15.0, P=0.006). The Undoolya birds were in poor condition due to a prolonged drought and they did not have access to as much seed and water as the 'town' birds that attended the baited trap at Alice Springs and who exploited irrigation. Even after the drought broke and grasses were seeding in the district surrounding the Research Centre birds still maintained a significant weight differential over the 'bush' birds and fewer stopped their moult in mid-cycle.

Of all the passerines studied to date only the Double-barred Finch (Schoepfer 1989) and the Masked and Long-tailed Finches (Tidemann and Woinarski 1994), moult primaries as slowly as the Zebra Finch. In a comparative study of moulting patterns in four species of sympatric estrildines in New South Wales, Schoepfer found that the Double-barred Finch took almost twice as long to moult its primaries as Red-browed Finches, Plum-headed Finches and Diamond Firetails, all of which moulted at conventional rates (125–150 days for all nine primaries). Nonetheless, all four species actively regrew remiges in the summer months, the normal nesting period. In the Northern Territory, the Gouldian Finch took 150–180 days to complete wing moult and waited until breeding finished before starting moult, whereas the Long-tailed Finch started moulting during its breeding season (Tidemann and Woinarski 1994).

Slow moult is believed to minimise the impairment of the aerodynamic qualities of the wing and spread the energy demands of feather replacement in a way that minimises interference with other physiological processes such as breeding (Ashmole 1962). The fact that many arid species of birds in Australia breed and moult simultaneously when favourable conditions arise, yet still manage to complete the moult within a relatively brief period (Keast 1968) suggests that physiological demands may not be the only factor in determining slow moult. Slow moult has been found in two other arid-adapted Australian species, the Budgerigar

Melopsittacus undulatus (Wyndham 1981) and the Squatter Pigeon *Geophaps scripta* (Crome 1976), although only the former overlaps successive cycles of moult as does the Zebra Finch. Slow moult allows maximum mobility to be maintained at all times. Ready mobility may be essential for survival for many arid-adapted and oceanic species, and this may be the factor that led to the evolution and maintenance of slow moult. Conceivably, slow moult evolved in the common ancestor of the Zebra Finch and the Double-barred Finch, but the requirement for mobility in the latter does not appear to be an essential trait for survival in the mesic habitats it currently occupies and may not be responsible for maintaining its existence.

Effects of captivity on wing moult

Providing they did not breed, the duration to moult all nine primaries in wild-caught Zebra Finches did not differ significantly from that of free-living ones. If allowed to breed freely they increased the duration by a mean of 68 days in males and 79 days in females (R. Zann unpublished observations). Among wild-caught birds there was a significant positive correlation in both sexes between duration of primary moult and the number of clutches attempted during that interval (males: $r = 0.67$, $t_{28} = 4.78$, $P<0.001$; females $r = 0.56$, $t_{37} = 4.11$, $P<0.001$). Therefore, in the aviary where conditions are assumed to be optimal, breeding still had a strong retarding effect on the moult process, with most birds slowing moult rather than stopping completely. Weekly inspections showed that all primaries regrew to full size within three weeks of loss of the predecessor except number 9 which took four weeks. This finding agrees closely with the estimates made on free-living birds, and since number 9 was not the shortest feather it means that it was simply slower to re-grow than the others. Each secondary feather also took three weeks to fully re-grow.

First-generation aviary-bred offspring of wild-caught birds lost their first primary at a mean age of 80 days, the same age as in free-living birds. However the duration to complete the whole cycle was, on average, 60 days longer. Aviary-bred birds also moulted in a continuous stepwise manner and there was no significant difference to free-living birds in the length of the interval between the first cycle and the second. However, free-living birds had significantly longer intervals between subsequent cycles ($F_{1,202} = 26.5$, $P<0.001$), that is, aviary-bred birds tended to overlap their cycles more. One effect of this increasing overlap was a progressive loss of the rigid order of replacement of primary and secondaries. The clear order of replacement of remiges in the first cycle matched precisely that described above for free-living birds, but in subsequent cycles there was an increasing tendency for several adjacent feathers to be lost and replaced simultaneously so that the centre of moult activity and its step-wise progression among the primaries became increasingly obscure.

This study of wing moult in wild-caught and aviary-bred Zebra Finches is instructive, for three reasons. First, it confirms all the main findings in free-living birds; second, it demonstrates the flexibility of the moulting processes in response to environmental conditions, especially the availability of food and the incidence of breeding, and third, it reinforces the need for caution in interpreting data obtained from captive subjects.

Summary

Lesser Sundas Zebra Finches are about 13–18% smaller than Australian Zebra Finches. Plumage of the females is identical, but Australian males have redder bills and a proportionally larger black breast band than Lesser Sundas males. Sexual dimorphism is more extreme in the Australian subspecies. Females are heavier than males in breeding and non-breeding seasons, but males have longer wings, and longer and deeper bills; there is no sex difference in tarsus length. Zebra Finches from central Australia were larger (head–bill, bill-depth, tarsus), but not as heavy as those from southeastern Australia. The harsher climate in central Australia may be responsible for the lower weight and poorer physical condition. Biochemical investigation of the Australian subspecies may reveal some degree of geographic differentiation.

Captive Zebra Finches have become highly domesticated and were first bred in Europe almost 130 years ago. There are about 30 main strains of colour morphs of the Australian subspecies in captivity and about 50 distinct combinations of these. In Europe, domesticated females are larger and heavier than offspring of wild-caught birds and males are heavier, although they are not larger. By contrast, in Australia, domesticated colour morphs are smaller and not as heavy as wild-caught birds and their first-generation offspring. Different selection criteria among bird keepers may be responsible for these differences. First-generation aviary-bred birds do not develop the red iris of free-living birds.

In free-living birds the secondary sexual plumage begins to replace the juvenile plumage around 35–40 days after hatching, and is complete by about 60–70 days and reaches the full colour intensity of adults by 90–100 days. The rate of development is not uniform and is slowest for young hatched in the autumn months. Free-living adults moult the wing feathers in a very slow, continuous step-wise manner and the whole cycle takes about 235–290 days. The time from loss to full growth of the replacement primary takes about three weeks. The next moult cycle may start before the previous cycle has finished, so that most birds are in a state of permanent primary moult, although a greater proportion of birds in central Australia stopped their moult during a cycle than did birds from northern Victoria. Young begin their first primary moult at a mean age of 80 days after hatching but the actual age depends on when young hatch. Adults may moult during a breeding cycle, but do so more rapidly outside the breeding season. Females moult more slowly than males. In captivity intense breeding activity slows wing moult and the order of feather replacement is gradually obscured. In unfavourable climates and during harsh seasons moult slows and may frequently stop until conditions improve. Among passerines only the Double-barred Finch, the Long-tailed Finch and the Masked Finch moult as slowly as the Zebra Finch. The main advantage of slow moult is its minimal impairment of flying ability so that birds maintain maximum mobility, an essential requirement for survival in arid and semiarid habitats for some species.

4 Feeding ecology

'It passes much of its time on the ground, and feeds upon the seeds of various kinds of grasses.'

J. Gould 1865.

All the Australian estrildines feed on the seeds of grasses, but no species concentrates on them to the same extent as Zebra Finches. Six quantitative studies of diet have been published since Immelmann's (1962a, 1965a) original observations, and despite differences in habitat and climate among the studies, they agree on two aspects, namely the rarity of insects and the rarity of seeds other than those of grasses. The picture of the diet is fairly complete in southeastern and central Australia but in other parts of the distribution information is scanty, especially that for breeding birds. The role played by half-ripe seeds during the breeding season would be a fruitful area of research for nutritional physiologists.

Diet

Methods

Three methods have been used to investigate the diet of Zebra Finches. Immelmann (1962a, 1965a) simply observed what birds were feeding on and collected a sample for identification. This technique is difficult with fallen seed because often nothing is found when the ground is examined. Zebra Finches have been shot in two studies (Davies 1977a; Morton and Davies 1983) and the contents of the crops identified. Identification can be a problem since the characteristics of the dehusked seed are different from those of whole seed. Consequently, uncertainties arise, especially when several congeneric species have been eaten, and extensive reference collections of dehusked seeds are necessary. The virtue of shooting is that it is simple and efficient, and the whole contents of the crop can be examined, but it is a bit drastic for the unfortunate bird. The third method is a non-destructive one devised by Zann and Straw (1984b): a sample of seed from the crop of mistnetted birds is extracted by means of a narrow plastic tube. This tube is gently pushed down the oesophagus into the crop's store of seed where individual seeds are pushed up into the bore of the tube by manipulating the wall of the crop with a finger.

Up to ten to fifteen seeds can be extracted and if the crop contains water, capillary action simply draws seed and water up. Schöpfer (1989) bevelled the tip of the tube to allow large seeds to enter. The tube method is not very aversive for the birds and it provides representative samples of the types of seeds in the crop, although it underestimates the abundance of insects (Zann and Straw 1984b). Where the diet is fairly well known, conspicuous types of seeds can be identified through the thin, almost transparent wall of the crop (Tidemann 1987), but samples are required for confirmation and reference.

Grass seed specialists

Zebra Finches eat grass seeds and rarely anything else. This is the consistent finding in all investigations, irrespective of whether samples were taken from birds that were shot (Davies 1977a; Morton and Davies 1983) or from those that had seed extracted from the crop (Zann and Straw 1984a, 1984b; Tidemann 1987; Schöpfer 1989; Zann *et al.* 1995). Seeds of a large variety of species are eaten, as one would expect of a species that occupies a broad range of habitats across a wide distribution. Sixty-one species of seeds have been identified in the diet, and all but three species were seeds of grasses (Appendix 2). Nine seed types were from introduced species of grasses. All seeds are dehusked and may vary in age from soft green seeds to old desiccated ones. A great range of sizes are eaten. The smallest dehusked seeds eaten were those of *Chloris truncata* (1.5 × 0.4 mm; weight 0.2 mg) and the largest those of *Echinochloa crus-galli* (1.7 × 1.4 mm; 1.7 mg) (Zann and Straw 1984a). Zebra Finches are capable of eating larger seeds, for example, Yellow Millet *Setaria italica* (1.9 × 1.1 mm; 2.0 mg) is one of the standard commercial seeds fed to captive Zebra Finches. Extra-large seeds, such as those of wheat and oats, which were available in northern Victoria throughout much of the year, were not eaten, and only a few fragments were found in crops; presumably, the whole seeds were too large. Although Immelmann (1962a) saw Zebra Finches feeding on the seeds of spinifex (*Triodia* spp.) they were not found in the crops of birds inhabiting country where spinifex was the predominant species of grass and so must be regarded as an unsuitable food; perhaps the seeds are too long to easily ingest. Seeds of dicots were only found in a few individuals and constitute a minute part of the overall diet.

Zebra Finches have been observed eating insects, especially termites (North 1909) but these are rarely found in crops. In northern Victoria, Zann and Straw (1984a) found that less than one percent of birds (n = 540) had insects (aphids and ants) in their crops, and these constituted less than one percent of items (n = 17,645); at two locations in the eastern arid zone Morton and Davies (1983) found only two homopterans in 34,221 items extracted from 97 birds; all the remaining items were grass seeds. Insects were extremely rare in the crops of 510 Zebra Finches shot by Davies (1977a) in arid western Australia. Immelmann (1962a, 1970) saw Zebra Finches taking flying ants and termites in mid-air in northern Australia. Birds would leave a lookout perch and make short fluttering hawking flights then swallow the prey on landing again. Remains of snails and ostracods have also been found in some crops but these were probably eaten for their mineral content rather than as food. Green material, often leaves and seed heads nipped from the extreme tips of grasses and succulents, is often found in the crops during the breeding season in both adults and nestlings (Zann and Straw 1984a). Fine gravel and particles of sand are also frequently found in crops.

No significant sex differences in diet were found in any of the eight quantitative studies (Table 4.1).

Seasonal changes in diet

No marked seasonal changes in diet were evident in the two populations studied by Morton and Davies (1983) in the eastern arid zone. However, there were three distinct annual changes in the diet in northern Victoria although twelve species of seed were eaten in total (Zann and Straw 1984a). The predominant grass seed in the diet was *E. crus-galli*, which was eaten throughout winter, but, by the start of spring, these seeds were scarce and in poor and damaged condition. Birds switched to the small green seeds of spring flowering annuals, mainly *Poa annua* and *Cynodon dactylon*, but by mid-spring all these seeds had shed and the finches switched to the wallaby grasses, the next grasses to seed. *Danthonia caespitosa* seeded first, then a month later *Amphibromus neesi*. By February, the new season's supply of *E. crus-galli* seed was beginning to ripen on the heads and the finches switched back to this. These large seeds were produced in abundance and if the plants were eaten by cattle they reseeded soon afterwards to produce another wave of ripening seed. These changes in diet are depicted diagrammatically in Figure 4.1.

Density of fallen *E. crus-galli* reached a peak in June but the dead vegetative parts of the plant lay on top of the seed and prevented the finches from fully exploiting it until July. Seed was in limited supply in late winter and early spring (August–September)—the start of the new growing season, which is often the time of food shortage for granivores in both tropical and temperate regions (Wiens and Johnston 1977). At this time of year, most *E. crus-galli* seed was buried in mud and the spring annuals had not yet seeded. The finches had to feed for longer periods in the afternoon, had to move farther between successive foraging patches and remained on those patches for a briefer period before moving on. This indicates that patches of seed were smaller and fewer at this time. Muddy bills and foreheads from digging for seeds was further indication

Table 4.1 Indices of dietary diversity $(B)^*$ in Zebra Finches from eight populations; the higher the value the more diverse the diet

Location	Birds Sampled (n)	Samples (n)	Seed Species (n)	Dominant Species	B mean ± s.d.	Source
Fowlers Gap, N. S. W.	45	5	12	*Enneapogon* sp.	1.67 ± 0.47	Morton and Davies (1983)
Sandringham, Qld.	52	5	14	*Panicum decompositum*	1.66 ± 1.02	Morton and Davies (1983)
Northern Victoria	413	15	12	*Echinochloa crus-galli*	1.41 ± 0.44	Zann and Straw (1984)
Burrendong, N. S. W.	14	1	7	*Danthonia* sp.	1.32	Schöpfer (1989)
Undoolya Bore, N. T.[†]	39	1	4	*Enneapogon cyclindricus*	1.06	Zann et al. (1995)
Alice Springs, N. T.[‡]	23	1	2	*Cenchrus ciliaris*	1.05	Zann et al. (1995)
Alice Springs, N. T.[§]	25	1	5	*Cenchrus ciliaris*	1.19	Zann et al. (1995)
Ooraminna Rock Hole, N.T.[¶]	18	1	7	*Digitaria brownii*	3.15	Zann et al. (1995)

$* \ B = 1/\sum_{i=1}^{n} p_i^2$ where p_i is the proportion in the diet of the ith seed type in the sample and n the number of seed types (Wiens and Rotenberry 1979).

[†] 31 km E of Alice Springs (23°45'S, 134°06'E).

[‡] Settlement Tanks, 6 km SW of Alice Springs (23°44'S, 133°50'E).

[§] Airport, 13 km S of Alice Springs (23°48'S, 133°52'E).

[¶] 30 km S of Alice Springs (24°02'S, 134°00'E).

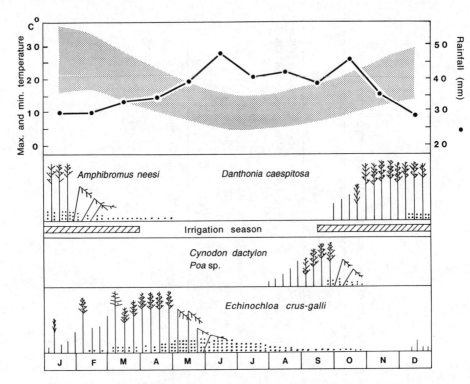

Fig. 4.1 Schematic portrayal of seasonal growth and production of five species of grass seeds heavily utilised by Zebra Finches in an irrigated region of northern Victoria. The top panel shows mean monthly rainfall and ranges of mean monthly maximum and minimum temperatures. (From Zann and Straw 1984a.)

of food shortage. Food shortages occurred in Alice Springs towards the end of a two-year drought in May 1986 (Zann *et al.* 1995).

Narrow diet

Despite the large range of sizes and species of seeds capable of being eaten by Zebra Finches, quantitative studies of diet at eight locations showed that they concentrated on seeds from a limited number of grass species. For example, in northern Victoria 66% of seeds eaten over a 15-month period were *E. crus-galli* (Zann and Straw 1984a) and in the eastern arid zone Morton and Davies (1983) found that 84% of seed eaten at Fowlers Gap (New South Wales) was an *Enneapogon* species and 74% eaten at Sandringham (Queensland) was *Panicum decompositum*. Mean dietary diversity *B*, which is an index of the breadth of the diet, ranged from 1.05 to 3.15 (Table 4.1). Indices for northern Victoria, Sandringham and Fowlers Gap were not significantly different (Zann and Straw 1984a) and are similar to that found at Burrendong (New South Wales) by Schöpfer (1989). The greatest diversity was found for the Ooraminna population near Alice Springs where $B = 3.15$. This population was in poor condition and few seeds were found in the crops, but they came from seven species (R. Zann and N. Burley, unpublished observations). Tidemann (1987) studied the diet of Zebra Finches, Gouldian Finches and Long-tailed Finches in the Top End of the Northern Territory. Unfortunately, her index of diversity was different from that used here so comparisons are difficult. Zebra Finches in her study area had the most diverse diet, eating the seeds of 11 species of grasses, whereas the Gouldian Finch ate only three species. Zebra Finches had the least dietary overlap when compared with the other species.

Morton and Davies (1983) were impressed by the dietary specialisation of the Zebra Finch populations they investigated, although they did not assess the proportions of seed available to determine if the diet departed from random. They compared the breadth of diet to that reported in the literature for a number of other species from different parts of the arid zone. Breadth of diet was not significantly different from that of Budgerigars (Wyndham 1980) which ate a smaller proportion of grass seeds than Zebra Finches, but breadth of diet was narrower than that of the Spinifex Pigeon *Petrophassa plumifera* which also ate seeds from many species of shrubs (Morton and Davies 1983). Larger granivores eat a more diverse diet than smaller granivores, presumably because larger bills enable them to eat a larger size range of seeds. The same conclusion has been reached by many others who have investigated the diet of avian granivores (e.g. Newton 1967; Willson 1971; Abbott *et al.* 1977).

Zebra Finches had a significantly narrower diet ($B = 1.32$) than that of four other species of finches studied by Schöpfer (1989) at Burrendong (Red-browed Finch, $B = 3.57$; Diamond Firetail, 2.66; Plum-headed Finch, 2.27; Double-barred Finch, 2.67). However, Zebra Finches were uncommon at this study site and all the samples ($n = 14$) was taken in the same month, hence the sample may not be representative and the diet not as narrow as thought. What was clear from Schöpfer's study was that Zebra Finches take fewer non-seed items than the other species of finches and do not increase their intake of insects during the breeding season.

Why only grass seeds?

The diet of the Zebra Finch raises a number of interesting points. First, why limit seed types almost exclusively to those of grasses; second, why concentrate on just a few species to make up the bulk of the diet; and third, why omit insects almost completely? Taking the first point, Morton and Davies (1983) emphasised that grasses are particularly abundant in the arid zone and provide a relatively stable and accessible resource in this environment. Perennial grasslands cover 40% of the arid zone and, in addition, form an important understory layer in the *Acacia* and *Eucalyptus* shrublands that form another 41% of the arid zone. These shrublands with their grassy understory also extend into many parts of the semiarid regions and thus allow granivores, such as the Zebra Finch, to occupy a wide range of country of differing aridity. Grass seeds are a surprisingly reliable source of food despite erratic rainfall in much of the arid zone. Davies (1977a, 1979, 1986) has shown that Zebra Finch populations at Mileura (Western Australia) manage to survive, even in poor seasons, by concentrating on limited parts of the landscape where grass growth and seeding is most likely. These microhabitats normally have some annual grasses germinating and seeding every year and allow some Zebra Finches to survive, but not necessarily reproduce. In good seasons, germination and seeding will be prolific and may also occur in the flood plains away from the watercourses and allow Zebra Finches to breed freely. Zebra Finches appear to specialise on those species of grasses that are adapted to such microhabitats. Not surprisingly, rainfall is patchy and difficult to predict, so when it fails completely in one area, Zebra Finches become vulnerable to starvation and seek seeds elsewhere. Such mobility is essential to survival in arid regions.

Grass seeds, in comparison to seeds of other species of plants, are a 'convenience' food for Zebra Finches. Most grass seeds are a convenient size for easy dehusking, they do not have hard protective or sticky coats, and they lack toxic chemicals (Morton and Davies 1983). Moreover, grass seeds may be produced in great abundance and can persist on the ground in good condition for many months, especially in a dry environment, and form a 'seed bank' of food that can be exploited when needed. The high (80%) carbohydrate content of grass seeds makes them an ideal food for arid-adapted species because they not only provide energy, but metabolic water as well (Chapter 5). Although granivores constitute only 17% of avian species in the arid zone they make up 44% of all individuals and are the most abundant group. However, in completely waterless regions, such as the Great Victoria desert they are almost absent (Fisher *et al.* 1972).

Foraging

Choice of seed

In northern Victoria, Zann and Straw (1984a) found that Zebra Finches preferred larger seeds over smaller seeds. *Echinochloa crus-galli*, the largest seed available, was preferred even when it was old, over small good-quality seeds, such as *Chloris truncata*. There was a significant positive correlation between the percentage of each species of seed eaten by the finches and weight of individual seeds. Laboratory tests on wild-caught finches confirmed this preference.

When domesticated Zebra Finches were raised exclusively on one type of seed, they showed a clear preference for that type when given a choice (Rabinowitch 1969). Similarly, in rigorous experiments Palmeros (1983) showed that domesticated adult Zebra Finches have an initial bias towards that type of seed with which they were most familiar. However, when forced to sample other types of seeds and then offered a choice of four seed types (White Millet *Panicum milateum*, Yellow Millet *Setaria italica*, Japanese Millet *E. crus-galli*, and Canary Seed *Phalaris canariensis*) in equal abundance and accessibility, they preferred the type that provided the highest net gain of energy per unit of foraging time. 'Profitability' was a function of the size of the seed and the ease it could be manipulated or dehusked, that is, minimum handling time. White Millet was preferred to Yellow Millet despite taking longer to dehusk and to swallow; yet its larger size (before husking 3×2 mm *vs.* 2.5×1.5 mm) allowed more grams of seed to be ingested per minute; consequently, the daily energy requirement was attained in a shorter time. Canary seed was the largest seed of all but it was difficult to dehusk. Costs and benefits of foraging choices in terms of rate of food intake is a widespread phenomenon in animals (Krebs and Davies 1993).

In their choice of seeds, laboratory Zebra Finches do not always optimise rate of intake of energy. In an early experiment on seed choice, Morris (1955) offered Zebra Finches White Millet, Yellow Millet and Canary Seed for two weeks, and in each case found a much greater proportion by weight of Yellow Millet was eaten (56–68% Yellow Millet, 21–41% White Millet and 2–4% Canary Seed) despite the fact that the Yellow Millet was smaller and harder to dehusk than the other two. This more natural and longer experiment may account for discrepancies with the findings of Palmeros who tested his birds for a maximum of three days and deprived them for several hours before the experiments began.

Feeding methods

Zebra Finches take seeds individually, mostly ripened ones that have fallen to the ground, but seed on the heads of standing grass is also eaten. Seed heads lying on the ground or within reach are pecked out individually but higher ones are reached by flying up and pecking out seeds one at a time or by perching on branches at the same level as the grass head. Birds may jump up, seize the head in the bill or occasionally grasped it with the feet, and pull it to the ground where it is clamped with a foot until all the seeds are picked off. I have seen birds fly at old grass heads, take a bill-full and land nearby and dehusk three or four seeds before flying up to grab another lot. Some birds, not necessarily pairs, seem to cooperate by taking turns to pull down seed heads and pecking out all the seeds together on the ground. When seed is scarce, Zebra Finches use their bills to dig into the soil or mud in search of buried seed. A sideways movement is used to move sand and husks away. All seed heads are checked and re-checked by hungry birds.

Crop samples show that every seed is dehusked before swallowing. Slow motion analysis of video recordings of domesticated birds shows that the mechanical aspects of dehusking are complex (Nelson 1993). On grasping the seed, the eyelids close for an instant. Next, the tongue positions the seed about a third of the way up from the tip of the bill on one side of the mandible. The tongue rotates the seed so that the margins of the lemma and palea, the two scales that form the husk, contact the edges of the upper and lower mandibles. While held in place with the tongue the two mandibles then close on the seed. Both mandibles have rapid movements with vertical and lateral components and act in synchrony so that resulting shearing forces first crack the seal between the husk and the kernel, and subsequently, further dexterous mandibulations force the edge of the upper mandible down between the lemma and palea and strip them away from the kernel. The husk is allowed to fall from the bill. Positioning of the seed and the mandibular movements vary according the type of the seed and its size.

The crop and the daily seed requirement

The crop, which is a diverticulum with two lobes extending either side of the vertebral column, is easily seen on each side of the back of the neck in Zebra Finches. A diurnal pattern of changes in crop volume was consistent throughout the year in wild Zebra Finches (Zann and Straw 1984a). Although hungry birds could fill their crop to bursting within two hours of dawn, this was not the pattern; rather, there were two peaks in crop volume, one in the late morning and one just before roosting. The afternoon peak was more pronounced in winter than in summer. Birds with full crops held overnight in captivity had empty crops by the next morning and there was also a significant loss in weight of around 1 g. Using dyed seed Cade *et al.* (1965) found that 87–106 minutes elapsed between ingestion and egestion depending on the amount of seed already stored in the crop and gizzard. They also found that between 100–300 mg of seed was normally stored in the crop and about 30–50 mg in the gizzard. Wild birds probably store about 600–1000 mg in the evening in winter.

The daily requirement of seed in domesticated Zebra Finches under standard laboratory conditions is about 3 g (Morris 1955; Calder 1964; Skadhauge 1981; Palmeros 1983; Skagen 1988; Lemon 1993 (2.4 g)), irrespective of day length (Houston *et al.* 1995a). Seed consumption falls to 1–2 g at temperatures between 30°–37°C (Lee and Schmidt-Nielsen 1971). Skadhauge and Bradshaw (1974) found that wild-caught birds ate 4 g per day, but this fell to 3 g when birds where deprived of drinking water. Estimates of digestive efficiency of domesticated Zebra Finches fed on commercial seed ranges from 78% (El-Wailly 1966) to 88% (Lemon 1993). Efficiency increases slightly when Zebra Finches are breeding (El-Wailly 1966) or moulting (Meienberger and Ziswiller 1990). Efficiency also depends on the quantity of food provided and its percentage of protein. When protein content is low, Zebra Finches (but not Bengalese Finches *Lonchura striata* or Java Sparrows *Padda oryzivora*) will conserve it and reduce the amount converted to energy (Meienberger and Ziswiller 1990). The standard metabolic rate of Zebra Finches during the active phase of their diurnal cycle in a metabolic chamber is from 0.80 kJ h^{-1} (Calder 1964) to 0.88 kJ h^{-1} (Cade *et al.* 1965) and the daily energy requirement of a bird in an aviary is 35.7 kJ day^{-1} (Lemon 1993).

In the wild, more seed is needed because of the greater oxygen consumption associated with low temperatures and vigorous activity. For conditions in northern Victoria, Zann and Straw (1984a) calculated that 5.3 g of seed (equivalent to 3,127 *E. crus-galli* seeds) was needed per day in winter; in summer, when a pair might be incubating, the male needed 3.65 g a day (6,643 *A. neesi* seeds) and the female 3.84 g (7,000 *A. neesii* seeds). There are no measures of rate of seed intake in wild Zebra Finches but Palmeros (1983) found domesticated birds ate 3 g of commercial *E. crus-galli* (about 1000 seeds) from a bowl in 30 to 50 minutes, depending on the individual. Lemon (1993) found that the time spent foraging for the daily seed intake (2.4 g) in his domesticated birds depended on searching time. Only 33.5 minutes were needed in the control situation when no husks were added to the supply of seed; however, when husks were added experimentally, foraging time increased dramatically (seed mass to husk mass ratio 1:1, 47.7 minutes; 1:2, 69.4 minutes; 1:3, 120 minutes). The minimum time for wild Zebra Finches to forage for the daily winter requirement of 5.3 g of wild *E. crus-galli* seed would be about 90 minutes. This assumes that patches of wild seed were as abundant as that in the laboratory and the seed itself of equivalent size and quality. Of course, this is not the case, and the calculation does not include search time for such patches. A realistic estimate of total foraging time in winter would be about 120–180 minutes. In summer, patches of *A. neesi* seeds would take longer to find than those of *E. crus-galli*, they would also take longer to dehusk and many more would be needed to obtain the daily requirement so, at a minimum, we are looking at about 180–240 minutes foraging time for this habitat in northern Victoria.

Choice of foraging patch

In northern Victoria, Zebra Finches ranged up to 1 km from the nesting trees in their search for seed (Zann and Straw 1984a), but at Alice Springs, birds came to a walk-in trap from many kilometres away (Zann *et al.* 1995). Seed was patchily distributed in northern Victoria and depended mostly on where moisture accumulated and produced the most luxurious growth of grass. Once a rich patch of seed had been found and exploited, the location may be frequently re-checked long after the bank of seed has been exhausted. For example, our walk-in traps may stand empty of seed and finches for months but a day or so after restocking they are quickly discovered again by many birds. Thus, flock feeding and colonial roosting and nesting may aid in the discovery of seed and feeding knowledge through local enhancement, although this has not been studied in Zebra Finches. In an interesting experiment using domesticated Zebra Finches, Beauchamp and Kacelnik (1991) found, paradoxically, that membership of a mated pair actually hindered the transfer of feeding knowledge of certain types from the skilled, or knowledgeable partner, to the naïve one, since the latter just exploited the knowledge of its partner rather than learning itself.

The size of seed and its density were not the over-riding factor in choice of a foraging patch. For example, wild Zebra Finches prefer patches of seed with fewer husks (Zann and Straw 1984a). To the human eye, the empty husks appear identical to whole seeds because the lemma and palea spring back after dehusking. The finches may experience momentary confusion in seed selection and this increases searching time so that foraging would be faster in patches with fewer empty husks. Naturally, the proportion of husks to whole seeds increase with feeding so that search time in a patch should gradually increase and rate of seed intake diminish. Subsequent laboratory tests on wild-caught Zebra Finches showed that they had a significant preference for feeding at bowls of *E. crus-galli* seed that had fewer husks to whole seed but they did not distinguish between bowls with 60% (by weight) whole seed and 40% whole seed.

Effects of foraging efficiency was investigated experimentally on domesticated Zebra Finches by Lemon (1991) and Lemon and Barth (1992). Those forced to feed inefficiently needed to forage longer each day, which lowered net energy intake and resulted in fewer offspring. These birds not only had smaller broods and longer intervals between successive broods, but shorter lifespans as well.

Nutrient quality must be important in choice of foraging sites because birds seek out green and ripening seed on the heads when dry seed on the ground is available. Moreover, at those study colonies where commercial *Setaria italica* was provided at baited walk-in traps birds tended to visit the feeder on their return from foraging for a 'top-up' (Zann and Straw 1984a). Inspection of the crops showed the recently eaten yellow-coloured commercial seeds lying on top of wild seeds of various of colours. Wild seed was preferred during the peaks of the breeding season when nutrient content of food may be a significant factor in growth of nestlings. At Alice Springs, parents preferred to feed their nestlings wild seed rather than dry commercial seed from the feeder even when the latter was close by and available in abundance (Zann *et al.* 1995).

Nesting activity and diet

Annual changes in breeding and flocking activity at the Cloverlea colony are shown in relation to the flushes of ripening seeds of the main food items found (Figure 4.2). Three main peaks of breeding were evident from September to the following May and coincide with the onset of flushes of seed production. The increasing size of the three peaks, from spring to autumn, corresponded with an increase in the size and abundance of the main species of grass seeds eaten. The first small burst of breeding in early spring coincided with the modest seed set by *Poa annua* and *Cynodon dactylon*, both of which have extremely small seeds (0.33 and 0.23 mg, respectively). In November, *Danthonia caespitosa* (0.44 mg) set seed but breeding fell slightly only to rise steadily as *Amphibromus neesi* seed (0.55 mg) ripened in December and January. The first seeds of *E. crus-galli* (1.69 mg) began to ripen in February but were not produced in abundance until March. The sharp peak in breeding in April

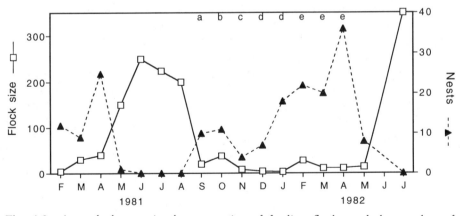

Fig. 4.2 Annual changes in the mean size of feeding flocks and the number of breeding nests by month at the Cloverlea colony, northern Victoria. No samples were made in June 1982. Birds were not fed at this colony. Error bars are not shown but were very small during the nesting period, but quite large outside it. Breeding nests had either eggs or young. The main flushes of ripening seed eaten by the finches are indicated by lettering: a, *Cynodon dactylon*; b, *Poa annua*; c, *Danthonia caespitosa*; d, *Amphibromus neesi*; e, *Echinochloa crus-galli*. (Adapted from Zann and Straw 1984a.)

resulted from a renewed seed set made in response to grazing by sheep.

Both El-Wailly (1966) and Houston *et al.* (1995a) failed to find any differences between the sexes in amount of seed consumed by domesticated Zebra Finches. However, during the laying period, El-Wailly (1966) found an increase in seed consumption by females, whereas Houston *et al.* (1995 a) failed to find any increase at all, although there was a conspicuous increase in the amount of calcium taken from cuttle bone. Skagen (1988) estimated the mean daily seed requirement of nestlings aged 1–14 days post-hatch: a nestling at 5 days requires 0.74 g; at 10 days, 1.83 g; and at day 14, 2.70 g. Thus, parents with five young about to fledge would need to provide 13.5 g of seed daily, in addition to their own requirements.

Feeding flocks

The size and behaviour of feeding flocks are a function of the abundance and distribution of seed and the level of breeding activity. Feeding flocks during the breeding season were much smaller than during the non-breeding season (Figure 4.2). This is partly due to the way breeding activity disrupts the freedom to flock, especially in the morning when incubation and provisioning duties are greatest. There was a significant negative correlation between mean monthly flock size and the number of active nests at the Cloverlea colony in northern Victoria (Zann and Straw 1984a). Mean size of feeding flocks was below 10 in November, December and January, but between 10 and 20 during other months of the breeding season. The large flocks in early spring may include non-breeding adults and those in the second half of the season may include young. Once breeding ceased in April (in 1981) or May (in 1982) there was a dramatic increase in the size of feeding flocks; mean monthly flock sizes ranged from 150–350 birds. These large flocks formed in two ways: a smallish feeding flock would be joined on the ground by other small flocks, or birds would assemble in dribs and drabs in a nearby tree until a large flock was formed whereupon they would descend to the ground and feed as a single flock. There was a significant negative correlation between mean monthly minimum temperatures and flock size during the winter months but none during the summer months.

The flight-distance of feeding flocks varied seasonally at Cloverlea. During the breeding season, small flocks could be approached to within 10 m before they fled, whereas large flocks could not be approached within 15–20 m; some fled when 40 m away. Larger flocks provide an increased level of vigilance over smaller flocks but experiments are needed to determine the size beyond which vigilance no longer increases. Selfish herd benefits would increase with increased flock size.

Competition for food may limit flock size and its dispersion on the ground. In winter, *E. crus-galli* was abundant in large, dense patches at Cloverlea and flocks were at their maximum, whereas smaller flocks existed during the breeding season when food was more scattered and in smaller patches. A similar thing was observed at Alice Springs during a prolonged dry spell in 1986. Flocks were small in most places as food was scarce although large flocks could form in areas distant from the roosting nests. Here they foraged in a coordinated way although they were widely dispersed over the ground (see below), except at one location, where irrigation had produced dense patches of Buffel Grass seed. Here, tight flocks of 60–160 would form, and once I caught 40 birds out a foraging flock of 80 in a single collision with a mistnet. Flock size is thus a function of both avoidance of predators and avoidance of foraging competition. It is interesting to note that large flocks still formed during August at Cloverlea by which time the bank of *E. crus-galli* seed was exhausted, hence competition must have been high. Perhaps large flocks were still an advantage because they enhanced the probability that some large patches of yet unexploited seed may be discovered; alternatively, even if birds were starving at this time they could still reduce their risk of predation by joining a large flock.

When feeding, Zebra Finch flocks are remarkably silent for what is normally a noisy bird and can cause one to startle when a flock is disturbed in long grass. They must be conspicuous from the air since small parties flying high overhead will suddenly change direction and dive down to join a feeding flock. Perhaps the white rump is a releaser. Flocks feeding on seeds in standing grass are conspicuous as birds jump up and pull down heads. A flock feeding on fallen seed move across the ground in one general direction, the speed depending on the density of seed. When high, the movement is smooth and gradual with the birds in front advancing as those behind catch up, but when seed is scarce the movement is fast, jerky and erratic, with individuals towards the rear taking to the air and landing in front of the leading birds. This 'roller feeding' goes in waves with intervals of 20 to 60 seconds.

An undisturbed feeding flock disperse gradually into smaller flocks from two to ten individuals that commute to the nearest trees or back to the nesting colony, or off in search of another patch of seed. If suddenly disturbed, the whole flock instantaneously takes to the air in a deafening din and rushes to the nearest tree or fence where they alight. First impressions suggest chaos, but the flock is highly cohesive and well coordinated with pairs and family groups keeping together so that there is little re-positioning of members of families and pairs after landing.

Mixed-species feeding flocks

In northern Australia, Immelmann (1962b) was impressed by the number of mixed-species flocks of finches that formed during the non-breed-

ing, dry season. These were genuine flocks as distinct from the aggrega-
tion of species around water-holes and rich patches of seeds. Members of
these flocks maintained their coherence and attraction for one another
beyond localised resources. He described three types of mixed finch
flocks: those whose members all had red bills, those with grey bills and
those with yellow bills. He believed that the common bill colour was the
optical releaser that permitted them to flock together. The red-billed
flock consisted of the Zebra Finch and the Star Finch; the yellow-billed
flock, the Long-tailed Finch and the Masked Finch; and the grey-billed flock
consisted of three species of mannikins: the Chestnut-breasted Mannikin,
the Yellow-rumped Mannikin and the Pictorella Mannikin. Strangely,
S. Tidemann (pers. comm.), who for a number of years studied the
Gouldian Finch in the same region where Immelmann made his observations thirty years earlier, failed to observe any mixed-species flocks of
finches, but only aggregations around water holes. Evans *et al.* (1985)
observed Long-tailed and Masked Finches forming mixed-species flocks
in the vicinity of waterholes in the Kimberley, but recorded that the
other species noted by Immelmann (above) maintained single-species
flocks. In eastern Australia it is not uncommon for Zebra Finches to be
seen feeding on dense patches of ripening grass with Double-barred
Finches, yet when alarmed the latter fly to heavier vegetation while the
former stay out in the open.

At the end of May 1986, I observed a mixed-species flock in the
rugged Macdonnell Ranges near Alice Springs. When 40 m up a rocky
slope in Heavitree Gap I heard distance calls of Zebra Finches and saw
about 50 birds flitting among the rocks and spinifex; approximately half
turned out to be Painted Finches. These were darker and more upright
than the Zebra Finches and gave a 'tidit-tidit' call every time they flew a
few metres. The Zebra Finches were more conspicuous; they perched on
tops of rocks and called more frequently. The flock was widely scattered
but appeared well coordinated and moved slowly down the slope for
more than 50 m feeding as they went. An Australian Magpie *Gym-
norhina tibicen* suddenly landed among the flock and they immediately
took to the air in a single dense formation and flew 30 m to a dead tree.
Some Painted Finches then flew off in their own small flocks but the
remainder stayed and gradually moved away with the Zebra Finches,
feeding as they went. Clearly, this was a genuine mixed-species flock,
and interestingly, both species had red bills. Members of both species
appeared to benefit from the relationship in terms of finding food and
avoiding predators, although the following week at the same location I
saw a similar number of finches of both species, but this time they were
in separate flocks.

Juvenile feeding flocks

Small all-juvenile flocks were commonly formed about a week after fledging, but were fairly inconspicuous at my colonies in northern Victoria and Alice Springs. Sometimes young from several families formed crêches in thick trees until called out for feeding by returning parents, but in most cases families kept separate. By the time young are between 30 to 35 days they can usually fly well enough to join their parents in feeding flocks. Young less than 30 days of age are often seen feeding with other colony members up to 400 m from the nearest nesting trees and on only two occasions have I seen all-juvenile feeding flocks away from the nesting sites.

Summary

Zebra Finches are grass-seed specialists and unlike most other estrildines rarely take insects, even when raising young. Seeds from 61 species of grasses are eaten; nine are exotic species, but Zebra Finches have not become crop pests. Seeds of a great range of sizes are eaten and each seed is dehusked before swallowing. There are no sex differences in diet. Food is limiting during drought in central Australia and at the onset of new growing seasons elsewhere. Dietary diversity is quite low with birds in most areas specialising on certain species of seeds; in northern Australia the diet was more diverse than that of other species of estrildines. In the arid zone grass seeds provide a rich and fairly dependable source of food for Zebra Finches. Size, familiarity and dehusking time are the important factors in choice of seed type, but nutrient content is also important. Most seed is taken from the ground, but half-ripe seed is taken from the heads of standing grasses. In captivity the daily requirement per individual is about 3 g of seed while up to 5.3 g is needed in the wild. Search time and seed abundance determine choice of seed patch in the wild. Net energy intake of domesticated Zebra Finches during foraging affects lifetime reproductive success. Half-ripe grass seeds are an important component of the diet of breeding birds. Outside the breeding season feeding flocks of up to 350 individuals may form but flocks of 10–20 are frequent in the breeding season. Mixed-species feeding flocks and all-juvenile feeding flocks are rare.

5 Drinking, water relations and temperature regulation

'The Chestnut-eared Finch is regarded by many bushmen as a good indicator of water. They are great drinkers, and soon after sunrise they resort to the nearest pool or well, where they spend the day, returning to camp towards sunset.'

G. A. Keartland (in North 1909).

Although Zebra Finches have become one of the symbols of the Australian arid zone, they are not considered among the true desert specialists that have evolved in the deserts and can live nowhere else (Frith 1973; Schodde 1982). Zebra Finches are among a number of species, including many granivores, that have more widespread distributions and have penetrated the deserts from surrounding semiarid regions. Selection in these species has tended to favour flexibility rather than specialisation, and this strategy is exemplified in the adaptations and pre-adaptations of Zebra Finches to the three main elements of desert environments—dryness, heat and cold.

Drinking

Tip-down drinking

Immelmann (1962a, 1966) observed that Zebra Finches swallow with the bill tip immersed in the water (Figure 5.1), while most other species of birds bring the bill tip up in order to swallow. The functional morphology of 'tip-down' drinking in the Zebra Finch and 'tip-up' drinking in the Bengalese Finch *Lonchura striata* has been studied in detail by Heidweiller and Zweers (1990) using film and radiogram analysis and their interesting results are briefly summarised below.

During each bout of drinking the bill is immersed in water for an average duration of one second and pulsating throat movements occur for about half a second later as the head is raised. Drinking actions of the Bengalese Finch are superficially similar except that the bill is only immersed for about one third of a second. Essential differences between the species occur during the immersion phase. In the Zebra Finch, the fleshy tongue with its arrow head shape is used to scoop up, then pull back a dose of water into the pharynx. The front of the larynx then immediately forces this dose into the oesophagus where peristalsis transports the water to the crop. This 'double-scooping' or pulling action

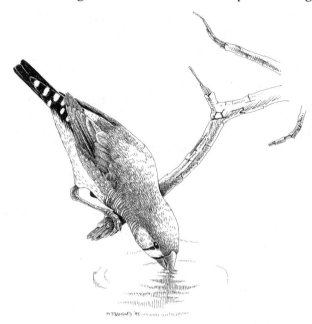

Fig. 5.1 A female Zebra Finch uses the tip-down method of drinking to draw up water from a high position.

occurs while the bill tip is still immersed in the water and the tongue scoops in and out in very rapidly (about 50 ms per cycle) so that 18 to 20 double scoops occur in one immersion of the bill. The Bengalese Finch also uses the tongue to scoop up water into the mouth and pharynx in the same way as the Zebra Finch, but can only do about five scoops before the pharynx is full, then it must withdraw the bill from the water and raise it so that gravity and the scooping action of the larynx transport the dose of water down into the oesophagus. Thus, both species have a two-scoop mechanism but the Zebra Finch integrates both into a continuous, uninterrupted sequence whereas the Bengalese Finch has two disjunct steps. Water enters the mouth of the Zebra Finch twice as fast as that of the Bengalese Finch and because the latter must raise the bill tip in order to swallow the former can get water into the crop about ten times faster. Heideweiller and Zweers (1990) found that in the laboratory, both species drank 1.5 ml per day but the Zebra Finch only needed a mean of 3.6 seconds while the Bengalese Finch needed 39.6 seconds to imbibe the daily requirement. Finally, both species could scoop up very small (0.01 ml) droplets of water equally proficiently. This thorough study also concluded that Zebra Finches do not suck up water by means of peristaltic movements of the oesophagus, as suggested by Immelmann and Immelmann (1967), nor do they drink in the same way as do pigeons. What appears to be sucking is really a continuous 'conveyer belt' of scooped doses entering the oesophagus.

Heideweiller and Zweers (1990) believe that the evolution of graminovory in estrildines provided the anatomical pre-adaptations necessary for tip-down drinking. The anatomical adaptations of the palates, tongue, pharynx and larynx, which enable rapid dehusking and swallowing of grass seeds, also play a crucial role in the evolution of two-scoop drinking. However, only seven Australian estrildines have made the next step to the double-scoop tip-down method, namely the Double-barred Finch, the Long-tailed Finch, the Black-throated Finch, the Masked Finch, the Star Finch, the Gouldian Finch and the Diamond Firetail. Tip-down drinking probably evolved independently several times in estrildines, but only in the Australian members (Immelmann 1962a; Chapter 12). Heideweiller and Zweers (1990) give three main advantages of tip-down drinking:

(1) minute quantities, such as dew drops can be exploited rapidly before they disappear;

(2) water can be drawn vertically upwards from sources difficult to access, and

(3) birds can drink much faster, and so reduce exposure to predators.

Daily drinking

Continuous observation of waterholes has established that most Zebra Finches come to drink fairly regularly throughout the day if water is close by (Fisher et al. 1972). When water is beyond about 5 km some observers have detected an increase in frequency of drinking in the hours around noon (Davies 1971; Evans et al. 1985; Schleucher 1993); this is the pattern of drinking for all other Australian estrildines, except for the Poephila which drink once in the early morning and once in the later afternoon (Immelmann 1965a). Some arid zone species avoid drinking around mid-day, presumably to avoid the high temperatures and intense solar radiation at this time, but these factors do not deter Zebra Finches.

Almost all keen observers of wildlife in the arid zone have remarked on the tightness of the link between the availability of surface water and the presence of Zebra Finches. Zebra Finches are usually found at watering points because daily drinking appears essential to their survival and large numbers congregate around permanent sources during dry spells. For example, Fisher et al. (1972) recorded 17,750 drinking visits in a single day by Zebra Finches at one waterhole north of Alice Springs in September 1967; this number of visits is probably close to the actual number of individuals coming into drink since temperatures did not exceed 23°C on the day the observations were made, so repeated visits by the same individuals are unlikely.

Zebra Finches have been used by thirsty humans to find water (Serventy 1971). The direction flocks take towards water can be followed and small, hidden rock holes can be located by listening for their distance calls. Immelmann (1962a, 1965a) and Fisher *et al.* (1972) maintain that Zebra Finches come to water hourly under summer conditions. This may occur when water is locally available to a breeding colony or feeding flock, but often water is some distance away so that frequent visits are impossible. When I mistnetted for whole days in summer around bores and rock-holes in the Alice Springs district in 1986 I never caught a same-day retrap from 250 captures, nor did I see banded birds return a second time to drink. Furthermore, I have observed hundreds of Zebra Finches coming to drink each day at bores and dams, but when I searched widely in the surrounding country I found few signs of them. Hence, they probably travel considerable distances to water during dry periods, but it is difficult to establish precisely how far they travel. Birds may be observed far away from the nearest known sources of water, but in rocky country it is possible that some minute water hole is hidden away and unknown to all but the finches. However, in sand country where rock holes do not occur, all bores and wells are well known to the local inhabitants so that some degree of certainty can be placed on distances birds must travel to drink. Thus, in the Tanami Desert, some 500 km northwest of Alice Springs, S. Morton (pers. comm.) has seen Zebra Finches roosting at least 25 km from the nearest source of water. Similarly, in the Simpson Desert, Badman (1979) has seen Zebra Finches 27 km from the nearest surface water at Purni Bore (26°17′S, 136°06′E). Finally, S. J. J. F. Davies (pers. comm.) observed Zebra Finches living in parts of the Sandy and Gibson Deserts where there was no water, which lead him to conclude that, providing they can find seed, they do not require daily supplies of drinking water .

Regular surveys of birds conducted by P. Taylor (pers. comm.) in the Simpson Desert around Cowarie Station (27°45′S, 138°15′E) in South Australia provide interesting evidence on the ability of Zebra Finches to move away from sources of surface water. Zebra Finches were found in most survey blocks (17 km × 17 km), but in five there was no surface water and the nearest sources were 12–25 km away. At some sites Taylor observed Zebra Finches 50–80 km from known surface water. The source of water in a number of blocks was not surface water but leaves of a succulent, the Large Pigweed *Portulaca intraterranea*, which is a ground cover species and is common in some areas. Zebra Finches were seen eating these leaves, as were five other species of granivores, namely the Peaceful Dove *Geopelia placida,* the Diamond Dove *Geopelia cuneata*, the Crested Pigeon *Ocyphaps lophotes*, the Galah *Cacatua roseicapilla* and the Little Corella *Cacatua sanguinea*. Presumably, these birds extracted enough preformed moisture to forego daily intakes of surface water. Eco-physiological investigations are needed to understand this phenomenon. Some species of African estrildines, although basically seed-eaters, get sufficient preformed water from their diet of insects to free them from the need to drink for many months (Immelmann and Immelmann 1968).

In central Australia, the crops of Zebra Finches I trapped around drinking sites were often full of water, almost to bursting point in some cases. The precise amount was difficult to determine but at least several millilitres must have been present. When a tube was inserted to extract seeds from the crop, water flowed out freely. I gained the impression that birds were loading up with as much water as they could in order to make a long return journey, and that more could be carried when the crop was not packed with seed. Mid-day drinking makes sense if birds have a long way to travel since an early morning feed would restore energy reserves depleted the previous night and provide fuel for a long trip to water. However, too much seed in the morning would be an unnecessary load on a long flight and would displace precious water on the return leg. Sometime after mid-day, birds would presumably arrive back at the feeding grounds in time to restore the supplies of seed for the night. Thus, it is likely that Zebra Finches in the arid zone tend to roost near the feeding grounds, not near the drinking points and observations confirm this. Of course, in good seasons there will be dependable supplies of both seed and water together, or nearby, so birds would not need to travel much, but as supplies dwindle daily, commuting becomes necessary. I observed the converse of this phenomenon near Alice Springs in May and June 1986. There had been almost no rain for the previous six months, and all but the most permanent sources of water were dry; the ground cover was dead and the earth bare. I observed Zebra Finches arriving at permanent bores and pools in their hundreds. For example, at Simpson's Gap near Alice Springs between 300–400 finches per hour drank from a shallow pool only 1.0×0.5 m. These thirsty birds were fearless and oblivious of the tourists that almost stepped on them as they drank. At the end of June 177 mm of rain fell over a 10-day period. As a consequence, I saw no finches at my dam watches or at Simpson's Gap for the next three months. Presumably, birds drank at temporary sources local to their roosting and feeding sites that had been recharged by the run-off. Casual observers at inland waterholes have the impression that Zebra Finches are resident on a permanent basis, but this could be misleading in most cases due to the continuous commuting traffic from the feeding and roosting sites. Grasslands surrounding permanent water points are usually the first to be exploited by grazing cattle and granivores, and are often in poor condition so that more durable sources of seed are often some distance from the water supply. Capture-mark-recapture studies are needed to determine how frequently Zebra Finches come to water each day and how this is affected by high temperatures.

Drinking and bathing sites

If possible, Zebra Finches avoid drinking at large expanses of water, preferring to drink from rain puddles or tiny shallow pools, but will drink anywhere if necessary. They prefer to drink at those places with gently sloping banks, where after drinking, they usually bathe for a minute or so. Exposed drinking sites are preferred to those surrounded by heavy vegetation, whereas other species feeding in the same area, such as Double-barred and Plum-headed Finches, usually prefer the latter. After bathing, a warm sheltered spot is chosen to dry off and re-oil the plumage. If necessary, Zebra Finches can reach down to water almost 6 cm below the lip of cattle troughs and draw it up even where it is too difficult for larger species of birds to reach. Often they perch on branches sticking out of the water when the sides of rock holes or wells are steep and smooth, but occasionally weakened birds slip in and drown.

Zebra Finches can drink drops from a leaking tap and Immelmann has seen them taking drops of dew from the tips of leaves, but they do not bathe in wet vegetation. On the Canning Stock Route in the Great Sandy Desert in northwest Australia, parched flocks have been observed squeezing through cracks in the covers of wells, flying 5–6 m down in the dark to drink black, putrid water then come streaming vociferously out the gap again a few seconds later (Gard and Gard 1990). Some travellers believe the birds hang around waiting for tourists to arrive to remove the covers from the wells so exposing access to the water below.

Before Europeans arrived, Zebra Finches would have drunk at natural sites such as waterholes in the beds of creeks and rivers, at lakes, ephemeral ponds, springs and at rock-holes formed in rocky outcrops in the ranges. Aboriginal wells and soaks dug into creek beds would have also been exploited. After European settlement more sites became available as the pastoral industry expanded. Initially deep wells were dug, but by the late 1800s the first bores were drilled deep down to tap the artesian water that exists under vast areas of the arid zone. Many bores form boredrains where water under high pressure rises up and flows away in a drain. Other bores have low pressure and water is pumped up by windmill to a tank, either metal or earthen, from where it flows to cattle troughs; surplus water flows to a dam. Some bores are simply a hole in the ground with a tap on top and so cannot be accessed by animals. The Great Artesian Basin in the eastern arid zone had at least 4,700 bores installed. Of 171 bore drains surveyed by Badman (1979, 1987) in northeast South Australia, the driest part of the continent, Zebra Finches were one of the most numerous and widespread of 175 species listed, and only exceeded by several species of migratory waders. Evidently, Zebra Finches travel vast distances to find and exploit new watering points, thus providing further support for the view that daily access to surface water is not a rigid requirement. Nevertheless, where there is no water or succulents across large areas, such as in the Great Victoria Desert, there will be no Zebra Finches (Ford and Sedgwick 1967).

Drinking flocks and predation

Zebra Finches are vulnerable to predators around waterholes and flock in response to this danger. During the non-breeding period around Alice Springs I found the size of flocks flying to water significantly larger than those flying away (median (interquartile ranges) (n): to water 12 (3–25) (15) *vs.* from water 3 (2–7) (51); Wilcoxon Rank Sum Test $z = 2.44$, $P = 0.015$). Several small flocks would often assemble in a nearby bush or tree until numbers had built up sufficiently to fly down to drink for a few seconds then they would dash away. Eight seconds was the longest I observed a bird drink. In the Kimberley, Evans *et al.* (1985) found Zebra Finches and other species of finches spent about half a minute on the ground although they do not say if the bird was drinking all this time. Flocks flying to water are swift and direct and I could usually find my way back to a bore by walking in the direction they took. Black Kites *Milvus migrans*, Brown Goshawks *Accipiter fasciatus*, Collared Sparrowhawk *Accipiter cirrhocephalus*, Black Falcons *Falco subniger*, Australian Hobbys *Falco longipennis*, Brown Falcons *Falco berigora* and Pied Butcherbirds *Cracticus nigrogularis* are the main raptor species that ambush Zebra Finches around waterholes in central Australia. When raptors are abundant, Zebra Finches and other small granivores, such as Diamond Doves and Budgerigars, can be forced to wait in trees around waterholes all morning before drinking in the hottest part of the day, when raptors are less active (Schleucher 1993). Snakes will also prey on drinking Zebra Finches and I have seen a Mulga Snake *Pseudechis australis* waiting for Zebra Finches coming to drink at a puddle beneath a tap. Dingoes will occasionally prey on drinking Zebra Finches (A. Newsome, pers. comm.). In the Kimberley, Brown Goshawks, Blue-winged Kookaburras *Dacelo leachii* and Pied Butcherbirds were the main predators (Evans *et al.* 1985).

Water relations

Daily requirements

While there are no studies of the amount of drinking water needed by free-living Zebra Finches a number of physiologists have examined the problem in wild-caught and domesticated birds. When fed on a standard mixture of air-dried seed and kept in temperatures of 22–23°C, domesticated and wild-caught birds consume 24–28% of their body weight in water over a 24-hour period (Calder 1964; Cade *et al.* 1965; Skadhauge and Bradshaw 1974). This is equivalent to about 3.0 ml of water for an average bird (12–13 g). Naturally, water consumption increases dramatically when temperatures increase and Cade *et al.* (1965) found that when birds were held at 40°C they drank 6–12 ml of water per day. While there was significant individual variation in the amount of water needed, no individual drank 100% of its body weight in one day, but this was the norm for another estrildine, the Black-rumped Waxbill *Estrilda troglodytes*, that was also tested by Cade *et al.*

Although Zebra Finches find fresh water the most potable they will drink water from bores with salinities from 0.15–0.3 M NaCl (Badman 1987; Skadhauge and Bradshaw 1974). Oksche *et al.* (1963) conditioned wild-caught Zebra Finches to tolerate salinities of 0.7–0.8 M NaCl; however, domesticated birds could tolerate only 0.6 M which suggests that the wild birds have a higher maximal renal concentrating ability (Skadhauge 1981).

Oksche *et al.* (1963) found that wild-caught birds could be conditioned through gradual reduction in drinking water to survive for at least several months on 0.5–1.0 ml of water per week at 22–24°C. Similar results were found by Priedkalns *et al.* (1984) and Vleck and Priedkalns (1985), again using wild-caught birds. These were kept in a room at 27°C with a relative humidity of 40% and fed on standard dry finch seed, and, while there was no loss of weight, dehydration caused a significant fall in testis size.

Laboratory birds have been deprived of water completely yet have managed to survive on a diet of dry seed only. Domesticated Zebra Finches held by Cade *et al.* (1965) lived for more than 250 days without water, a finding replicated by Lee and Schmidt-Nielsen (1971) who concluded that their Zebra Finches could survive almost indefinitely providing there was no heat stress. More than half of Sossinka's (1972a,b, 1974) birds survived a total deprivation experiment that ran for 513 days. There was no significant difference in survivorship between the sexes or between domesticated and wild strains of Zebra Finches. Body weight initially fell on deprivation, then increased and later stabilised. Feather condition deteriorated and the bills of a few grew so long that feeding became impossible, and they starved.

Although these laboratory studies in no way suggest that Zebra Finches in nature can exist without regular access to water, they nevertheless demonstrate an extraordinary ability to endure extremes of thirst and dehydration that must enable them to survive long periods without drinking. This reserve capacity may be essential to locate new sources of water when old ones dry up or when seed stocks become depleted. Given the erratic nature of the climate over much of the arid and semiarid zone this capacity should ensure survival.

About a dozen species of granivorous birds have demonstrated the capacity to survive in the laboratory without drinking water providing seed was abundantly supplied. These species have a body mass of less than 30 g and are mostly fringillids or estrildines (MacMillen 1990). This phenomenon puzzled a number of comparative physiologists who set about researching the water economy of desert birds using wild-caught and laboratory Zebra Finches in many instances.

Water economy

Water balance or water flux in birds is an outcome of water intake and water loss and research into this field has been the subject of a number of detailed reviews (e.g. Dawson 1981; Skadhauge 1981; Webster 1991). Birds gain water in a 'preformed' state via drinking and via the moisture content of the ingested seed. The water content of commercial air-dried seed ranges from 7–10% by mass (Morris 1955; Calder 1965), but would be much higher for green and half-ripe seeds and less for old, desiccated ones. Water is also made by the bird when it metabolises seed. Metabolic water production (MWP) is highest when carbohydrate is oxidised. Consequently seed, which is rich in carbohydrate and relatively low in protein and fat, not only supplies the energy needs of the bird but can produce significant amounts of water. Metabolic rate and MWP depend on ambient temperature and the range of temperatures where metabolism in the Zebra Finch is at its minimum (the thermal neutral zone) is 30–40°C (32–40°C, Calder 1964; 36–42°C, Cade et al. 1965). This range is one of the highest recorded in birds and means that when ambient temperatures fall below about 36°C metabolic rate and MWP are progressively increased to the extent that below 23°C dehydrated laboratory Zebra Finches do not need to drink at all because MWP equals the amount lost (MacMillen 1990). If Zebra Finches are not dehydrated through gradual water deprivation, ambient temperatures must drop to 12°C, or lower, before MWP offsets that lost from the body. Thus, the ability of individual Zebra Finches to meet their water needs from MWP depends on their drinking histories. Tolerance to dehydration in Zebra Finches may vary clinally from arid habitats to mesic and humid ones, as well as seasonally within habitats. Serventy (1971) has observed 'drinking clines' in other arid species of Australian birds.

Zebra Finches lose water mainly through evaporation (64%), but also by excretion and egestion (36%) (MacMillen 1990). Water-deprived birds are more efficient at reducing losses than non-deprived birds. The Zebra Finch kidney is highly efficient at extracting water from urine and produces one of the most concentrated urines found in birds (2.8 times more concentrated than blood plasma) due to much of the nitrogen being excreted as crystalline uric acid (Skadhauge 1981). A Zebra Finch dropping is rod-shaped with white uric acid at the anal end of the faeces, all of which is surrounded by clear liquid urine (Lee and Schmidt-Nielsen 1971). The faeces are relatively dry due to resorption of water in the colon and cloaca. The dry weight of faeces produced daily by Zebra

Finches is 0.3 g but water content (% mass) varies from 80% in individuals not deprived of water (Calder 1964) to 55% in deprived ones (Lee and Schmidt-Nielsen 1971); these values are among the lowest found in birds (Skadhauge 1981). Evaporative water loss in the Zebra Finch occurs mainly through the skin (63%, Bernstein 1971; 48% Lee and Schmidt-Nielsen 1971); the remainder is lost from the lungs via expired air. Rates of evaporation relative to oxygen consumption are low in comparison to other species of birds, and Zebra Finches deprived of water can make even greater reductions in evaporative water loss by reducing that lost through the skin but not that lost through pulmonary evaporation (Lee and Schmidt-Nielsen 1971). A lipid barrier in the skin of nestling Zebra Finches prevents, almost completely, the passage of water through the skin; this impermeability is gradually lost after 10 days of age so that by adulthood permeability to water is about 20 to 30 times greater than that of young nestlings (Menon *et al.* 1988). However, if adults are deprived of drinking water for up to six weeks the intracellular lipids are re-mobilised and the impermeable barrier is partly restored so that water loss through the skin is reduced by at least half; the exceptionally low levels found in the naked nestlings are never attained (Menon *et al.* 1989).

Thermoregulation

Mean cloacal temperatures of laboratory Zebra Finches rise with increasing ambient temperatures and range from 38° to 44°C (Calder 1964; Cade *et al.* 1965; Bech and Midtgård 1981; Marschall and Prinzinger 1991). This demonstrates a tolerance of about three degrees to hyperthermia. If body temperatures reach 45–46°C death occurs within one hour (Calder 1964; Cade *et al.* 1965). At high air temperatures, Zebra Finches become completely immobile and initially dissipate heat passively by sleeking the plumage and holding the wings out from the body to allow radiation from the thinly feathered sides and axillar region (Figure 5.2). Evaporative water losses also increase above 37°C. When body temperatures reach 42–43°C, all water conservation measures are abandoned and heavy panting maximises evaporative cooling (Cade *et al.* 1965). Schleucher (1993) observed that heavy panting began in some wild populations when ambient temperatures reached 38°C, and in northern Victoria, birds will even begin panting when shade temperatures exceed 32°C. Zebra Finches have an extraordinary ability to dissipate heat through evaporative cooling and can lose up to 1.4 times as much heat as is produced; this capacity is greater than that known for any other species of passerine (Calder and King 1963). However, water losses are so great (0.6 g of water per hour at ambient temperatures of 43°C; Calder 1964), that birds already hardened by long periods of

Fig. 5.2 Panting posture of a heat-stressed Zebra Finch. Note the extended legs, the sleeked plumage, the open gape and raised carpels.

water deprivation soon become exhausted. Under these conditions, access to drinking water is critical to restore water balance.

In the arid zone during summer, daytime air temperatures recorded in the shade may frequently exceed 40°C and temperatures of 49°C are not exceptionally rare (Serventy 1971). In most instances, these maxima are transitory and birds seek shade and water where they rest and bathe until temperatures fall (Schleucher 1993). For example, over four years at Mileura (Western Australia) Davies (1986) found that no period remained above 37°C for more than 12 hours, even though temperatures ranged from –1°C to 48°C. However, on occasion in large parts of central Australia searing temperatures continue unrelenting for days on end so that birds become exhausted in their attempts to keep their bodies below lethal temperatures with disastrous consequences. The most graphic descriptions of massive deaths come from the record heat wave in January 1932 in northern South Australia. For 16 consecutive days the temperature ranged from 47–52°C, and after the tenth day 'the birds appeared to have become so stupefied by the heat that they dashed straight into the waters, became wet and were unable to rise again' (McGilp 1932). Tens of thousands of bodies of Zebra Finches and Budgerigars were cleaned out of dams and tanks, and more bodies were found beneath trees in surrounding areas. When the zoologist, H. H. Finlayson (1932), arrived by train at Rumbalara (25°20′S, 134°29′E), 600 km north of where McGilp had made his observations, the train was invaded by scores of gasping Zebra Finches, some of which hit the ceiling fans and were killed or maimed; others sought shade under the carriages. More than a thousand birds fluttered about the station and scores were dying everywhere. About 80% were Budgerigars and Zebra

Finches. When water was placed out for them few attempted to drink and Finlayson concluded that temperature, and not dehydration, was the cause of their demise.

Zebra Finches may be more susceptible to extreme heat than some species of birds because their brains may not be kept sufficiently cool. Beche and Mitgård (1981) found that the brain of Zebra Finches is only about 0.2°C cooler than other parts of the body whereas nine other species that have been investigated to date maintained a differential of around 1°C. Correspondingly, the vascular heat exchange mechanism, which is located in the head, is poorly developed in the Zebra Finch by comparison with other species. However, it should be pointed out that the Zebra Finch is the only passerine that has been examined in this respect, so the possibility remains that this is characteristic of all species in the order.

Summary

Zebra Finches are dependent on daily access to drinking water, but some arid populations can live long distances from watering points, and may not need to drink every day. Small pools in open areas are preferred sites for drinking, but Zebra Finches will drink at any location and have exploited the numerous bores drilled by Europeans that tap artesian water across much of the eastern arid zone. Zebra Finches do not drink by sucking, but use a complex 'double-scoop' mechanism that enables them to hold the bill tip down in the water when swallowing. The main advantage is high drinking speed which reduces their vulnerability to predators; it also enables them to exploit minute quantities of water, and to exploit sources of water difficult to access. Fresh water is preferred but Zebra Finches have high tolerances to saline waters in the laboratory and in the wild. Laboratory birds drink about 28% of their body weight each day, but this increases steeply with increased air temperatures. In the laboratory they can survive almost indefinitely without any water providing there is no temperature stress and seed is unlimited. Sufficient water is obtained from dry seed where small amounts exist in a pre-formed state and the remainder is produced from the metabolism of carbohydrate found in the seed. Laboratory work suggests that wild Zebra Finches have an extraordinary reserve capacity to withstand thirst and dehydration that is only drawn upon during periods of extreme stress. Zebra Finches have exceptionally efficient means of conserving water; evaporation through the skin is very low and the kidneys and cloaca produce extra-dry uric acid and faeces. A high thermal neutral zone pre-adapts Zebra Finches to high ambient temperatures. Above 43°C heavy panting maximises evaporative cooling and heat is dissipated very efficiently, but access to drinking water is essential to restore the water balance. Normally, extremes of heat are of short duration and cause no major stress, however, during heat waves of exceptional duration and intensity, evaporative cooling for long periods may cause exhaustion, and death may result when body temperatures reach 45–46°C. Massive deaths across whole populations may occur under these circumstances despite the availability of water.

6 Coloniality and breeding ecology

'Chestnut-eared Finches (Taeniopygia castanotis) *are nesting. Nests containing four or five eggs, everywhere, high and low.'*

H. Simpson, Oodnadatta, 1932.

The behaviour and ecology of breeding Zebra Finches has been the principal focus of field studies to date. Investigations of various duration and intensity have been made in most parts of the range, except for the northeastern region.

Coloniality

Ten Zebra Finch colonies have been studied in detail to date (Table 6.1). Most studies have concentrated on nesting activity and breeding seasonality, but only my studies in northern Victoria have monitored seasonal changes in the size of free-flying flocks.

Nest dispersion

Zebra Finches use nests for breeding and roosting and are one of ten Australian species of estrildines that build a nest specifically for roosting purposes (Immelmann 1962a). A resident pair have a regular roosting nest in a colony which they strongly defend from others in the evening, but after dark they will eventually allow a desperate individual to enter and share it. The breeding nest, in contrast, is always defended. Dispersion of breeding nests varies according to location, availability and type of sites, and density of breeding birds. The majority of potential nesting sites tend to be non-uniform in distribution in most parts of the range. Consequently, it is unreasonable to make estimates of nesting density for every location. However, in four colonies, nesting trees were fairly uniform in distribution and densities could be estimated (Table 6.1). The maximum density found was 76.6 nests per hectare at York, Western Australia, where a small colony bred in a large thicket (150 m × 20 m) of hakeas (*Hakea preissei*), a prickly bush (Immelmann 1962a). The lowest density estimated was 0.7 nests per hectare for a medium-sized colony nesting in a low woodland of Long-leafed Corkwood (*Hakea suberea*) at Alice Springs, central Australia (Zann *et al.* 1995).

Some pairs of Zebra Finches nest away from the main colony. At

Table 6.1 Size of ten Zebra Finch nesting colonies during breeding and nonbreeding seasons[a]

Source	Location[b]	Season	Number of active nests (ha^{-1})	Number of adults Breeding	Non–breeding
Frith and Tilt (1959)	Griffith	1953–54	16 (3.3)	–	–
Immelmann (1962a)	York	1959–60	14 (76.6)	–	–
	Kununurra	1960	24 (28)	–	–
Kikkawa (1980)	Armidale	1961–62	32	–	200[c]
		1962–63	18	–	200[c]
		1963–64	20	–	–
		1964–65	18	–	180[c]
Zann (unpublished)	Shepparton	1975–76	24	33[d]	116[d]
		1976–77	36	63[d]	23[d]
		1977–78	23	–	–
		1978–79	19	37[c]	–
		1979–80	15	–	–
		1980–81	13	–	–
		1981–82	11	–	–
	Bunbartha	1981–82	11	–	70[c]
Zann and Straw	Padgett	1976–77	35	54[d]	27[d]
(1984a)		1977–78	26	27[c], 45[c]	28[d]
		1978–79	13	76[d]	77[d]
		1979–80	11	52[d]	–
		1980–81	15	59[d]	66[e]
		1981–82	7	51[e]	48[c], 60[f]
		1982–83	12	68[d], 112[f]	49[c], 84[f]
	Cloverlea	1980–81	24	36[d]	300[c]
		1981–82	36	48[d]	350[c]
Zann (1994a)	Danaher	1985–86	30	66[e]	20[e]
		1986–87	15	24[e]	67[e]
		1987–88	23	61[e]	64[e]
		1988–89	47	123[e]	94[e]
Zann et al. (1995)	Alice Springs	1986	27 (0.7)	229[e]	130[e]

[a] Breeding season size is based on (i) the number of active nests (eggs or young) during the peak month of the season and (ii) the number of adults captured that month. During the nonbreeding season colony size is based on the maximum number of adults observed in winter feeding flocks near the nesting colony or captured in nets or traps at the colony. In four colonies dispersion of nesting sites permitted estimates of nesting density (nests per hectare). Dash means no data collected.
[b] See Figures 2.4 and 3.3.
[c] Visual estimate of feeding flocks.
[d] Mistnetted.
[e] Walk–in trap.
[f] Estimated from capture, mark, resighting.

York, Immelmann found one pair nesting alone 800 m away from the others and similarly, at Kununurra, pairs nested on their own from 70 m to 100 m from the main colony. At the Danaher colony the majority of breeding birds nested in 12 to 16 small African boxthorn bushes that grew in a ragged hedge about 190 m long. These formed the main colony, but other pairs nested in four to seven satellite colonies of one to two bushes each that grew at distances of 100 to 1,870 m away from the main colony (Zann 1994a). A few pairs nested on their own, hiding them in small bushes several hundred metres from their nearest neighbours. Despite the dispersion of nests at Danaher all the pairs formed one social group, and there was continuous commuting of members between all parts of the colony.

Both Immelmann (1962a) and Kikkawa (1980) reported that breeding pairs at their colonies preferred, where possible, to nest in their own exclusive bush. Pairs vigorously defended the bush against other conspecifics, especially those searching for a nesting site. The ground around the nesting bush was not defended. At Armidale up to 220 separate bushes and trees were used for nesting by one colony over the four years it was monitored by Kikkawa (1980), yet the most breeding nests used in any one month of the breeding season was only 32. By contrast, at my Danaher colony, multiple nesting in the same small bush was the norm rather than the exception (Zann 1994a). The majority of nests were one to two metres apart with up to four or five active nests in the one small bush being quite common. Pairs defended the nest itself and the nest-approach perch but persistent newcomers were allowed to build in the same bush after some initial skirmishes. Incubating and brooding birds regularly interrupted their duties to chase trespassers. Occasionally, two occupied breeding nests would be in physical contact, but the entrances faced different directions thus reducing provocation. In the main colony at Danaher, most nesting occurred in only a third of the bushes while apparently equivalent bushes nearby remained empty much of the time. Favoured nesting bushes tended to change inexplicably from one season to the next. Multiple nests also occurred in the small bushes that formed the satellite colonies. Thus, nesting bushes were not a limiting resource at the Danaher colony and this cannot account for the high level of nesting sociality.

Multiple nests in the one bush have been reported for many locations in the arid zone where suitable nesting sites are often in short supply at some locations. However, rarely is evidence provided on this aspect. McGilp (1944) claimed that 21 nests, all of them active, were found in one *Acacia victoriae* bush at Oodnadatta, northern South Australia. In northwestern Australia, Whitlock (1948) found 13 active nests in one bush and, in central Australia, Immelmann (1962a) reported nine active nests in a single bush. By contrast, the most nests I found in a single corkwood tree at Alice Springs in 1986 was three although these trees were numerous throughout the woodland, and many apparently suitable trees had unoccupied nests or none at all.

Nest predation

It appears that resolution of the forces of attraction and repulsion between breeding pairs of Zebra Finches clearly differ within and among colonies across the distribution. Pressure from nest predators appears to the most important determinant affecting nesting sociality in Zebra Finches. Unfortunately, there are few data on the species of predators involved, their density and seasonal abundance although most predation occurs in the breeding season. Nest predators at the colonies in northern Victoria included dragon lizards *Amphibolurus* spp., Tiger Snakes *Notechis scutatus*, Brown Snakes, rats *Rattus rattus* and mice *Mus musculus*; Australian Ravens *Corvus coronoides* and Brown Goshawks were seen near nests and presumably were responsible for the destruction of those nests torn apart. At Alice Springs, nest predators included Grey-crowned Babblers *Pomatostomus temporalis*, Yellow-throated Miners *Manorina flavigula*, Little Crows *Corvus bennetti*, Torresian Crows *Corvus orru* and Pygmy Mulga Monitors *Varanus gilleni*. At York, Immelmann (1962a) found that the Singing Honeyeater *Lichenostomus virescens* was the chief nest predator while unspecified species of snakes, goannas and carnivorous marsupials and rodents were the major predators at his Kununurra study site in the Kimberley. Barn Owls *Tyto alba* will take roosting adult Zebra Finches; for example, in central Queensland 65 Zebra Finch skulls were found in 25 regurgitated pellets at one roost (P. Woolley, pers. comm.).

At the Danaher colony rates of nest predation were found to be significantly lower at bushes in the main colony than at equivalent bushes in the satellite colonies (Zann 1994a). Re-nesting pairs were significantly more likely to move to a new bush more than 20 metres away if their first attempt suffered predation than if it did not; however, they were conservative in this respect, with many simply re-nesting in the same bush, some even on the same site.

Zebra Finches rarely attempt to repel potential predators; they have never been observed mobbing in the wild, but something resembling it has been observed in captivity (Lombardi and Curio 1985a,b). Zebra Finches attempt to reduce predation by making the nest difficult to find and/or difficult to reach. If this is not possible their only recourse is to nest socially in order to exploit the 'selfish herd' effect. This only appears to be effective when predation levels are moderate. When vegetation of the nesting bush or tree is dense, nests tend to be placed in inconspicuous positions and may be dispersed fairly uniformly throughout a colony thereby making it difficult for predators to find. This strategy appears to operate at the York, Armidale, Padgett and Cloverlea colonies (Table

6.1). Where nesting trees are thinly vegetated, as is the case at Shepparton and Danaher, and in many parts of the arid zone, nests are almost impossible to hide and are frequently clumped loosely together, presumably to exploit the dilution effect should they be discovered by a predator. It is unlikely that colonial nesting would saturate large local predators.

Irrespective of whether Zebra Finches attempt to hide their nests or not, all prefer thorny or prickly bushes as a nesting site if these are available. This hinders but probably does not prevent nests from being robbed. Where rates of predation are high, Zebra Finches exploit exotic nesting sites that predators avoid or cannot access. In northern Victoria, for example, Zebra Finches at several colonies have forsaken the formidable yet still vulnerable defence provided by boxthorn bushes to nest in hollow, steel crossbeams that support insulators on power poles along roadsides (Zann and Rossetto 1991; Zann 1994a). In addition, most nest predators appear to avoid places of human habitation, such as farmhouses and outbuildings, and the vicinity of large raptor nests where they themselves may fall prey. Both sites appear to attract breeding Zebra Finches and suggests that they confer lower rates of nest predation (Zann and Runciman 1994). Immelmann (1962a) thought that the tendency of Zebra Finches and several other species of estrildines to nest in sites over water and in the vicinity of nests of stinging wasps is a further attempt to make nest predation more difficult.

Establishment of breeding colonies

Immelmann (1962a) witnessed the establishment of a breeding colony at York at the start of the 1959–1960 breeding season. During the winter he found that birds slept in roosting nests scattered throughout clumps of hakea bushes growing in a sheep paddock. The roosting colony was normally deserted during the day, but as temperatures rose with the approach of spring, some pairs left the feeding flock some kilometres away and returned to the roosting colony where they searched for a suitable breeding site, usually the foundations of an old nest. Once a site had been chosen they would confirm it with nest ceremonies (Chapter 9). The new nest was constructed in the following days; sometimes a pair would renovate the roosting nest while still constructing the breeding nest, despite the fact that the two nests could be at different ends of the colony. When the breeding nest was completed the pair would spend their first night there, but only after extensive nest ceremonies on the first evening. The first breeding nests were built as far away from other breeding pairs as possible but inter-nest distances gradually fell as new pairs settled at the colony. In northern and central Australia, Immelmann found that breeding nests were built more rapidly and more synchronously than those at York. These birds had one nest only and simply converted the roosting nest into a breeding nest.

Duration of colonies

In the non-arid parts of Australia, most Zebra Finch colonies are permanently occupied throughout the year. Numbers, of course fluctuate, but in the non-breeding season nests are used for roosting and some breeding pairs are found nesting throughout the long breeding season. Colonies may persist for a number of years. Kikkawa's (1980) colony at Armidale existed for at least four years and several of my colonies in northern Victoria have been continuously occupied for at least 10 years.

Size of colonies

The number of pairs breeding in a colony varies throughout the season but the number of active nests (eggs and/or young) found in the peak month of breeding indicates the maximum size of the breeding colony (Table 6.1). This varied from seven to 47 nests in the ten colonies studied to date. The number of breeding pairs does not seem to be determined by supplies of seed bait used at those colonies with walk-in traps. The number of adult birds roosting at a colony is difficult to determine during the breeding season since feeding flocks are fragmented and birds may roost on skimpy platforms or other nest remains that are easily overlooked. Nevertheless, trapping data show that many more adults occur than there are breeding nests and suggests that non-breeding birds may be present at the breeding colonies. These may be dispersing birds in transit, immigrants about to make a breeding attempt, or visitors from distant colonies. The seed bait provided at the trap may be the stimulus that attracts and keeps some of these birds. At Danaher there was no relationship between the number of pairs nesting each season and the number of adult and young that arrived from other colonies (Zann and Runciman 1994).

After the breeding season, most colonies increase in numbers of adults; some almost tenfold whereas a few colonies fall in numbers. There appears to be no consistent relationship between trapping method and numbers of adults in the non-breeding season (Table 6.1).

Annual changes in colony composition

Numbers of Zebra Finches found at colonies, and their age class, change significantly over the course of a year. The pattern was fairly consistent in the two colonies in northern Victoria that were monitored monthly for a number of years (Figure 6.1). Nestlings were found from spring to autumn, but few of those in the spring months reached the free-flying

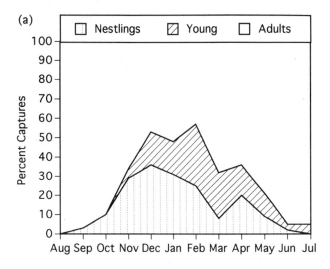

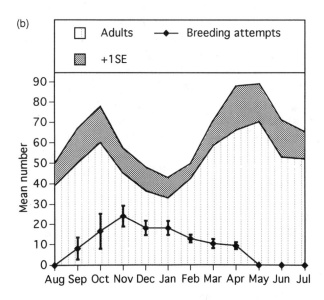

Fig. 6.1 Annual changes by month in the composition of Zebra Finches at the Danaher colony over the period 1985–1989: (a) age structure (*n* = 1,585); (b) mean number of resident adults (*n* = 547) and breeding attempts (at least one egg: *n* = 467 attempts). Residents were those recaptured ≥28 days after initial capture. Nestlings: 12–20 days of age; young: 20–80 days; and adults >80 days. Birds were trapped in walk-in traps at least once every month except for winter 1986. Nests were monitored weekly in the breeding season and monthly in the non-breeding season.

stage. There were three peaks in numbers of young over the breeding season: early and late summer and early autumn. Peaks and troughs in numbers of nestlings and free-flying young coincide, rather than being offset as one would expect as cohorts move through the age classes. This is probably due to the fact that the peaks of breeding differ slightly from one year to the next so that sequences of cohorts are obscured.

The number of adults peaked in mid-autumn, probably due to the maturation of young of the season. This number fell in late autumn and early winter due to dispersion and mortality, but peaked again in mid-spring due to immigration; numbers fell again in the middle of the breeding season, presumably as fewer immigrants took up residence in the previous months due to breeding commitments.

Roosting nests

A typical roosting nest is approximately spherical with an outer diameter ranging from 8–15 cm with walls and roof seldom exceeding a thickness of 1–2 cm. The side entrance is wide and may have a slight overhang or awning, but in many cases only half the roof is present. Visitors may be forced to overnight on an old nest floor or platform. Nests are built of grass stems or twigs, whatever is locally available. Most roosting nests are simply the remains of old, dilapidated breeding nests in which the walls and roof are partly renovated. Breeding nests in a reasonable state may be modified for roosting by destruction of the entrance tunnel and enlargement of the entrance itself (Kikkawa 1980). Purpose-built roosting nests are smaller and more flimsy than breeding nests with the number of stems involved in the construction ranging between 180–350 (Immelmann 1962a). Structure of roosting nests varies throughout the range, for example, neither Immelmann nor Kikkawa found roosting nests lined with feathers or wool but this was the norm at my colonies in northern Victoria; furthermore, Kikkawa could distinguish un-mistakenly a roosting nest from a breeding nest by its wide entrance, but this was not possible at my colonies in northern Victoria and central Australia.

Micro-climate of roosting nests

Roosting nests may reduce heat loss of sleeping Zebra Finches and could be an important adaptation to desert environments where cold stress is often more a problem than heat stress (Davies 1986). At Armidale, Kikkawa (1980) found that roosting nests increased in number as mean minimum monthly temperatures fell and diminished as temperatures rose. At the Danaher colony in the winter of 1989 we found that ther-mocouples placed in the air space just below the roof of the nest chamber of occupied roosting nests registered overnight temperatures 1–7°C higher than ambient temperatures in the bush (Zann and Rossetto 1991). The roof and walls of the nest, thin as they are, must reduce radiation loss from the roosting occupants. Vleck (1981) calculated that laboratory Zebra Finches roosting in nests save on average 18% of energy used when roosting without a nest. Air temperatures in the nest chambers of roosting nests were also significantly higher than those of breeding nests in the same bush despite the fact that the chamber walls and roof were thinner. This suggests that more heat escapes from birds roosting together than from breeding birds incubating alone. One advantage of roosting in nests over roosting on perches may simply be that the nest allows more individuals to huddle together and thus conserve body heat by exposing less surface area to cold air (White *et al.* 1975). Most adults nest in pairs or family groups and up to seven fully grown Zebra Finches may roost in the same nest, although this is exceptional.

Daily pattern of colony activity

The structure of colonies and the daily pattern of activity observed in northern Victoria largely conforms to that described by Immelmann (1962a) for his two Western Australian colonies. Around sunrise, and well after most other species of birds have left their roost, distance calls (Chapter 10) suddenly erupt in volleys in the scattered roosting and breeding nests across the still colony, and, after a minute or so, birds emerge and fly in pairs and small groups to a tall, dead or sparsely vegetated tree in the centre of the colony that serves as the 'assembly point' for departure from, and arrival at, the colony. After noisily congregating for a few minutes they depart in groups of two to ten individuals after which the colony falls silent again.

Time spent away from the colony depends on the season and weather. On cold days in the non-breeding season in northern Victoria a colony could remain empty until about one hour before sunset (Zann and Straw 1984a). After about two hours of feeding, the flock usually breaks up into small groups to rest, preen, sing and court in nearby trees and bushes for a variable period after which they re-form into a large flock and foraged uninterruptedly for the last three to four hours before sunset. On fine, sunny days small groups may return to the nesting colony during the mid-morning break where they nest-build, court, sing, rest, etc. On cold, windy days they spend longer away from the colony in the morning but return earlier in the evening for roosting. Flocks land in the assembly trees and quickly and quietly retire to their roosting nests although they may not enter these until shortly after sunset.

In the breeding season, pairs incubating eggs or brooding young may be away with the feeding flock for less than an hour or so after leaving the roost. On cold mornings egg temperatures may drop significantly when nest attendance is low (Zann and Rossetto 1991). Nest-building pairs and those searching for a nesting site also return early from the feeding flock. Pre-laying pairs may delay their departure with the feeding flock and fly to special 'courting' trees or bushes where they court and mate in relative privacy (Chapter 9). Some young and non-breeding adults may stay in the feeding flock for most of the day, but after several hours of feeding there is continual traffic between the colony and the feeding flock. In the afternoon, social activities predominate to the extent that nests with eggs or young may be unoccupied for an hour or so, but eggs with thermocouple implants failed to show noticeable falls in temperature at this time (Zann and Rossetto 1991). The colony usually has one or two 'social' trees or bushes that serve as centres for resting, preening, allopreening, singing etc. In southern Australia these sites are sheltered from the cold wind and trap the late afternoon sun. The branches of the social trees have fewer leaves than typical nesting bushes and allow for comfortable perching of many flock members. Our telescopes were often trained on the social tree in order to determine mated pairs among colour-banded individuals. Immelmann (1962a) also observed regular bathing and drinking sites with favourite trees for post-bathing preening, but I have not observed these in northern Victoria.

Breeding pairs

Over four breeding seasons at the Danaher colony, breeding attempts of 144 unique pairs (122 males and 123 females and 21 re-pairings) of colour banded birds were followed. Only 19% of males and 28% of

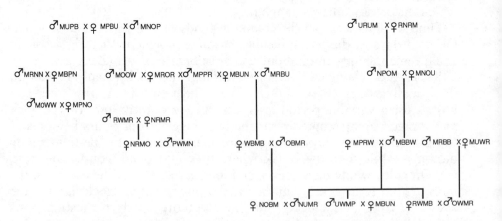

Fig. 6.2 A genealogical tree showing four generations of Zebra Finches at the Danaher colony, Victoria, over the period 1985–1989. Birds with more than one partner only re-mated after their first partner had disappeared. Female MBUN, a resident between 1987 and 1989, appears twice in the tree, having mated with her grand-daughter's 'brother-in-law'.

females were hatched at Danaher, the remainder having immigrated from elsewhere (Zann 1994a). With such an outbred population few family pedigrees could be established, despite the potential for rapid production of generations. Parents of both partners were known in only six pairs and in 12 pairs both parents of one partner were known. A sample pedigree is illustrated in Figure 6.2.

In one pair, parents were brother and sister from the same clutch. They hatched in autumn 1988 and made several unsuccessful breeding attempts with unrelated partners in the spring, but these disappeared, and the following autumn brother and sister made two breeding attempts together; the first clutch suffered predation and only one egg of four in the second attempt hatched. The one nestling fledged successfully. DNA fingerprinting showed that both brother and sister were full sibs (Chapter 9). When brother and sister paired it is unlikely that it was a case of mistaken identity. If cues for kin recognition exist they could have been learnt since both siblings were raised in the same clutch; moreover, the brother had faithfully learnt his father's song and Distance Call (Zann 1990) so that these diagnostic family traits would have been available to the sister at re-pairing.

Pair bonds

Except when incubating and brooding, pairs of Zebra Finches are inseparable in both breeding and non-breeding seasons. At Danaher, 65% of the 144 uniquely colour banded pairs identified making breeding attempts formed initial bonds, that is, the first bond observed for either partner; bonds of the other 35% of pairs were either subsequent bonds for both partners, or a combination of initial or subsequent bonds (Zann 1994a). Breeding pairs that immigrated to Danaher after the age of pair formation (68% of immigrants) may have had previous partners at previous colonies.

Fifty re-pairings were detected during the Danaher study. These only happened when a partner disappeared from the colony and, presumably, had died. No bird was ever observed to have more than one mate at any one time, therefore, social monogamy and divorce by death was the rule. In 23 cases females re-paired, and in 27 cases males re-paired. These findings are the first confirmation of the long-held belief that Zebra Finches are monogamous and pair for life (Immelmann 1962a, 1965a).

Most birds (85%) had one only partner during their term at the Danaher colony; a few had two or three and one female had four. There were no significant differences in the number of mates acquired by the two sexes (Zann 1994a). Replacement mating could be more common than this, for in 97 pairs both male and female disappeared simultaneously. Both may have died together or, more likely, emigrated together, so nothing is known about how their pair bonds terminated. Since wild

Zebra Finches may live up to five years (Chapter 8) there is ample time for multiple mates, given the fact that high mortality frequently ruptures bonds.

Age at pair formation and first breeding

Joint breeding attempts, mutual and exclusive allopreening of one another, and defence of the partner were evidence of a pair bond in wild Zebra Finches. The earliest bonds form between 50 and 60 days of age, shortly before full adult plumage is attained in males. The youngest bonded female observed was 51 days old and the youngest male 60 days old. In 47 pairs, where the age of one or both partners was known or could be estimated accurately, the male was oldest in 31 pairs and the female in 16 pairs ($G_1 = 2.44$, n.s.). The age difference at pair formation could be large: in one pair the male was a mature adult (at least 100 days of age, or older) and the female only 51 days; in another pair the female was a mature adult and the male only 65 days of age.

In a pair formation study of domesticated Zebra Finches Schubert et al. (1989) found that females preferred older males regardless of experience, and argued that older males may possibly have an advantage over younger males because they have more 'polished' courtship displays, and would be more competent foragers and nest builders. The Danaher data show a weak, non-significant trend in this direction, but availability of mates and other factors blunt its expression.

The age at which Zebra Finches made their first breeding attempt at Danaher depended on whether they hatched in the first or second half of the breeding season (Zann 1994a). Fifty-four per cent (37/69) bred in the same breeding season in which they themselves hatched and the remainder waited until the following season (Table 6.2). The youngest female laid her first egg at 62 days of age; the youngest male was 67 days old

Table 6.2   Age of Zebra Finches in days on the date of the first egg of the first breeding attempt; after Zann (1994a)

Season of first attempt		Males	Females	Wilcoxon Two Sample Test z (P)
Current	median	95	92	0.41 (0.68)
	range	67–139	62–162	
	n	18	19	
Following	median	300.5	266	2.37 (0.02)
	range	229–362	197–373	
	n	14	18	
Total	median	129	162	0.57 (0.56)
	range	67–362	62–373	
	n	32	37	

when his partner laid her first egg; the median age was approximately 90 days in both sexes. There was no significant difference between proportions of novice males and females that bred in the current season and the proportion that waited until the next season. For birds breeding in the season of hatching, there was no significant difference in age of first breeding between males and females, yet among those that held over to the next breeding season for their first breeding attempt, females bred at a significantly younger age than males (Zann 1994a).

Precocial breeding of Zebra Finches has long been known among aviculturalists, but only isolated instances have been reported in the wild. Immelmann (1962a) mentions a colour-banded female at York breeding when 86 days old; at Armidale, Kikkawa (1980) found a male and a female that hatched in spring making breeding attempts in the following autumn. During the second half of the breeding season at Danaher, precocial breeders comprised 44% of all pairs making breeding attempts and thus contributed significantly to the breeding effort of the colony. Only four other avian species are know to breed in their season of hatching. The earliest known age of breeding in birds is 37 days post-hatch in the Zitting Cisticola *Cisticola juncidis* (Ueda 1985) where precocial female breeders make a vital contribution to breeding output at the end of the season. In years of exceptional rain, two species of Darwin's Finches, the Medium Ground Finch *Geospiza fortis* and the Cactus Finch *Geospiza scandens*, were found breeding at 81 and 89 days of age respectively (Gibbs *et al.* 1984). Occasionally in Europe, Quail *Coturnix coturnix* that hatch in early summer are believed to breed in the early autumn (Cramp 1980) and it is well known among captive breeders that young of this species become sexually mature at around six to eight weeks after hatching (Kovach 1975, Shephard 1989).

While both domesticated and wild Zebra Finches become sexually mature and may begin to breed between two to three months of age, other species of Australian estrildines appear to wait until at least six months of age (Chapter 12). There are no data on age at first breeding in wild estrildines other than Zebra Finches, but the avicultural literature gives age of maturation and first breeding of six to twelve months (Queensland Finch Society, 1987).

Not all maturing Zebra Finches attempt to breed before the end of the season. Seventy-two per cent (33/46) of Zebra Finches that hatched before 1 January, the mid-point of the eight-month-long breeding season, attempted to breed before the end of the current season (Zann 1994a). Only 21% (5/24) of those hatched after this date attempted to breed. December was the pivotal month: 8 of 21 that hatched in December bred in the current season, the remainder waited until the next season (Figure 6.3). There was no significant difference in month of hatching between the sexes for those that attempted to breed in the season of hatching.

Clearly hatching date is an important factor in determining whether a

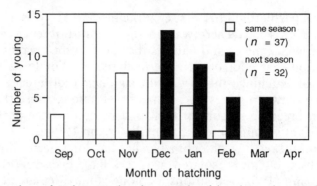

Fig. 6.3 Number of Zebra Finches by month of hatching that made a breeding attempt in the season in which they were hatched (same) or held over their first breeding attempt to the following breeding season (next). (After Zann 1994a.)

youngster will attempt to breed in the current season or defer to the next season. There will be the classic trade-off between costs and benefits of current reproduction versus survival of the parent and its offspring to the next breeding season. Foraging skills and seasonal abundance of the wild seed necessary to feed nestlings should be important considerations in the decision to defer breeding to the next season or not. At locations where there was a superabundance of wild seed in late autumn, such as at Cloverlea, one would predict that the pivotal month for deferring breeding would be later than that found at Danaher where wild seed was less abundant. There should be strong selection for parents to breed as early as possible in the breeding season to enhance the possibility of producing at least two generations per season, and thus increasing their reproductive value.

'Dispersal' distance

Sixteen males and 24 females banded in the nest made their first breeding attempts in the Danaher colony. One male and two females made the attempt in the same bush in which they were hatched while 15 males bred in bushes ranging from 165–1,815 m (median 653 m) away; 22 females bred from 105–1,815 m (median 505 m) from their hatching bush. Differences in dispersal distance between the sexes were not significant.

Nesting

Nest structure

Zebra Finches build a typical estrildine breeding nest: a spherical, domed egg chamber of dead grass stems with a side entrance to which is

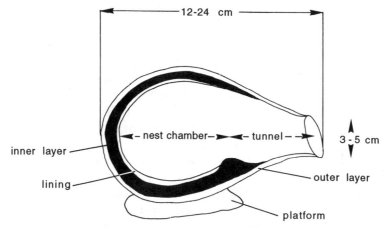

Fig. 6.4   Structure of a typical Australian Zebra Finch breeding nest.

attached a short horizontal tube or entrance tunnel (Figure 6.4). Nests tend to be on the untidy and superficial side although there is enormous variation within and among colonies in dimensions, construction materials and robustness. Immelmann (1962a) summarised the dimensions and construction of 84 breeding nests from different parts of Australia and found that the basic structure was similar to that of the Double-barred and Long-tailed Finches. Length varied between 12–24 cm and the outer diameter of the egg chamber ranged between 12–20 cm; the length of the entrance tube varied from 0–8 cm. The internal diameter of the tube was 3–5 cm. A raised lip separates the egg chamber from the floor of the entrance tunnel and prevents eggs from rolling out (Figure 6.4). The egg chamber rests on a foundation, usually an old nest, although occasionally new foundations will be laid for nests constructed *de novo*; here numerous, short, stiff stems are criss-crossed over horizontal branches. The walls, which range in thickness from 1–3 cm, consist of an outer structure of long (15–20 cm), stiff, coarse grass stems, and an inner structure of numerous shorter (5–10 cm), soft, fine stems; some individuals use small twigs or leaves for the outer wall, but coarse leaves are not used. Between 180–500 separate components may be used in the nest chamber and tunnel. The floor and lower walls of the egg chamber are lined with feathers, mostly white, plant 'wool'; sheep's wool is used where available, but this can cause problems when it gets wet and may dampen the nest to such an extent that it kills eggs or young or is deserted by the parents. In central Australia, Immelmann (1962a) found breeding nests smaller and more flimsy than elsewhere, but I found all sizes and qualities there.

Nest sites

Given their extensive distribution and range of habitats it is not surprising that Zebra Finches are the least specialised of all the Australian estrildines in their choice of nesting sites. Where possible they prefer to anchor the nest in densely branching thorny shrubs and small trees, siting the nest in vertical forks or between horizontal branches on the periphery. In many parts of the range suitable nesting sites are limiting and almost any site is used, including bizarre ones such as clumps of dry seaweed on a beach, rabbit holes, holes in cliffs or even the bleached skulls of cattle; nevertheless, they rarely, if ever, nest in grass.

Unless cited specifically the following descriptions of nesting sites are based on my own observations and those extracted from the RAOU Nest Record Scheme. In the more settled parts of southeastern and eastern Australia, Zebra Finches frequently nest in introduced species of trees and shrubs including fruit trees, (especially *Citrus* and grape), thorny weeds (boxthorn, briar rose, gorse, hawthorn, blackberry, brambles and thistles), and ornamentals (e.g. pinus, cupressus, tamarisk, schinus, cotoneaster). Of the native shrubs, Melaleucas provide the most common nesting sites, especially those dense forms, such as M. *armillaris*, M. *linearifolia*, and M. *styphelioides*, which are planted as windbreaks in farmland or along roadsides. Thorny species such as *Hakea* spp. and native blackthorn *Bursaria spinosa* are also used for nesting. In swampy areas, lignum, *Muehlenbeckia* spp. often provides suitable sites and, in rare instances, cumbungi *Typha* spp. will do. In north Queensland, Zebra Finch nests are frequently found in two exotic thorny weeds, *Zizyphus mauritaniana* and *Parkinsonia aculeata*, and a thorny native, *Acacia farmesiana*. In southwest Western Australia dense clumps of the prickly *Hakea pressii* were the only sites where Immelmann (1962a) found Zebra Finches nesting.

The flimsily made nests are not suitable for most species of eucalypts because the sparse branching provides poor anchorage for the foundations and they quickly get blown to the ground. Nevertheless, eucalypts are used if the nest can be lodged in a hollow or in dense clumps of fruiting bodies or in parasitic mistletoe (Family Loranthaceae). In the Kimberley region Zebra Finches nest in trees of the following species *Eucalyptus papuana*, E. *camaldulensis*, E. *dichromorphloa*, E. *microtheca*, *Bauhinia cunninghami*, *Acacia bidwilli*, and *Carisa lanceolata* (Immelmann 1962a). Immelmann believed that Zebra Finches nest higher above the ground in northern Australia than in other parts because of the greater abundance of nest predators such as snakes and goannas.

Across much of the arid zone the preferred nesting sites are small prickly trees of three species, *Acacia victoriae*, A. *tetragonaphylla* and *Hakea suberea*. In addition, *Capparis mitchelli* is commonly used in the

Alice Springs region, and *Eremocitrus glauca* and *Apophyllum anomalum* in western New South Wales. *Acacia aneura, A. cambagei, Casuarina decaisneana, Atalaya hemiglauca, Santalum lanceolatum*, and *Melaleuca glomera* are the principal non-spiny species in the arid zone where nests are found. Zebra finches will also nest in dense, or prickly, chenopod clumps including *Atriplex nummularia, Maireana* spp., *Chenopodium nitrariaceum, Rhagodia* spp. and *Bassia quinquecuspis*.

Where suitable sites are not available Zebra Finches will use those made by other species including Welcome Swallows *Hirundo neoxena*, Yellow-rumped Thornbills *Acanthiza chrysorrhoa*, Southern Whitefaces *Aphelocephala leucopsis*, European Goldfinches *Carduelis carduelis*, and Australian Ravens *Corvus coronoides*. At Kulgera, in the Northern Territory, Zebra Finches established a small colony of four nests on the remains of old mud nests of Fairy Martins *Cecropsis ariel* placed high up on the piers of a bridge in a region that provided few suitable nesting places. Zebra Finches will also exploit nesting hollows excavated in termite mounds by species, such as the Red-backed Kingfisher *Todiramphus pyrrhopygia*. In some cases Zebra Finches will not wait until the rightful owners have vacated their nests and may encounter resistance, with protracted battles ensuing (Nielsen 1959).

Zebra Finches are renowned for nesting in the foundations of raptor nests even where suitable nesting bushes are available nearby and while the owners are engaged in their own breeding attempts. Nests have been found in nest foundations of the following species: Black-shouldered Kite *Elanus axillaris*, Black Kite, Whistling Kite *Haliastur sphenurus*, Black-breasted Buzzard *Hamisrostra melanosternon*, Wedge-tailed Eagle *Aquila audax*, Little Eagle *Hieraaetus morphnoides*, Spotted Harrier *Circus assimilis*, and Nankeen Kestrel *Falco cenchroides*. In southwest Queensland, 11 Zebra Finch nests were found in a nest of a Wedge-tailed Eagle, and at least four were active.

It is not uncommon for nesting Zebra Finches to be attracted to busy human settlements even when natural sites are available nearby. For example, nests are often found beneath verandahs of houses and sheds, where they are located in hanging plant containers, in roof spaces and rafters; nests are also found in hollow fence and gate posts, in pulley blocks on moored fishing trawlers, in insulator cross-bearers on power poles, up irrigation pipes, and in tractors, old car tyres and car bodies. Even the fuselage and engine cowlings of newly parked aircraft on airstrips have provided nesting sites. At all these sites the birds fearlessly go about their business, especially during the building stage, and tend to ignore most human presence. North (1909), for example, describes a pair nesting successfully amid the din of a busy blacksmith's shop. Of course, they become furtive and wary should anyone attempt to make detailed observations.

Micro-climate of breeding nests

When ambient temperatures were mild and there was no wind, temperature probes placed in the nest chamber of occupied breeding nests at the Danaher colony did not detect any significant difference between nest temperature and from that measured in the bush in which the nest was located (Zann and Rossetto 1991). However, in late winter and early spring when temperatures were low, the air in the top of the chamber of the breeding nest was two to three degrees higher than that of the bush. Furthermore, temperatures would be much higher on the floor of the nest where insulation is thickest. Over much of the range, the enclosed nest may prevent solar radiation from increasing temperatures to lethal levels since many nesting trees have thin foliage and provide little shade for the eggs and young.

In a constant temperature chamber, Vleck (1981) found that a grass nest inside a metal can was three degrees warmer for a range of temperatures from 8–35°C and led to considerable metabolic savings.

Clutch size

Zebra Finches lay a modal clutch of five eggs, but range in number from two to seven (Figure 6.5). Only three studies of clutch size have been made to date. At Griffith, New South Wales, Frith and Tilt (1959) found a mean of 4.7 ± 1.1 (*s.d.*) eggs in 221 clutches examined over three breeding seasons. Clutch sizes were significantly smaller at the start and

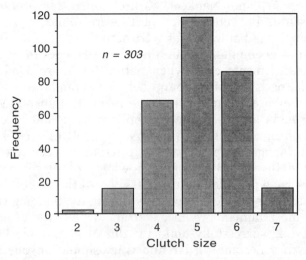

Fig. 6.5 Number of eggs laid per clutch in Australian Zebra Finches at the Danaher colony in northern Victoria over four breeding seasons.

end of the breeding season, presumably when conditions were not optimal for breeding. At the Padgett colony in northern Victoria, I found a
mean of 4.85 ± 0.83 eggs (n = 112), which was not significantly different
from that at Griffith. These Padgett clutches came from the second half
(January to April) of the 1976–1977 breeding season; unfortunately, no
nest inspections could be made in the first half of the season. However,
at the nearby Danaher colony the mean clutch size over four breeding
seasons (1985–1986 to 1988–1989) was 5.04 ± 0.98 (n = 303), and was
significantly larger than those found at Griffith and Padgett (Zann
1994a). Clutch sizes in the autumn months (March and April) were significantly smaller than those for the preceding six months of the season.
A seasonal decline in avian clutch size, such as this, is typical of bird
populations and three non-exclusive hypotheses have been proposed (see
review by Rowe *et al.* 1994):

(1) sub-optimal parental condition;

(2) a decline in offspring value due to a lower probability of survival of
    late-hatched young, which in turn, leads to

(3) an increase in the cost of reproduction.

While it is unlikely that hypothesis (1) would apply to experienced
breeders it is possible that it could apply to novice breeders, who themselves hatched earlier in the season (see below).

Birds at Danaher were trapped at a walk-in trap baited with seeds *ad
libitum*, so it is conceivable that the extra seed was responsible for the
enlarged clutch size—it is known in other species that experimental supplementation of food can improve female condition and lead to an
increase in clutch size (Högstedt 1981; Newton and Marquiss 1981;
Dijkstra *et al.* 1982). Nevertheless, no significant differences were found
in the size of clutches laid in the second half of the season among the
Danaher, Padgett and Griffith birds (Zann 1994a). Thus, *ad libitum* food
may not necessarily inflate the clutch size, at least in the second half of
the season in northern Victoria. Brood size (and presumably, clutch size)
was not significantly affected by experimentally induced differences in
net energy gain among breeding domesticated pairs of Zebra Finches
(Lemon 1993; Lemon and Barth 1992).

In experimental manipulations of diet of domesticated Zebra Finch
nestlings, Haywood and Perrins (1992) discovered that clutch size was
permanently affected by the level of nutrition experienced by a female during her first month of life. The amount of food, rather than its type or
quality, caused the effect (see Chapter 7). Thus, female nestlings that had
high rates of food consumption grew faster than those with low rates, and
subsequently laid larger clutches of eggs even though both groups of
breeding adult females were kept under identical conditions of *ad libitum*
food. Although samples sizes in this experiment were small (n = 12), this is

the first demonstration in birds that environmental conditions experienced early in life can permanently affect the size of clutch a female can lay.

The Danaher study also found significant variation in clutch size among the four breeding seasons despite the availability of seed from the trap (Zann 1994a). Although supplies of wild food were not monitored at Danaher it was assumed that the regular irrigation of the pastures during dry periods in summer would reduce shortages. Furthermore, re-nesting pairs did not differ in the size of first, second or third clutches laid in the one season. In the second half of the breeding season at the Danaher colony some young that hatched a few months earlier attempted to breed. Nevertheless, there was no significant difference between the proportion of small clutches (< 4 eggs) laid by these novices and the proportion laid by older more experienced females hatched in the previous year, or in earlier years (Zann 1994a).

The modal clutch size in domesticated Zebra Finches varies according to breeding stock. For example, Sossinka's (1970) birds at Bielefeld had a modal size of four eggs (4.3 ± 1.25, $n = 69$), which did not differ significantly from that laid by offspring of wild-caught birds (4.2 ± 1.63 $n = 74$). However, Birkhead's stock at Sheffield had a modal clutch size of six (6.05 ± 0.87, $n = 40$) (Birkhead *et al.* 1989) and was larger than that of first-generation aviary-bred offspring held under identical conditions (Birkhead *et al.* 1993).

Control of clutch size

Haywood (1993a) conducted extensive investigations into aspects of the mechanism that controls clutch size in domesticated Zebra Finches. First, he established that Zebra Finches belong to those bird species classified as indeterminate layers (Haywood 1993b), namely those that are capable of increasing their clutch size in response to (experimental) egg removal. Second, by systematically removing and replacing eggs at different times over the laying period he established, in a meticulous series of experiments, that tactile contact with only a single egg on the second or third day of the egg-laying period is sufficient to trigger the cessation of laying, although under natural conditions, most females have already laid three eggs by this time. Tactile contact with an egg on the second or third day of laying causes disruption of ovarian follicular growth sometime between 13:00–15:00 hr on the third day of laying. Third, the timing of follicular distruption could be experimentally advanced or delayed by adding or removing eggs, respectively. Consequently, the onset of tactile sensitivity of the brood patch to eggs present in the nest is not related to the onset of egg laying itself but to stimulation provided by the first egg; this sensitive period lasts less than two days. Fourth, variation in the size of clutch within and among females was not due to variations in the timing of follicular distruption, but to differences in the number of growing follicles present at the time distruption was triggered, since these advanced follicles were able to complete their rapid-growth phase and ovulate. Therefore, those females with only one yolky follicle present at distruption layed a clutch of four eggs, those with two, a clutch of five and those with three layed a clutch of six eggs. Apparently, it takes five days to make a Zebra Finch egg—one day for a follicle to be 'recruited', three days for addition of yolk ('yolky follicle' stage), another day for ovulation from the ovary, and another for fertilisation and the addition of albumen and shell before laying. Removal of a clutch on the fifth day of laying releases inhibition on follicle growth so that the first egg of a replacement clutch occurs on the fifth day after clutch removal. Therefore, it takes a minimum of ten days to replace a modal clutch of five eggs.

Laying times

In common with most species of birds Zebra Finches lay one egg each day, mostly around sunrise. To determine this, we inspected new nests at the Danaher colony every two hours, from before sunrise until 15:00 hr Eastern Standard Time during January and February 1991. Of 31 eggs from 16 clutches, 27 (87%) were laid in the two-hour period starting just before sunrise. No eggs were laid after 13:00 hr. Domesticated Zebra Finches also lay their eggs in the early morning, and, coincidentally, Haywood (1993a) also found that 87% of eggs were laid within the first two hours after the cage lights were turned on.

The first egg is frequently laid before the nest is complete. Frith and Tilt (1959) found that the average time between the onset of building and the first egg was six days with building being completed in another seven days. In both central Australia and northern Victoria, the first egg may be laid with only the floor and lower walls of the nest chamber standing and this may only take two days to construct. In most instances, the nest is fully constructed before the first egg is laid, but lining is not added until incubation begins.

Eggs

Eggs are matt white in colour when newly laid but become polished with age. At the narrow end faint greyish streaks or mottling may occur in eggs laid by certain females. Only preliminary studies of egg size have been made on the Australian Zebra Finch. Dimensions and shape of eggs laid by the Australian subspecies vary considerably, even when food is not limited. In my aviaries, eggs freshly laid by second and third generation offspring of wild-caught Zebra Finches weigh about one gram (1.0 ± 0.10, range 0.75–1.25 g, $n = 147$) and laying sequence had no significant effect on weight ($F_{1,145} = 0.44$, $P = 0.51$), nor was there any difference between fertile and unfertile eggs ($t_{145} = 0.27$, $P = 0.79$). However, there was a significant female effect (Welch test, $F_{26,31} = 9.49$, P<0.001) indicating that egg weight tends to be consistent within individual females. Zebra Finch eggs are 21% smaller than expected (1.0 g *vs.* 1.26 g) for estrildines of their body size according to the allometric equation (Egg Mass (g) = 0.18 $B^{0.76}$, where B = female body weight (g)) calculated on 52 species by Rahn *et al.* (1985). Eggs represent 8–10% of the female's body weight (13 g), which is also less than expected (13%) for a bird of their size according to the allometric equation for passerines (Rahn *et al.* 1985).

Eggs (*n* = 286) at the Danaher colony averaged 15.10 ± 0.72 mm (range 13.12–17.35) long and 10.94 ± 0.491 mm (range 9.57–12.7) wide. Egg volume (V = 0.51 × length × width^2; Hoyt 1979) was 927 ± 115 mm^3 (mean ± *s.d.*) and varied significantly among females ($F_{55,229}$ = 26.04, P<0.0001), but not with laying sequence within a clutch ($F_{5,30}$ = 0.5, P = 0.77), nor between successive clutches of the same female within a season ($F_{1,54}$ = 1.8, P = 0.18).

Fresh eggs from the Padgett colony averaged 1.0 g and had a relatively high proportion (26.9%) of yolk compared to 17 other altricial Australian species, which led Lill and Fell (1990) to conclude that nestlings probably use the high levels of residual yolk to offset costs of slow growth.

Unfortunately, no data are available on what factors affect the predisposition of wild Zebra Finch females to lay eggs of a particular size, but this was a consistent finding in both aviary-bred and wild birds, despite the fact that high levels of egg dumping at Danaher, and presumably in aviary birds as well, would tend to increase the variance within clutches. Consistency in size and appearance of eggs within females has been established in other species of birds, and found to be greater in colonially breeding species, and Møller and Petrie (1991) suggest that it is possibly an adaptation against nest parasitism. Experimental feeding with supplementary protein during the laying period will produce larger eggs in domesticated adult Zebra Finches (T. D. Williams, pers. comm.). Consequently, when food is limiting, females may lay smaller eggs than those measured here since seed was provided *ad libitum* in the aviary and at Danaher. It has also been shown that malnutrition during the first month and a half of a female Zebra Finch's life has permanent detrimental effects on the size of eggs she subsequently lays, even when food becomes abundant, before and during laying (Haywood and Perrins 1992).

The effects of geographic variation, female age and 'quality' on egg size needs to be investigated. Sossinka (1970) found that domesticated European Zebra Finches lay larger eggs than wild-caught ones (948 mm^3 *vs.* 875 mm^3), but the eggs constituted a smaller proportion of body weight.

Re-nesting

The mean (± *s.d.*) number of clutches per season for males at the Danaher colony was 1.9 ± 1.16 (range 1–6, *n* = 127) and 1.7 ± 0.91 (1–4, *n* = 127) for females. Within a breeding season the mean interval in days between the dates of the first eggs of successive clutches was significantly shorter when the previous attempt failed (33 ± 11 days, $n = 35$) than when it was successful (52 ± 16 days, $n = 48$) (Zann 1994a). The shortest time between re-nestings was estimated at 34 days, when a female was found incubating a full clutch while her mate cared for two fledglings about 22 days old. Within a single breeding season at Danaher, 54% of females ($n = 123$) laid a single clutch only, 25% laid two clutches, 14% three, 4% four, 1% five and 1% six. These are minimum values because some attempts may have gone undiscovered, and attempts made at other colonies before, or after the breeding pair arrived at Danaher, were likely.

Incubation and hatching

Incubation in free-living and wild-caught Zebra Finches was investigated by means of temperature probes placed in infertile eggs located in the centre of the clutch; temperature readings were made every five minutes and stored in a data logger (Zann and Rossetto 1991). Both sexes incubate, but only the female develops a typical brood patch. El-Wailly (1966) mistakenly took the bare apterium of male Zebra Finches for a brood patch. When it forms, the few down feathers on the female's ventral apterium are lost, the skin becomes thick and loose, and turns a pale yellowish-white, but there is no increased vacularisation commonly found in most species of birds. The brood patch forms before the clutch is complete and disappears a few days after hatching so does not appear necessary to heat the brood. The absence of a brood patch does not affect the ability of the male to warm the eggs (Zann and Rossetto 1991).

Diurnal bouts of incubation do not differ between the sexes in free-living and wild-caught birds (Zann and Rossetto 1991). In domesticated birds both El-Wailly (1966) and Delesalle (1986) found that female attentiveness was significantly greater overall than that of the male, although there was considerable variation among pairs. The median duration of day-time bouts of incubation at Danaher was 35 minutes for males and 40 minutes for females, but ranged from 7–81 minutes. In Western Australia, Immelmann (1962a, 1965a) found that the average incubation bout during the day was 92 minutes and the longest 175 minutes. At our Alice Springs study site, the median bout was 47 minutes (range 10–120 minutes). Conceivably, distance from nest to foraging areas may influence the duration of incubation bouts. The artificial sources of seed located in the nesting colonies at our study sites may have allowed the non-incubating partner to forage closer to the nest. Consequently, it was able to relieve the incubating partner more frequently. This trend is exaggerated in captivity where wild-caught birds in our large aviaries had a median incubation bout of 20 and 18 minutes for males and females respectively.

Contrary to Immelmann's (1962a) statement, females appear to do all the incubation at night (Zann and Rosetto 1991).

During incubation, both parents may leave the nest unattended for long periods, especially in the afternoon, and many incubation bouts during the day are interrupted by social activities and by disturbance from potential predators and trespassers. The Double-barred Finch, by contrast, rarely leaves the nest unattended (Immelmann 1962a). Ambient temperatures also affect attentiveness of incubating parents. When the shade temperature exceeds 38°C Immelmann (1962a) found that Zebra Finches stop incubating and leave the eggs unattended for periods up to 16 hours duration. Constant temperatures above 40°C are lethal to most avian embryos (Drent 1975); however, Vleck (1981) found that domesticated Zebra Finches in constant temperature chambers continued to incubate when egg temperatures exceed 40°C by attempting to conduct heat away from the eggs through panting and feather erection to increase evaporative cooling. Zebra Finch embryos do not develop properly if artificially incubated at temperatures of 35°C (Vleck 1981) although embryos of most avian species fail to develop under 25–27°C (Webb 1987).

Once full incubation began the average temperature of the eggs for a number of both day- and night-time readings from six pairs at Danaher was 36.1 ± 1.18°C, but, taken separately, night time egg temperatures were significantly higher than daytime ones despite much lower ambient temperatures (Figure 6.6). Similar day–night differences occur in incuba-

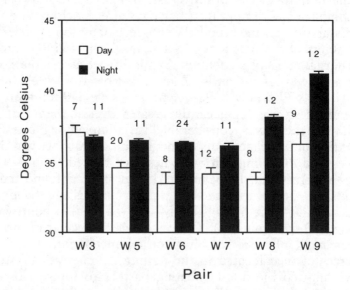

Fig. 6.6 Day and night egg temperatures during incubation for six pairs of free-living Zebra Finches at the Danaher colony in northern Victoria. Bars show means and one standard error. The number of hours with 12 five-minute readings is given above the bars. (After Zann and Rossetto 1991.)

tion temperatures of wild-caught Zebra Finches (n = 7 pairs); daytime ones were significantly lower than those at night but were higher than daytime temperatures of free-living birds. Nest attentiveness accounts for these differences—it is greater at night than during the day and less among free-living Zebra Finches than wild-caught ones. We found no significant differences in egg temperatures between early and late stages of incubation, unlike what has been found in other small passerine species (Webb 1987). Temperatures were also maintained or even increased after hatching.

Vleck (1981) found that domesticated Zebra Finches had a mean egg temperature during incubation of 37°C whereas El-Wailly (1966) obtained a lower temperature of 35.2°C. The incubation temperature of most species of birds is around 35°C (Kendeigh *et al.* 1977).

Onset and duration of incubation

Warming of the eggs begins gradually after the first egg is laid as both partners spend more and more time just sitting in the nest, either together or alone (Immelmann 1962a; Birkhead *et al.* 1988a). Initially, the female spends more time than the male, but by the third egg attentiveness is about the same. Temperature probes show that eggs may be well above ambient levels for short irregular periods, but they are still cold to touch for much of the day, especially in the afternoons. This warmth probably does not come from the brood patch of the female, or the male's apterium, but is due to the birds squatting over the eggs. If the clutch is five or more, incubation proper beings after the fourth egg is laid, while it begins after the last egg is laid if the clutch is less than five (Zann and Rossetto 1991). Daytime nest attentiveness suddenly increases and egg temperatures increase about 15°C and are sustained around 35–38°C, varying according to the pair (Figure 6.7). Enhanced attentiveness alone is not responsible for the temperature increase since night-time temperatures at the third egg stage reach only 33°C despite 100% attentiveness; sustained contact between the brood patch/ventral apterium must be responsible. Although wild-caught pairs also show the sudden increase in egg temperature after the fourth egg it is not as large as in free-living birds, and some embryonic development may already occur in some pairs. Presumably, absence of foraging restraints may allow captive birds time to increase their level of attentiveness sooner than free-living birds.

In wild-caught pairs (n = 13) the incubation period ranged from 11–15 days, with a median of 14 days. In free-living Western Australian birds the incubation period ranged from 12.5–16 days (Immelmann 1962a, 1965a). Eggs of domesticated European Zebra Finches hatched in a minimum of 11.25 days when warmed in an incubator set at an optimum temperature of 38.6°C (Ziswiller 1959). Ziswiller (1959) also found that during incubation, Zebra Finch eggs were remarkably resistant to

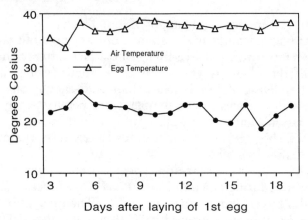

Fig. 6.7 Mean daily egg temperatures before, during and after incubation in a wild-caught pair of Zebra Finches. Each data point is a mean of 24 × 12 readings logged every five minutes. Full incubation began after the fourth egg was laid (day 4) and full incubation temperatures were not reached until the fifth day. The first egg hatched on day 14. (After Zann and Rossetto 1991.)

cooling: most eggs still hatched despite being cooled down to 18°C for periods of 10 hours and were much more resistant to cooling than other species tested (estrildines, fringillids and ploceines).

Energetics of incubation

By measuring oxygen consumption over ambient temperatures ranging from 8° to 28°C, Vleck (1981) showed that an incubating Zebra Finch expends about 20% more energy than a bird roosting in the same nest. The energy cost increases with a drop in the ambient and nest temperatures to compensate for the rate of heat loss from the eggs and the greater amount of heat that must be transferred from the incubating birds. Above 28°C, which is towards the lower critical temperature for Zebra Finches (Calder 1964; Cade *et al.* 1965), the energetic cost of incubation is negligible because heat for incubation can come from the bird's normal heat production. Vleck's (1981) estimates of the cost of incubation are significantly lower than those estimated by El-Wailly (1966) who used a less accurate method based on increases in food consumption.

A major energetic cost of incubation is re-warming eggs that have cooled during absences. Incubating birds quickly sense the temperature of the eggs, and if they are below 35°C adjust their position and contact with the eggs and begin to elevate heat transfer with a commensurate increase in oxygen consumption. During the short re-warming phase the metabolic rate may increase threefold (Vleck 1981). Biparental incubation with high rates of attentiveness prevents cooling of eggs and provides considerable energy savings when ambient temperatures are low. Absences are usually limited to later in the day when temperatures are higher.

Hatching

Hatching of the clutch normally spans two days in wild Zebra Finches, but is quite variable. Weekly nest inspections made late in the day at Danaher over four breeding seasons yielded 144 nests in which the first young were believed to have hatched on inspection day. In some clutches all six eggs were estimated to have hatched on inspection day, that is total synchrony, but the majority (47%) of clutches had only 50% of eggs hatching on inspection day (Figure 6.8). During a subsequent study of egg dumping, we made daily inspections of 30 nests at Danaher and found that all clutches hatched within two days and that 60% had more than 50% of the clutch hatch the first day. This proportion is not significantly different from that estimated from weekly inspections ($G_1 = 1.63$, *n.s.*). These findings are consistent with those of Immelmann (1962a) who found that, in his colonies in Western Australia, two young hatched in the morning, another one or two around midday, and depending on the size of the clutch, another one or two the following day. Thus, hatching in wild Zebra Finches normally spans more than one day, which, by definition, is considered asynchronous hatching (Amundson and Slagsvold 1991).

Hatching in most pairs of wild-caught Zebra Finches is less synchronous than that of free-living birds. Eggs tend to hatch one each day,

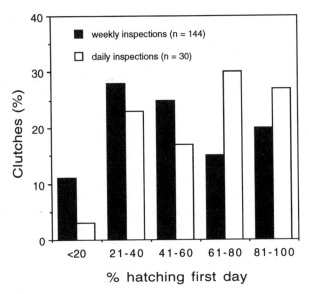

Fig. 6.8   Hatching synchrony in free-living Zebra Finches at the Danaher colony in northern Victoria when inspections were made on a weekly and daily basis. (Modified from Zann and Rossetto 1991.)

in order of laying. Asynchronous hatching here is probably a consequence of early nest attentiveness in which sub-optimal levels of heat applied to the first three eggs triggers some embryonic development, and the differential is maintained throughout the incubation period, leading to asynchronous hatching. If wild-caught pairs keep egg temperatures below 30.5°C before full incubation begins on day four, then all eggs hatch on the one day (Zann and Rossetto 1991).

In an experimental investigation of the effects of asynchronous hatching on growth rates and survival of young under conditions of food shortage, Skagen (1988) found that clutches of domesticated Zebra Finches hatched over a two-day period, and were considered asynchronous when the difference in hatching weight exceed 0.5 g. Artificially created synchronous and asynchronous broods did not differ in nestling survival under food-limited and food-abundant conditions. However, in clutches with asynchronous hatching there was a hierarchy of nestlings by size in which the heaviest nestlings were fed preferentially by the parents so that they grew more quickly. This finding partly supports Lack's (1968) brood reduction hypothesis which posits that asynchronous hatching of altricial birds is an adaptive response to unpredictable food shortage where the youngest are starved in order that more of the larger siblings might survive. However, contrary to predictions that asynchrony would not be disadvantageous under conditions of food abundance, the older siblings still grew faster and had a higher survivorship. Skagen considered this a maladaptive response of asynchrony to food abundance. Overall, asynchronous hatching in these experimental Zebra Finches is superior to synchronous hatching under both good and poor laboratory conditions. This trend is opposite to that found in most other studies (Amundson and Slagsvold 1991), consequently there is a need to replicate the study in free-living Zebra Finches, and if possible, follow young through to breeding age.

Development of young

Mouth-markings

Like all estrildines, Zebra Finch nestlings have a conspicuous species-specific pattern of black markings in the mouth, which are prominently displayed during begging when the gape is opened wide (Figure 6.9). Whereas the mouth markings of the Lesser Sundas Zebra Finches have not been described, those of the Australian subspecies have been studied in detail. There are five sets of marks in the Zebra Finch:

(1) twin marks that form a chevron in the tip of the premaxilla;

(2) five roundish spots that form a 'domino' pattern in the centre of the upper palate;

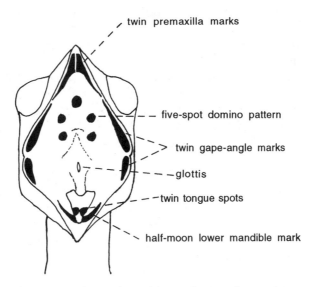

twin premaxilla marks

five-spot domino pattern

twin gape-angle marks

glottis

twin tongue spots

half-moon lower mandible mark

Fig. 6.9 Typical pattern of mouth-markings of Australian Zebra Finch nestlings. This pattern was drawn from an aviary-bred nestling whose grandparents were wild-caught from northern Victoria.

(3) two lateral tongue spots towards the tip of the tongue;

(4) a 'half-moon'-shaped mark in the sublingual region of the lower mandible; and

(5) twin tear-shaped marks each side of the gape flanges.

The background colour of the upper and lower palate is off-white with yellowish tints in some individuals. The tip of the tongue is white, the remainder, a pink flesh colour. A swollen white-pigmented flange forms outside around the beak angle and contrasts with the twin tear-shaped marks when the gape is open wide. There is considerable variation in the darkness and shape of the marks, but less within than between broods. All marks begin to fade around fledging time. This is fastest with the pre-maxilla marks and slowest with the domino spots which may persist beyond 50 days of age and even after day 100 faint smudges can be seen.

Domesticated strains of Zebra Finches often fail to develop mouth markings. While the wild-type has markings similar to those of wild birds, the white-morph has no black marks in the mouth at all and the beak flange lacks the white pigmentation. The completeness of markings in other colour morphs varies between these two extremes (Muller and Smith 1978). Experiments showed that parents of both wild-type and white morphs prefer to feed wild-type nestlings, which led Immelmann *et al.* (1977) to conclude that the mouth markings were primarily responsible for stimulating the feeding response in the parents. Moreover, they predicted that discrimination would be stronger among free-living birds where food would be limited so that young with mouth patterns that deviated even slightly from the species-specific pattern would receive less food and suffer higher rates of mortality. Indeed, Skagen (1988) found that nestlings of domesticated Zebra Finches without palate markings had significantly slower growth rates than those with markings under conditions of food shortage, but there was no difference when food was abundant. It would be interesting to test whether Australian and Lesser Sundas parents can discriminate young in mixed broods.

The mouth pattern in the Double-barred Finch is similar to that of the Zebra Finch, except that the upper three spots of the domino are elongated into thick lines and the pigmentation between the two lower domino spots is more intense. In the Long-tailed and Black-throated Finches the three upper domino spots are elongated still further so that they are fused together to form one mark of a 'horseshoe' shape. Steiner (1960) and Wolters (1957) believe that the five-spot domino pattern is the most primitive form, occurring in most African estrildines and in more than half of the Australian ones, whereas the horseshoe pattern is the derived form. Immelmann (1965a) pointed out that this evolutionary transition appears to have occurred independently in both African and Australian species among distantly related groups. Inter-specific brood parasitism is postulated as one factor that has led to the origin and maintenance of mouth patterns in the Estrildinae (Nicolai 1964; Kunkel 1969). Furthermore, it is feasible that intraspecific brood parasitism, which can occur at fairly high levels in Zebra Finches (below) may be responsible for maintaining variation in mouth patterns between broods and possible discrimination by parents. Experiments are needed to test this idea.

Morphological changes

My observations on nestling development of free-living Zebra Finches agree with Immelmann's (1962a) description. On hatching day (day 0) nestlings are naked and pink-skinned with a few pale-grey down feathers on the head and back, but by day 2 slate grey pigmentation appears in the middle of the back and quickly spreads laterally covering the flanks, thighs and top of the head. Pigmentation of the dorsal surface of the wings begins by day 3, and the whole dorsal surface, flanks and outside thighs are pigmented by day 5 or 6. The bill of newly hatched nestlings is a light-horn colour, but on day 2 and 3 dark pigment begins to spread on upper and lower mandibles and by 10–12 days of age pigmentation is complete and both mandibles are uniformly black. In the white morph young do not develop black pigments on the bill, but by fledging time the horn-colour gradually changes to a pale pink (Immelmann 1959).

The eyes begin to open around 6–7 days of age, forming a narrow slit at first and become fully open by day 10–11. Thin hairs precede the first pins of the remiges which break through the skin on day 6 and grow to about 3 mm by day 10, thereafter increasing by about 3 mm per day until day 18 when they are about 28 mm. The sheath of the pin splits open around day 11 exposing the dark vanes of the feathers. The first pin of the rectrices appears on day 9 reaching 5 mm in length by day 12 and 13 mm by day 18. Growth of the rectrices and remiges is faster than that of the Long-tailed Finch, but slower than that of the Crimson Finch (Immelmann 1965a). Pins of the contour feathers appear on the dorsal tracts around day 7 and those of the ventral tracts appear the next day leading to complete coverage of the body by around day 14. Down feathers are gradually lost, first from the shoulders, then the back, the thighs and lastly the back of the head.

In the wild, nestlings are remarkably resistant to cold and exposure (Immelmann 1962a). Nestlings as young as four or five days will survive without food and endure near-freezing temperatures for periods of up to 36 hours. Immelmann believes that they enter some form of torpor where digestive processes cease and food remains in the crop; warmth restores them.

On hatching day, the crop of the newly hatched nestling is empty as parents normally do not feed them until the next day, but some give small amounts of fine, green seeds which sit on the right side of the crop, which fills first. After day 3 the left side of the crop also stores seed, but in smaller amounts than the right side; equal amounts are found in both sides by day 5 if large quantities need to be stored.

Growth rates

Unfortunately, growth rates of nestlings have not been studied in free-living Zebra Finches, but have been investigated in domesticated birds in a number of studies. Sossinka (1970) found no significant differences in the growth of aviary-reared nestlings of domesticated and wild-caught birds. Like most estrildines investigated to date, Zebra Finch nestlings are slow developers by comparison with other granivorous species: they spend longer in the nest and take longer to reach adult weight than fringillids and ploceines (Ziswiller 1959; Lack 1968; Sossinka 1972b).

Nestlings of domesticated Zebra Finches weigh between 0.6–0.9 g at hatching and gain on average 0.66 g per day for the first 10 days; they weigh on average 4.6 g at day 7 and 8.0 g at day 12 (Skagen 1988) and 9–10 g at fledging in both sexes (Burley 1986c). Newly fledged Zebra Finches at the Danaher colony weigh around 10 g and reach their adult weight at a median age of 51 days (range 26–129 days, n = 100), with those attaining a heavier adult weight taking longer. In domesticated females, Sossinka (1980a) found that length of body, wing and bill reach adult size between 27 and 29 days of age.

The energy demand of a Zebra Finch nestling increases steeply each day from hatching to day 8, where it peaks at 8.8 kJday^{-1}, after which demand decreases as increasing feather cover assists thermoregulation (Lemon 1993).

The quality and quantity of the nestling diet affects growth rates and the size at adulthood. Boag (1987) found that a high protein diet increased the growth rate of most morphological characters he measured, affecting weight and tarsus length the most, while bill and plumage were affected least. Weight at adulthood was also affected by the quality of the diet with differences reaching 7–10%. Skagen (1988) found that the growth rate of nestlings was significantly faster when food was abundant than when it was limiting, and that variation in rate of growth was greatest between nestlings where food was limited. Nestling weight at age 12 and adult weight was positively correlated for both males and females. This effect of food limitation on nestling growth rates and final adult weight was confirmed by Haywood and Perrins (1992) who also found permanent detrimental effects on egg and clutch size.

When food is not limiting, but the net energy intake each day is reduced by high foraging costs, weight of nestlings at fledging is significantly reduced and the growth rate slowed significantly (Lemon 1993). The task remains to see if similar relationships can be established in wild populations and in particular, the relationship between egg size, nestling size, and size at first breeding.

Sexual maturation

Sossinka (1970, 1972b, 1975, 1980a) investigated sexual maturation of both sexes of Australian Zebra Finches. There were no significant differences between the sexes in body growth, changes in colour of the bill, and onset of post-juvenal moult. Consequently, these changes do not appear to be controlled by sex-specific factors. However, development of the gonads proceeds at a different rate—growth is logarithmic in males and linear in females.

Testis volume increased dramatically, reaching adult levels around 70 days—its rate of growth was not affected by day length (LD 8:16 *vs.* 16:8). Spermatogenesis proceeded with corresponding swiftness, the earliest mature sperm appearing at 59 days, and by 70 days most individuals had fully developed sperm. Testosterone production matched testis growth. There was a brief juvenile refractory period between day 18–34 where testis size, spermatogenesis and testosterone production leveled off, forming a plateau on the steep growth curves (Pröve 1983). This rate of sexual development is rapid, even for small tropical species. For example, testis development in the Spice Finch *Lonchura punctulata* took around seven to eight months and the juvenile refractory period lasted over three months (Sossinka 1975). Sexual maturation is even more rapid in offspring of wild-caught Zebra Finches raised under identical conditions. By 44 days of age the testes were almost twice the volume of those of domesticated birds, and the stage of spermatogenesis significantly more advanced (Sossinka 1970). Sexually dichromatic plumage marks were also less advanced in domesticated Zebra Finches, but there were no significant differences in body size.

Sexual maturation in domesticated females is slower than that of males. Sossinka (1980a) measured the weight of the ovary and the largest follicle from day 10 to day 100. The diameter of the largest follicle increased linearly until 100 days of age when it reached 1.0–1.2 mm, after which it ceased growing. It only started to grow again if breeding conditions arose, increasing exponentially for a period of about two weeks, during which ovulation occurred and, ultimately, the first egg laid. Sossinka termed the first growth period the 'stage of primary development', and the second, the 'stage of stimulated development'. If breeding conditions arose during primary development there was no resting stage and eggs could be laid as early as 90 days of age. However, maturation must be faster in free-living birds since females as young as 62 days laid their first egg in the Danaher study (Zann 1994a). Sossinka (1980a) also found that only a few follicles reached the resting stage in the first 100 days of life after which other follicles began to undergo primary development. Possibly, young females rush through the development of a few follicles in order to exploit any breeding opportunities that might arise before the ovary is fully developed, since the weight of the ovary did not reach that of adults until at least 250 days of age. No differences in clutch size were found between young females and adults in the Danaher study, so it is conceivable that more follicles undergo primary development in the first wave in free-living females than in domesticated ones. Another interesting finding made by Sossinka was that females held under short day length (LD 8:16) took significantly longer to grow ovarian follicles during primary development than those held under longer day lengths (LD 16:8).

Intra-specific brood parasitism

In February and March 1988, we collected blood from 25 colour banded Zebra Finch families (16 pairs of parents and 92 offspring) breeding at Danaher for investigation of the incidence of extra-pair paternity by means of DNA fingerprinting. There was a mismatch of fingerprints with the rearing parents in 12 nestlings of which extra-pair paternity accounted for two (2.4% of offspring in 8% of broods). However, we unexpectedly found that 10 nestlings (10.9% of offspring in 36% of broods) did not match either of the rearing parents, having been 'dumped' as eggs in their nest sometime before incubation began (Birkhead *et al.* 1990). The identity of the parasitic female was not determined but analysis of the bands indicated she was not closely related to the rearing mother. Only one brood contained more than one parasitic nestling and this may have arisen accidentally through a change in ownership between two females that had contested the same nest. Consequently, the adjusted level of true dumping is 8.7% (8/92) of offspring and 32% (8/25) of broods. Parasitised clutches were significantly larger than non-parasitised ones (6.0 ± 0.82 *vs.* 5.0 ± 0.95) which were closer to the population mean (5.03), suggesting that the parasite simply added her egg without removing one of the host's at the same time.

These relatively high levels of intra-specific brood parasitism had not been suspected because we only checked nests weekly, which precluded the discovery of more than one egg being added to the clutch on the same day. In their study of breeding of Zebra Finches at Griffith, Frith and Tilt (1959) on no occasion found two eggs laid on the same day, although they do not state how many clutches they followed through the laying period. During careful observations of unbanded birds at two nests near Inverell, northern New South Wales, Baldwin (1973) saw two adult females emerge from the same nest during the egg-laying period, and in another nest found two eggs laid in the morning on the same day. Despite many hours spent at the Danaher colony watching pairs during egg-laying, on only one occasion was a strange female seen entering a neighbour's nest and an extra egg subsequently found (Dunn 1992). During trapping sessions we occasionally had females laying eggs in the holding box quite late in the morning, but we did not realise at the time that these individuals may have been on dumping runs when trapped.

The incidence of intra-specific brood parasitism in Zebra Finches is still far from clear. Moreover, almost nothing is known about strategies of host and parasite and naturalistic observations are difficult to reconcile. For instance, Immelmann (1962a) stated that strange pairs had an absolute inhibition ('absolute Hemmung') about entering unguarded nests containing eggs or young, although he noticed that this inhibition had disappeared in domesticated birds. Our observations at Danaher and central Australia suggested that strange pairs were interested in the nests of others, but were tentative about getting close and none was seen to enter. Yet, we also observed how, at times, owners guarding their nests became particularly vindictive if they detected other pairs inspecting the entrance and long chases ensued. Overall, these observations, on the one hand, suggest that there is a potential risk of egg-dumping during egg-laying and that the laying pairs guard against it. On the other hand, observations show that the nest is unoccupied and unguarded for long periods during egg-laying while the pair go off to feed with the flock.

During the laying of the first four eggs in three pairs at York, Western Australia, Immelmann (1962a) found that the nest was unguarded for an average of 40% of the time, while at the Danaher colony Birkhead *et al.* (1988a) found it unguarded for an average of only 20%. Differences in guarding time may possibly be due to differences in distances pairs need to forage from the colonies. Hence, guarding is not continuous throughout the day; it appears more intense in the morning, when laying occurs and the risk of brood parasitism is greatest.

In an unpublished study we found that the detected frequency of egg-dumping at Danaher as revealed by checking for the appearance of two eggs in a nest on the same day was less than half that detected by DNA fingerprinting. We inspected nests daily after 14:00, by which time most laying for the day had been completed. Only 13% (7/54) of nests monitored during egg-laying in the 1991–1992 seasons had two eggs deposited on the same day, and conversely only 9% (5/54) of nests had gaps in the laying sequence on days eggs were found dumped. Both values are believed to be underestimates. In three of the seven instances the parasitic egg was added on the day the first egg of the host was laid, in two instances it was on the day the second host egg was laid, and the other two on the day the third egg was laid. We did not include the days the fourth and fifth host eggs were laid in our analysis in order to avoid the situation where the host only laid a small clutch and the subsequent eggs laid by a parasite could have been mistaken for those of the host. With potential parasites, gaps appeared in the laying sequence on the day of the second egg in three cases and on the day of the fourth in another two. Where pairs missed a laying day they continued to lay the modal clutch (five eggs) so if they had in fact dumped the missing egg their clutch would have had six eggs.

The extra-egg method for detecting dumping is direct, cheap and, theoretically, large samples can be collected quickly because wastage of data is slight since only the five laying days need be monitored during which nest failures can occur, whereas a nest must survive at least 23 days longer before blood can be collected from nestlings for DNA fingerprinting. However, a parasitic egg would be detected by DNA fingerprinting, but not by the direct method, if the parasite laid her egg the day before the host laid her first egg, or the day after the host laid her last egg. Although the plain white eggs have few individually distinct marks they do differ in dimensions and, theoretically, it may be possible to detect the egg laid by the parasite from those of the host.

On the ground below some nests at Danaher we found fresh whole and broken eggs and suspected that they may have been parasitic eggs ejected by the hosts. However, when we experimentally added single marked eggs to host nests (*n* = 14) before, during and after egg-laying none were ejected.

In summary, intra-specific brood parasitism in Zebra Finches can be quite high since colonial breeding provides the opportunity for parasites to find suitable hosts, and the high levels of nest predation would make it advantageous for the parasite to spread this risk. Analysis of breeding success (below) shows that there is no penalty in fledging rate associated with clutches enlarged to six or seven by the addition of a dumped egg. Defence against brood parasites does not appear well developed in Zebra Finches: guarding is spasmodic, eggs are uniform and unmarked, and there is no ejection of dumped eggs. Intra-specific brood parasitism has not been recorded for any other Australian estrildine to date, but given their tendency towards colonial nesting and pressures from nest predators, it is reasonable to expect that it will be found.

Breeding success

Pooled data over four years showed that 35% of breeding attempts, 41% of incubated eggs and 54% of incubated clutches produced at least one 'fledgling' (Zann 1994a). These estimates are slightly inflated because the number of fledglings produced was based on the number of nestlings of 12–14 days of age, not fledglings observed outside the nest. It is possible that some may have fallen prey before they could fledge. However, young are quite mobile after this age and may escape nest predators. For example, when forced to fledge, young explode out of the nest amid a series of loud identity calls and some can fly 10–20 metres before hitting the ground, whereupon they freeze and are difficult to locate. Younger birds flutter to the ground beneath the nest and move rapidly to any dark nook or crevice at the base of the bush and freeze.

Breeding success at Danaher was similar to that found by Kikkawa (1980) at Armidale, however, but fewer clutches produced young than in the population studied by Frith and Tilt (1959) at Griffith (Table 6.3). Success of breeding attempts at the Danaher colony was low by comparison

Table 6.3 Breeding success of Zebra Finches at three locations in eastern Australia based on four seasons' data

Location	Source	Percentage success	n
Griffith (34°17′S, 145°02′E)	Frith and Tilt (1959)	74.4[a]	172
Armidale (30°30′S, 151°40′E)	Kikkawa (1980)	43.5[b]	382
Wunghnu (36°09′S, 145°26′E)	Zann (1994a)	53.6[a]	300
		41.4[b]	1526

[a] Per cent of completed clutches that fledged at least one young. Frith and Tilt (1959) are not explicit (p. 291): 'Of 221 clutches begun in the four years of study, 172 were completed, and of these 128 ultimately produced young.' I assume these young left the nest.
[b] Percentage of eggs laid that resulted in a fledgling.

to that of eight species of Australian passerines recently summarised by Ford (1989).

The number of successful breeding attempts and incubated clutches at Danaher differed significantly across the four breeding seasons due to the larger number of failures due to predation in the 1988–1989 season. Overall, breeding success was significantly greater in the second half of the breeding season than in the first half (Zann 1994a).

Of 626 young banded just before fledging, 197 (31.5%) were recaptured at the walk-in trap as free-flying individuals ≥ 35 days of age. This is after the age of nutritional independence and is presumably the number of fledglings that have survived to independence, since limited flying ability prevents dispersal before this age. At 80 days of age, the time of first breeding, 137 young were recaptured (69.5% survival), but some would have dispersed and would not have been available for recapture. Thus, at least 9.0% (137/1526) of incubated eggs produced young to breeding age (Figure 6.10).

Effects of clutch size

When predation was excluded from the analysis, breeding success at Danaher varied significantly according to size of clutch (Table 6.4). The percentage of eggs that produced nestlings to banding age was lowest for clutches of five (66%), the modal number of eggs laid in the colony, and highest for clutches of two and three eggs (100%). Of eight clutches of seven eggs, the maximum number laid, 81% of eggs reached banding age, consequently this is the most productive, or optimal clutch size for Zebra Finches at this colony. It is not uncommon among birds for larger clutches to be more productive than the more commonly found clutches in a population (Boyce and Perrins 1987). Causes of sub-optimal clutches are found in life history trade-offs, for example, trade-offs between clutch size and offspring survival and reproduction, and between clutch size and parental survival and subsequent reproduction (Stearns 1992). Experiments would be needed to determine which are the important trade-offs with Zebra Finches.

Effect of age of parents

Age of breeding pairs did not significantly affect breeding success of Zebra Finches at the Danaher colony. By limiting data to the second half of the breeding season it was possible to compare breeding success of three type of pairs:

(1) pairs in which both partners were novices, having hatched in the first half of the breeding season and were between two to four months of age;

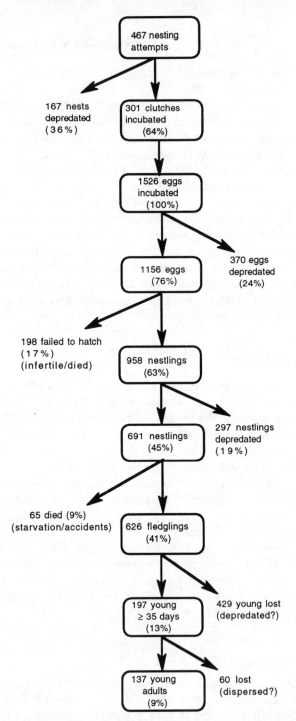

Fig. 6.10 Schema summarising breeding success of Zebra Finches at the Danaher colony in northern Victoria based on four seasons data (1985–1989). A breeding attempt was defined as one where there was at least one fresh egg. The number of fledglings was based on the number of nestlings banded in the nest around 12–14 days after hatching.

Table 6.4   Breeding success of Zebra Finches by size of clutch in nests that did not suffer predation (from Zann 1994a)

Clutch size	Number of clutches	Number of eggs	Number of eggs hatched	Per cent eggs hatched	Per cent hatchlings banded[a]	Per cent eggs banded[b]	$\overline{X}$ fledged per breeding attempt[c]
2	2	4	3	75	100	75	1.5
3	9	27	26	96	100	95	3.0
4	52	208	185	89	92	80	3.3
5	93	465	369	79	91	66	3.8
6	66	396	326	82	87	67	4.6
7	8	56	49	88	97	81	5.8
n	230	1156	958				

[a]   Based on number of nestlings banded 12–14 days after hatching.
[b]   The proportion of eggs that reached banding age from clutches of different size did not differ significantly ($G_4 = 18.03$, $P = 0.001$; clutch sizes 2 and 3 pooled).
[c]   Calculated on the number of clutches less those that suffered predation during the nestling stage.

(2) pairs where both partners were experienced, having hatched in an earlier breeding season and were at least eight months old or older; and

(3) pairs in which one partner was a young novice and the other an older bird.

However, no differences were found in four measures of breeding success: clutch size, per cent of eggs hatched, per cent of young that reached banding age and number of young per brood that reached banding age (Zann 1994a).

These results show that young novice parents are exceptional in their ability to breed as efficiently as older and more experienced Zebra Finches, since in most species of birds studied to date, older parents consistently produce more young than novices, at least until senescence sets in (Newton 1989). This precocial breeding proficiency may not necessarily be a consequence of the *ad libitum* supplies of dry commercial seed in the trap at the Danaher colony since wild seeds are the essential resource needed for rearing young. Wild grass seeds were not as abundant at Danaher as they were at Cloverlea.

Nest failures

Sixty-six per cent of all attempts (310/471) failed because of nest predation. Early in the nesting cycle, 36% (168/471) of attempts failed before the clutch size could be determined and another 47% (142/303) of incubated clutches were subsequently killed before young could be banded (Figure 6.10). In some seasons (e.g. 1985–1986) there were higher rates of predation in the first half of the season than in the second, whereas the reverse was true in other seasons (e.g. 1987–1988). Overall 17% (198/1156) of incubated eggs not taken by predators failed to hatch because of inadequate parenting. Failures were greater in the first half of the eight month breeding season than in the second half. Nine per cent (65/691) of nestlings failed to reach banding age due to starvation or accidents and the percentage of these failures varied significantly over the course of the study and increased progressively from the first season to the last.

Summary

Australian Zebra Finches nest in colonies throughout the year; these range in size from just a few pairs up to 40–50 pairs. Nests are located in thick, usually thorny, bushes close to feeding sites. There are pronounced annual changes in abundance and age of individuals present in colonies. Nesting density varies according to habitat and geographic location and ranges from 0.7 to 76 ha^{-1}. Territorality is confined to the nest itself, but an entire bush may be defended initially. In some colonies breeding nests may rest in contact. Nesting sociality reduces predation when predator levels are low to medium, but provides no protection when rates are high. Corvids, raptors, snakes, goannas, rats and mice are the main nest robbers. Zebra Finch colonies have a degree of functional structure with regular social sites, bathing sites, and arrival and departure sites.

Zebra Finches roost in nests and under some conditions this reduces loss of body heat. Breeding nests are slightly larger than roosting nests, and have a long side-entrance tunnel; they are made of grass stems or twigs, and are highly variable in dimensions, and construction materials. Breeding nests are lined with feathers and provide an improved microclimate during cold weather in southern parts of the range, and probably insulate eggs and young from excessive heat in summer. Eggs are laid one each day, just after dawn. Intra-specific brood parasitism (egg dumping) ranges from 13% to 32% of broods and may be a means of reducing the effects of nest predation. The modal clutch size is five eggs, with smaller clutches in the last months of the breeding season. Clutch size does not vary geographically, nor among successive clutches within females, nor between clutches laid by experienced and novice females.

Both sexes incubate, but only the female develops a brood patch and incubates at night, nonetheless, the male is equally competent as the female in heating the eggs. The average incubation temperature is 36°C and begins after the fourth egg is laid. Incubation bouts range from 35 to 190 minutes and may depend on commuting times to and from foraging sites. Eggs take 11–15 days to hatch, depending on levels of incubation attentiveness by the pair; the minimum incubation time is 11.25 days.

Hatching of the clutch spans around two days in the wild, but in captivity one egg may hatch each day so that young are of different sizes. Young Zebra Finches have species-specific palate markings that aid the release of the parental feeding response. Quality and quantity of rearing food affects growth rates of nestlings, final adult size, and future egg and clutch size. Laboratory males attain testes of adult size and can produce sperm around 70 days post-hatch and females can ovulate before 100 days of age, but wild birds mature even earlier.

Zebra Finches are socially monogamous, and pair for life, but re-pairing is frequent due to high mortalities. In northern Victoria, young hatched in the first half of the breeding season make up 44% of pairs making breeding attempts in the second half of the season. For birds breeding in the season of hatching the median age of first breeding is 95 days in males and 92 days in females. There were no significant differences in breeding success between novice and experienced pairs. High nest predation limits breeding success: 35% of nesting attempts, 41% of incubated eggs and 54% of incubated clutches produce at least one fledgling. Only 9% of eggs produce young to breeding age.

7 Breeding periodicity

'Why this hardy little finch should be so much more abundant than its congeners is a puzzle; but no other is so prolific, for I do not think that at any period of the year nests might not be found if searched for.'

F. Lawson Whitlock, Western Australia (in Cayley 1932).

The Australian Zebra Finch is renowned in the literature on avian breeding periodicity for its irregular aseasonal breeding, and for its ability to respond immediately to good falls of drought-breaking rain irrespective of the season. However, across the extensive range of distribution seasonal breeding tends to prevail, although timing varies among populations according to climatic region.

Unfortunately, there are almost no data on the timing of breeding of the Lesser Sundas Zebra Finch. On Flores there are no breeding records for Zebra Finches but Verheijen (1964) found that six other species of estrildines found there had a breeding season from February to July with a clear peak in March, April and May. Zebra Finches are likely to breed at these times as well. On a visit to Roti, southwest of Timor, Verheijen (1976) found Zebra Finches breeding in March and April. The climate of Timor and the more eastern islands of the Lesser Sundas is similar to that of Darwin and the northern extremes of Western Australia in having a typical wet–dry monsoonal seasonality. The monsoons stimulate a new cycle of plant growth which produces the ripening grass seeds necessary for breeding from about December to the following May, and this is the period when Lesser Sundas Zebra Finches would be expected to breed. Good sets of breeding data from different parts of the archipelago are needed to confirm this.

By contrast, there are reasonably good sets of data on breeding periodicity for the Australian Zebra Finch. Months of breeding for twelve localities and/or regions are shown in Figure 7.1. The best data come from five studies of breeding colonies where long-term monitoring of reproductive activity was undertaken with the specific aim of determining the timing and duration of breeding. In addition to these studies, less-rigorous data were extracted from the RAOU Nest Record Scheme and the Field Atlas of Australian Birds, which both compiled records from numerous scattered localities covering many years. I have combined these data into seven avifaunal regions identified by Blakers *et al.* (1984). These data were collected by volunteers along with information on many other species of birds. In some cases monitoring was systematic and localised, but most was opportunistic and dispersed, consequently months with no breeding records, or very few, may mean that no nests were found, or simply that none were checked. The size of these data sets are often small, especially for northern localities and additional data are badly needed for most northern regions of Australia.

Geographic variation in breeding seasonality

Inspection of the graphs in Figure 7.1 shows that most localities had peaks and troughs in breeding activity. Except for Alice Springs, there was a distinct reduction in nesting activity in the winter months which led to the complete cessation of breeding at higher latitudes for periods of up to four months duration (northern Victoria, 36°S). All 12 localities had a strong spring surge in breeding activity following a winter lull. Spring was the peak time for breeding at ten of the 12 locations. This spring surge began in August in northern and inland localities, and began a month later in more southern ones. In five locations there was a resurgence in breeding activity in the autumn months and this exceeded the spring surge at Griffith and Armidale.

Breeding patterns for Alice Springs, York, Wunghnu, Griffith, and Armidale (Figure 7.1: graphs (a), (d), (f), (g), and (i) respectively) summarise four to seven consecutive years of monthly data and are reasonably representative of the breeding season over that period, but the same cannot be said for the patterns shown for the other locations.

The annual patterns in breeding seasonality of Zebra Finches across the Australian continent are closely coupled with pulses of maximum plant productivity predicted by Nix's (1976) plant response model. This is based on climatic inputs of solar radiation, temperature, precipitation and evaporation. Broadly speaking, all avian species in the south and east of Australia breed in spring and those in the north breed in summer; breeding in the arid inland is irregular. The onset and breeding peak of granivores, such as the Zebra Finch, lag behind those of species that eat insects, nectar or fruit, since seeds are produced later in the growth cycle. In the following descriptions of breeding periodicity for the 12 locations and regions, most of the uncited climatic data come from Walter *et al.* (1975).

Alice Springs—central ranges (Figure 7.1a)

The climate is arid with an average annual rainfall of 263 mm, of which 70% falls in the summer months (Millington and Winkworth 1978). The incidence of precipitation is extremely unpredictable (Stafford Smith and Morton 1990), and there is no predictable growing season on a yearly basis. Slatyer (1962) estimated that, on average, rainfall at Alice Springs was only sufficient for two short growth periods each year. Data

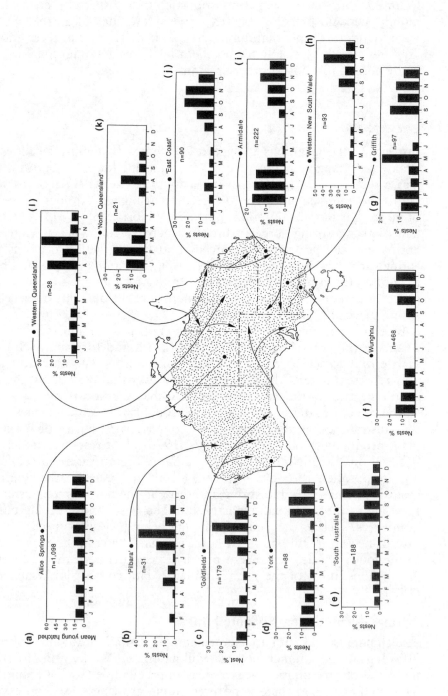

(a) Alice Springs n=1,098 — Mean young hatched; J F M A M J J A S O N D; 60 45 30 15 0

(b) 'Pilbara' n=31 — Nests %; J F M A M J J A S O N D; 40 30 20 10 0

(c) 'Goldfields' n=179 — Nest %; J F M A M J J A S O N D; 30 20 10 0

(d) York n=88 — Nests %; J F M A M J J A S O N D; 30 20 10 0

(e) 'South Australia' n=188 — Nests %; J F M A M J J A S O N D; 30 20 10 0

(f) Wunghnu n=468 — Nests %; J F M A M J J A S O N D; 30 20 10 0

(g) Griffith n=97 — Nests %; J F M A M J J A S O N D; 30 20 10 0

(h) 'Western New South Wales' n=93 — Nests %; J F M A M J J A S O N D; 50 40 30 20 10 0

(i) Armidale n=222 — Nests %; J F M A M J J A S O N D; 20 10 0

(j) 'East Coast' n=90 — Nests %; J F M A M J J A S O N D; 30 20 10 0

(k) 'North Queensland' n=21 — Nests %; J F M A M J J A S O N D; 30 20 10 0

(l) 'Western Queensland' n=28 — Nests %; J F M A M J J A S O N D; 30 20 10 0

Fig. 7.1 Timing of breeding of Australian Zebra Finches. Graphs show annual breeding at five locations (a, d, f, g, and i) based on monthly monitoring of specific populations. Other graphs (b, c, e, h, j, k, and l) are compilations of regional breeding data taken from the RAOU Field Atlas of Australian Birds (1977–1981) and the RAOU Nest Record Scheme (1957–1992). The y-axis shows the per cent of nests with eggs discovered each month except for graph (a) which shows the mean number of young hatched each month based on backdating of hatching date from aging of free-flying young caught at a walk-in trap. (a) Alice Springs—central ranges: 1986–1992 (From Zann *et al.* 1995); (b) 'Pilbara': locations between 20°–25°S and 116°–126°E; (c) 'Goldfields': locations between 25°–33°S and 115°–126°E (excluding York and the wheatbelt district of Western Australia); (d) York (31°52′S, 116°47′E): 1959–1964 (Davies 1979); (e) 'South Australia': locations between 27°–35°S and 133°–140°E; (f) Wunghnu: 1985–1989 (Zann 1994a); (g) Griffith: 1954 (Frith and Tilt 1959); (h) 'Western New South Wales': locations between 30°–35°S, 139°–145°E; (i) Armidale: 1961–1965 (Kikkawa 1980); (j) 'East Coast': locations between 21°–33°S, 145°–153°E; (k) 'North Queensland': locations between 17°–21°S, 139°–147°E; (l) 'Western Queensland': locations between 20°–26°S; 139°–145°E.

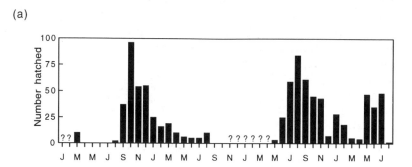

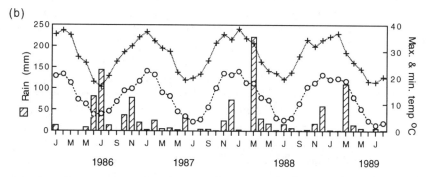

Fig. 7.2 Breeding activity of Zebra Finches (a) and climatic data (b) at Alice Springs for the period January 1986 to August 1989. Breeding activity is based on backdating the date of hatching of trapped young. No data were available for those months with a question mark. (Modified from Zann *et al.* 1995.)

summarised in Figure 7.1a were collected from 1986 to 1992 during which the average annual rainfall was 243 mm.

The graph gives the impression that breeding was more regular than it really was. Breeding was more erratic at this location than any other despite mean monthly breeding records shown for every month of the year. On the contrary, it was highly discontinuous and non-annual and only four breeding episodes were detected over the seven years (Zann *et al.* 1995). On the one hand, no breeding was detected in 24 monthly samples including continuous periods of 12 months duration, while on the other hand, breeding continued uninterrupted for intervals up to 15 months (Figure 7.2). The relationship between significant falls of rain and breeding was clearly established, even when rain fell in winter months, and this confirms Immelmann's (1963a,b, 1965a, 1971) well-known assertion that Zebra Finches are opportunistic breeders in central Australia. Zann *et al.* (1995) found a lag between the onset of drought-breaking rain and the onset of breeding that ranged from two to four months in winter to one to two months in summer. These lags are directly related to prevailing temperatures and their effects on the growth response of the annual and perennial grasses. In either case the hatching of the first clutches coincided with the first flushes of ripening grass seeds. Once a breeding episode began, pulses or surges of breeding activity corresponded with falls of follow-up rain with the appropriate seasonal time-lag. Clearly, Zebra Finches in central Australia have the ability to exploit for breeding purposes significant falls of rain at anytime of the year.

Northern Australia

No detailed breeding records are available for the northwestern part of Australia. This part of the wet–dry tropics forms the most northern part of the range of distribution of the Australian Zebra Finch. Here the northwest monsoons begin around November and continue to April or May with most precipitation occurring in January, February and March. No rain normally falls in the winter months from June to September. Immelmann (1963a, 1965b) spent October–December 1959 and February–April 1960 at the Kimberley Research Station near Kununurra (15°42′S, 128°36′E), about 100 km inland from the coast during which he made extensive observations of breeding behaviour of most species of birds in the area. This pioneering study was one of the first ever made in the wet season by an ornithologist in northwest Australia. There were three main types of vegetation in his study area: riverine forest, eucalypt savanna and grasslands. Grasses die back in the dry season and seeds can germinate at the start of the wet season and within a few weeks flower and set seed. However, after heavy deluges, extensive areas of low-lying land can remain flooded for many weeks so that germination and seed set can only happen when things dry out towards the end of the wet season.

According to Immelmann, the breeding period of Zebra Finches is limited to the start (November and December) and end (March and April) of the wet season. Rainfall is essential to germinate the annual grasses for the production of the half-ripe seeds, which he considered to be the prerequisites for breeding. However, the heavy falls of rain in the middle of the wet season interrupt breeding and drive Zebra Finches inland (Chapter 2). Therefore, the number of months in which they breed in this region depends on how abrupt the wet season starts and stops—the more protracted the light, scattered showers that precede and follow the main storms of the monsoon, the longer the breeding period. By comparison, nine other species of estrildines that remained at the study site throughout the wet season did not begin to breed until well into the second half of the season (Immelmann 1963a).

Immelmann (1965b) contends that essentially a similar breeding pattern prevails in other parts of northern central Australia but he presents no evidence. It is reasonable to assume that the Zebra Finches breed soon after the wet season begins in order to exploit the first of the ripening grass seeds. According to Nix's (1976) model for the monsoonal wet–dry tropics the peak of granivore breeding is from December to May although unseasonal rains may permit breeding in spring. Few Nest Record Scheme data are available for Zebra Finches from North Queensland (mainly Townsville and Mt Isa; Figure 7.1k). The pattern of late wet season breeding with some spring breeding, and the absence of breeding in November and December, is probably real, rather than an artefact of inadequate sampling as most data in this set were collected by local residents.

Southeastern Australia

Patterns of breeding in southeastern Australia reflect the changing climate where, with increasing latitude, winters become progressively more severe and rainfall switches from a predominantly summer regime to a winter one. Scattered Nest Record Scheme data covering a large area of the dry inland of southeastern Queensland and northeastern New South Wales show a pronounced breeding peak in late winter and early spring with low levels in other months (Figure 7.1j). This is mostly a region of moderately wet summers and dry, reasonably mild winters, but is subjected to long dry spells that become increasingly frequent towards the inland.

Armidale—Eastern Highlands (Figure 7.1i)

The good data set compiled over four seasons by Kikkawa (1980) for Armidale, show the combined effects on the breeding season of fairly severe winters (due to the elevation—1090 m) together with a reasonably high precipitation (mean annual rainfall 792 mm) that peaks in summer. This produces a reasonably long, slightly bimodal, growth period over all but the winter months. Correspondingly, Kikkawa found Zebra Finches breeding in all months except June and July, and in three study seasons there were two breeding peaks, one in spring (October or November) and one in autumn (January, February or March). In the fourth season there was no spring peak—possibly a consequence of lower mean minimum monthly temperatures in September and October. There was no strong relationship between monthly rainfall and the breeding activity for the current or the previous month. On the contrary, Kikkawa described a unique situation where there was a strong breeding surge in March 1965 during a prolonged dry period that killed grass and left soil exposed. Some nesting attempts were successful despite the absence of grass although Kikkawa does not mention whether the crops of nestlings were devoid of green or half-ripe seeds. It is possible that the finches may have found a cultivated source of half-ripe seeds that allowed them breed during the drought. Nevertheless, the observations are anomalous and stand in contrast with those made earlier at Armidale and those made in other parts of Australia.

Griffith (Figure 7.1g)

The first study of breeding seasonality of Zebra Finches was made from 1953 to 1956 by Frith and Tilt (1959) at Griffith in southern New South Wales which, because of its latitude and location on the inland plains, experiences mildly wet, but fairly cold winters and long, hot, dry summers. The region is semiarid (mean annual rainfall 403 mm), with about six months of frost, and has a bimodal season of maximum plant growth with the peaks in spring and autumn. Zebra Finches nested in a citrus grove and had a very regular breeding season. They bred in all months except June and July when low temperatures prevented breeding. No details on feeding ecology were given by the authors but breeding patterns correspond to the predicted availability of seeding grasses. The impact of irrigation on breeding of Zebra Finches in semiarid regions, such as this, is not to prolong the breeding season since this appears to be limited by low winter temperatures; rather, irrigation tends to dampen variation within and between seasons by reducing the troughs in breeding activity by compensating for irregular rainfall once the season has begun. Precipitation during the breeding season still appears to exert an effect since the spring and autumn peaks of breeding coincide with the maximum period for plant growth in non-irrigated areas.

Northern Victoria (Figure 7.1f)

The third detailed study of breeding in eastern Australia was conducted in northern Victoria by Zann and Straw (1984a) from 1976 to 1982 and by Zann (1994a) from 1985 to 1989. The semiarid climate is similar to that at Griffith, except that it is slightly wetter (450 mm mean annual rainfall) and the winter slightly longer being two degrees further south.

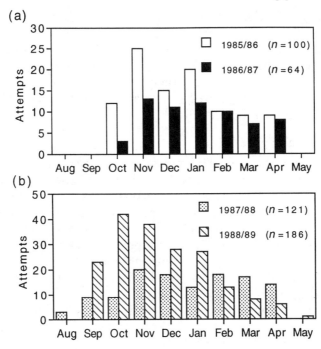

Fig. 7.3 Breeding attempts (≥ one egg laid) of Zebra Finches at the Danaher colony over four breeding seasons by month (a) 1985–1986 and 1986–1987, (b) 1987–1988 and 1988–1989. (From Zann 1994a.)

Although summers are not particularly dry, the region is irrigated from about October to April to compensate for the irregularity in rainfall and the high evaporation. Data shown in Figure 7.1f are from one small (Danaher) colony at Wunghnu and show breeding in all but the four winter months. Each of the four seasons data show a spring peak in breeding, but the timing and extent varied from year to year (Figure 7.3). In three of the four breeding seasons there was a significant positive correlation between the number of breeding attempts made each month of the season and the mean minimum temperatures for the current month (Zann 1994a).

Monthly breeding activity at Padgett and Cloverlea, 3.7 and 6.7 km west of Wunghnu, respectively, showed more pronounced autumn peaks in breeding. These were as large as those in spring, or larger (see Figure 4.2). A detailed study of diet and breeding conducted over 15 months at Cloverlea in 1981–1982 showed that each surge in nesting activity over the breeding season corresponded with a new flush of ripening grass seeds. The more abundant and larger the species of seed the greater the breeding response. The timing of the seed flushes depended on the prevailing temperature and the seeding phenology of each species.

The pattern of breeding in northern Victoria corresponds well with

Nix's (1976) plant growth response model calculated for Deniliquin, a region 80 km north-west of Wunghnu. This model has a bimodal distribution with a peak in spring and a smaller one in autumn. The indigenous grasses have comparable seeding phenologies to those of the exotic pasture grasses that have largely replaced them at this location so that breeding from early spring to late autumn would have also been possible before European settlement.

South Australia (Figure 7.1e)

Southeastern parts of the State of South Australia have a humid Mediterranean-type climate with wet, cold winters and dry summers. Spring is the time for maximum plant growth. As one progresses northward, precipitation decreases rapidly and its occurrence becomes increasingly episodic and irregular such that arid conditions prevail; consequently, there is no regular annual growth period. The 188 breeding records from the Nest Record Scheme come principally from the more populated eastern half of the state below latitude 30°S. They show a pronounced peak in spring that occurs in August and September in the arid northern parts and in October in the semiarid southern region.

Southwestern Australia (Figure 7.1d)

Breeding seasons of birds in the southwestern region of the State of Western Australia have attracted interest from a number of early workers because of instances of out-of-season breeding. Davies (1979) provided the most recent and definitive account, including an analysis of five seasons of monthly data collected by Serventy between 1959 and 1964 from a Zebra Finch colony breeding in the wheat-belt at Grass Valley, York, 100 km east of Perth. This colony was also studied by Immelmann for three to four months in the spring and summer of 1959–1960 (Immelmann 1962a). York is semiarid with a mean annual rainfall of 445 mm, but precipitation falls almost exclusively in winter due to moisture-bearing southwesterlies from the Indian Ocean, so there is a hot dry period of three to four months duration in the middle of summer. This asynchrony of maximum temperatures and maximum precipitation results in a bimodal plant growth pattern with a pronounced spring peak followed by a deep summer trough that rises to a smaller peak in autumn (Nix 1976). The smaller autumn peak is not only a response to the onset of early winter rains, but also to heavy episodic rain caused by cyclonic activity in late summer. Serventy's pooled data for Zebra Finches breeding at York correspond well with this pattern. There is a non-breeding period over the winter months followed by a nine-month breeding season that peaks in October and November, declines in the summer months, and peaks again slightly in April. Davies could find no direct correlation between the number of breeding nests started each month and rainfall figures for that month, nor could he find any correlation with mean monthly temperatures, although mean minimum monthly temperatures are probably more relevant. Immelmann (1962a, 1965a) was impressed by the sudden inhibitory effects of low temperatures on the onset of spring breeding by Zebra Finches; 50 per cent of clutches were abandoned during a cold snap in mid-September 1959. Unfortunately, no dietary studies were conducted during these surveys to determine the availability of grass seeds, which may have been responsible for much of the variation across seasons.

Arid Western Australia (Figure 7.1b,c)

Information on the timing of breeding by Zebra Finches in the arid regions of Western Australia comes from two sources: scattered Nest Record Scheme data across a vast area and an intensive three year study conducted at one location by Davies (1977a). Data from the former can be pooled into two regions: 'Goldfields' and 'Pilbara'. The Goldfields graph (Figure 7.1c) includes sites in the inland subdivision of the southwest below 25°S, and the Pilbara graph (Figure 7.1b) includes locations between 20° and 25°S. Mean annual rainfall is less than 250 mm in the Goldfields and is irregular throughout although more seasonal in the north (mainly occurring in autumn and early winter) than in the east and south. In the Pilbara region, mean annual precipitation is greater and more seasonal than Goldfields with most falling in late summer. The graphs show that it is possible for Zebra Finches to breed in most months of the year in both regions but there is a distinct peak in September and October. There is a smaller peak that occurs in early autumn in the Goldfields and in late autumn–early winter in the Pilbara. Data for the Pilbara are few and must be interpreted cautiously but they are consistent with Carnaby's (1954) observations that birds in this region nest predominantly in winter and spring but can nest at any time of the year if there are sufficient rains.

In 1973, 1974 and 1975 Davies (1977a, 1979, 1986) monitored breeding activity of Zebra Finches for 25 continuous months at Mileura Station. The mean annual rainfall for Mileura is 198 mm and occasional heavy falls can occur in summer and winter but not every year (Davies 1986). He estimated the month of hatching by back-dating the age of subadults from among 20–30 birds collected each month. The degree of ossification of the skull was used to age birds up to 120 days post-hatch (Serventy *et al.* 1967), and so provided a means of determining when breeding occurred up to four months previously. Although the number of young hatched each month is difficult to extract from Davies' (1977a) Figure 1, my analysis shows that in 1973 hatching occurred in every month with a slight peak in August and September. In 1974 hatching occurred in January, November and December with a few hatching in February, May and October. Data were also available for seven non-continuous months in 1975 when breeding was backdated to April,

May, June and August; none were estimated to have hatched in January, February, March and July. Despite strong variation among years, the data, taken together, indicate that Zebra Finches at Mileura have the ability to breed in any month and in some years may breed continuously at low levels, while in others there may be no breeding at all for seven months or longer. However, they breed slightly more in spring and early summer.

The relationship between rainfall and breeding at Mileura is not straightforward, and my analysis of the data differs slightly from that of Davies. Rainfall in winter 1973 was followed by a good burst of breeding the following spring, but good rains in winter 1974 only resulted in weak early summer breeding. The effects of rainfall on seed production is the crucial factor. Davies (1977a) found three main species of annual grass seeds (*Aristida contorta, Eragrostis australis* and *Eriachne* spp.) in the crops and believes that the availability of plentiful supplies of these seeds was responsible for the breeding of Zebra Finches at Mileura. Insects were rarely found and can be assumed to be of no importance in the diet. J. Mott (1972), a plant ecologist, found that these species of grasses at Mileura normally germinated after heavy falls of rain in summer and they flower and set seed in autumn. They can also germinate after winter rain and seed in spring if insufficient rain has fallen the previous summer, but the rate of germination and the extent of seed set is much lower than those produced by summer rains. Therefore, ripening seed should normally be available in abundance in late summer and early autumn in most years but can also be found in the spring of some years. Thus, the spring breeding in 1973 must have resulted from germination of seed in winter caused by the heavy winter rains. Further rains in October and November 1973 probably resulted in the germination of more seed and extended the breeding season into January and early February. The failure of the March and April rains to produce a new burst of winter breeding is probably a consequence of poor germination after the good summer germination (Mott 1972). Subsequent breeding episodes in 1974–1975 appear to be preceded by rainfall events with variable lag periods. The tightness of the relationship between rainfall and breeding can best be established by examining the crops of nestlings for seeds and relating these to the seed set and germination of the grass species concerned.

Eastern Australia (Figure 7.1l,h)

The two arid regions in inland eastern Australia for which there are Nest Record Scheme data show pronounced peaks in breeding during the spring months. The spring peak occurs earlier in western Queensland (Figure 7.1l) than in western New South Wales (Figure 7.1h), presumably because of the more prolonged winter in the latter region. The Queensland data are relatively few and a probably biased towards the winter and spring months when the area is more frequently traversed by amateur data collectors, whereas the data for New South Wales are more substantial and reliable since systematic samples were gathered by residents in the Broken Hill area. Rainfall is low (mean annual rainfall 244 mm) and aseasonal in distribution.

Length of breeding periods

Data summarised in Figure 7.1 indicate that Zebra Finch populations have the capability for very long periods of breeding at all localities throughout the range. In western Queensland, central Australia (Alice Springs) and the Goldfields region of western Australia breeding was possible in any month of the year, and in some years at Alice Springs and Mileura breeding occurred continuously for periods exceeding 12 months. However, when data across many years are combined, they show that some months were higher in nesting activity than others. A standard index of nesting activity is needed to compare the length of the breeding season in the different areas. This is the 'Equally Good Months (EGM)' breeding index devised by MacArthur (1964). Calculation of the EGM requires a good set of breeding data and I have confined its use to the five dedicated studies of breeding seasonality where evidence for the absence of breeding is reliable. The number of EGMs ranged from 7.6 at the most southerly study site (northern Victoria) up to 10.6 at the most northern site (Alice Springs) with intermediate values for intervening sites (Table 7.1). This trend is consistent with MacArthur's finding that birds at low latitudes have longer breeding seasons than those at higher latitudes. Wyndham (1986) found that latitude for latitude the breeding seasons of birds from African and Australia are longer than those from other parts of the world although there was no difference in the slope of the line of best fit when latitude was regressed against EGM. At each of

Table 7.1 Length of the breeding season expressed as the number of 'equally good months for breeding' (EGM)[a] for five Zebra Finch populations for which at least 12 months continuous data were available

Locality	Source (years)[b]	Latitude	Longitude	Expected EGM[c]	Actual EGM
Alice Springs	Zann *et al.* (1995) (3)	23°45′S	133°52′E	7.6	10.6
Armidale	Kikkawa (1980) (2)	30°30′S	151°40′E	7.0	8.9
York	Davies (1979) (5)	31°52′S	116°47′E	6.9	8.0
Griffith	Frith and Tilt (1959) (1)	34°00′S	146°0′E	6.6	8.6
Wunghnu	Zann (1994a) (4)	36°09′S	145°26′E	6.5	7.6

[a] EGM is calculated from the formula, $\exp(-\Sigma\, p_i \ln p_i)$, where p_i is the proportion of nests started in the ith month (MacArthur 1964). Values range from 1, when all nests started in the same month, to 12, when an equal number of nests started each month.
[b] Number of years of continuous monthly breeding samples.
[c] Based on Wynham's (1986) regression of length of breeding season at ten locations in the Australian region.

the five study sites where Zebra Finch breeding was monitored, the EGM was greater than that predicted by Wyndham's line of best fit for the Australian region. This means that Zebra Finches have longer breeding seasons than most other Australian species from equivalent latitudes. The breeding period for Alice Springs was longer than reported by Wyndham for any location in the world.

In further analysis Wyndham (1986) failed to find any correlation between mean annual rainfall and EGM in the Australian region, nor any differences in EGM between Africa and Australian localities, and concluded that irregular rainfall in Australia did not prolong the breeding season in birds of the arid zone as predicted by Immelmann (1963a,b). This was supported by the finding that in Western Australia there was no difference in EGMs for a suite of species from coastal, more mesic localities, and those from inland arid ones. In light of this, Ford (1989), in his review, concluded that opportunistic breeding in Australia was exaggerated and that seasonal breeding prevailed in the arid inland as elsewhere. However, it must be pointed out that all Wyndham's localities came from southwestern Australia between latitudes 24° and 32°S, and could in no way be considered representative of the whole arid zone, consequently, his three conclusions must remain tentative at this stage until more representative data become available. The exceptionally long breeding season of Zebra Finches at Alice Springs is more consistent with Immelmann's hypothesis than that of Wyndham's. This may be related to the fact that seasons are less predictable in central Australia than they are in southern Western Australia (Nix 1976).

Opportunistic breeding

Although breeding seasonality is low in all populations of Zebra Finches studied to date there is a consistent trend for stronger breeding in the spring months. This does not mean it is an annual event, rather the pattern only emerges in some locations when records are summarised over a number of years. Moreover, in the five studies where breeding activity was monitored each month for a number of successive years, it was clear that no two years were alike, presumably, because of differing environmental conditions. Taken together, these observations emphasise the highly flexible nature of breeding in the Zebra Finch, with, nevertheless, a predisposition for breeding in the spring months. The data, however, are incomplete for the northern part of the range, and a different pattern may emerge in this region with good sets of nesting data.

Before any long-term breeding data were available, Zebra Finches were widely regarded as opportunistic breeders in arid parts of Australia. Short-term and incidental observers in the arid zone were impressed by the ability of Zebra Finches and other arid species of birds to breed in response to sporadic rainfall, no matter what the season, in order to exploit favourable conditions that arose shortly afterwards. Observers included Carter (1889), Carnaby (1954), Serventy and Marshall (1957) and Harrison and Colston (1969) in arid regions of Western Australia; Keast and Marshall (1954) and Immelmann (1963a,b, 1965b) in central Australia; and McGilp (1923) in the Lake Frome region of South Australia. The effect of rainfall on breeding was especially impressive when observers, such as Immelmann (see below), witnessed birds starting to court and nest-build while rain was still falling. Similarly, Carnaby (1954) was also impressed by the dramatic impact of localised rainfall on plant growth and breeding by birds in the Pilbara of Western Australia. Here winter rains from locally restricted thunderstorms produced small patches, or 'oases', about two by seven kilometres in one case, of green lushness in which breeding activity was prolific, and outside of which nests were far less common.

These incidental observations would indicate that breeding by Zebra Finches and other similar species always follows rainfall; it is implied that rainfall at any time in the arid zone always produces a supply of ripening grass seeds for the provisioning of young. However, our observations at Alice Springs (see below) and those by Maclean (1976) in western New South Wales and Davies (1977a) in Western Australia show that in certain circumstances there may be a delay between rainfall and breeding, and the response may be weak. Naturally, the key factor is whether rain always stimulates a growth cycle of grasses that results in ripening seeds and whether Zebra Finches always exploit these for breeding every instance they are available. Growth and seeding of arid zone grasses is complex and varies regionally according to climate, topography and species composition.

Slatyer (1962) found that the timing of grass growth and seeding in central Australia not only depends on the timing of rainfall and the prevailing soil temperatures and available nutrients, but it also depends on what rainfall has occurred in previous seasons. Normally, to promote plant growth in this region a certain quantity of 'initial effective rain' is needed to germinate annuals and to stimulate the regrowth of perennials, but this must be followed by 'effective carryover rain' to continue vegetative growth to a stage where flowering and seed set can be achieved. Slatyer (1962) found that the prevailing temperature, through its effects on evaporation rates, determined what quantity of rain would be effective: 7–16 mm of initial rain was needed in winter and 22–29 mm in summer; carryover rains needed to be double these amounts. A single large fall of rain could sustain moisture levels so that it could serve both to initiate and carry over a cycle of growth. Another important factor is that the seeds of different species of plants germinate at different temperatures. In general, the seeds of most grasses are sensitive to frosts and remain dormant during the frost period although there are some short-lived annuals (Five-minute grass *Tripogon loliiformis*, Armgrasses

Brachiaria muliiformis and *B. gilesi*, and some species of the Nineawns *Enneapogon* spp.) that will respond to effective winter rains, but most species will not germinate or re-grow without the warm rains of summer.

In contrast, herbs, mostly daisies, germinate and grow when effective rainfall occurs in winter. I encountered this phenomenon in 1986 when I stayed at Alice Springs for eight months. In May 1986 we began monitoring the breeding activity of Zebra Finches towards the end of a prolonged dry period in which only 143 mm of rain had been recorded for the previous 17 months; all grass was dead and consisted mostly of overgrazed 'butts' (Zann *et al.* 1995). Hungry Zebra Finches were digging into the soil for scarce seeds and large flocks were coming to the few remaining drinking sites that still held water. Although many nests were checked, all were empty, and no subadults were seen among the large flocks inspected through binoculars or captured in mistnets—indicating that no breeding had occurred for at least three to four months previously. Between 19–23 June, 25 mm of rain fell and another 151 mm fell a week later. Within a few weeks this exceptional winter rain caused the rapid germination and growth of forbs and soon resulted in the renowned 'carpet of daisies' across the landscape. The exotic perennial, Buffel Grass *Cenchrus ciliaris*, the dominant species of grass in the district and the mainstay of local Zebra Finches, did not begin to regrow until early September, and seeds did not ripen until early October. Not surprisingly, Zebra Finches did not breed with the rain but waited until the food supply for their young was available. The first eggs were found on 21 August and the first young had green *C. ciliaris* seeds in their crops in mid-September (Zann *et al.* 1995). Thus, the Zebra Finches had found green seeds several weeks before we did and displayed the typical spring bout of breeding although not in the frenzied burst that we had expected from descriptions of earlier authors. This was probably a consequence of the fact that the grasses were not growing with the expected vigour and abundance despite heavy falls of follow up 'grass rains' in the warm months of September, October and November. Local plant ecologists had never experienced the situation where heavy winter rains were followed by heavy summer rains. The poor growth of perennial grasses in summer was thought to be due to the fact that the heavy layer of forbs produced by the winter rains had locked up the necessary soil nutrients and their shade perhaps prevented or retarded the germination of annual grasses.

Despite our experiences in 1986, subsequent monitoring of Zebra Finches at Alice Springs showed that not only was winter breeding possible, but it could be prolific. We believe that heavy rains in March 1988 were responsible for the strong breeding activity recorded for the following June, July and August. Similarly, in the following year, another burst of breeding in May, June and July appeared to be a consequence of heavy rain in the previous March (Figure 7.2). Unfortunately, the seeding phenology of the grasses was not monitored after 1986 so nothing can be concluded about the availability of the of grass seeds that, no doubt, made breeding possible. One must conclude therefore, that in the central arid zone not only is rainfall spasmodic and unpredictable, but its effects on the growth of grass and the subsequent production of seeds are also fairly unpredictable. In order to exploit every available opportunity for breeding Zebra Finches must track the occurrence of ripening grass seeds in both time and space and respond to any environmental cues that might predict the impending production of this resource.

Proximate causes of breeding

Most species of birds must make physiological and behavioural preparations before breeding is possible and the mechanism responsible for this has two components: an endogenous annual cycle and the ability to react to those environmental changes that predict the oncoming favourable season and entrain the cycle (reviewed by Immelmann 1971).

Gonadal activity

One possibility that should be considered first is that male Zebra Finches may not need much preparation time for a breeding episode because their testes could be held in a permanent, semi-activated state by a unique hypothalamo–hypophysial system that maintains a tonic level of gonadotropins (Farner and Serventy 1960; Farner 1967). Under this hypothesis the males would be ready to breed soon after inhibitory factors, such as low temperatures or dehydration, are removed, so that no proximate stimulating factors would be needed. Sossinka (1974, 1975) found evidence consistent with this hypothesis from aviary-bred Zebra Finches—testes of most males became fully functional by 70 days after hatching and remained active despite various regimes of water and light. Females, by contrast, developed ovarian follicles to a resting pre-egg stage by 70–90 days after hatching, but regimes of short day-lengths prolonged development (Sossinka 1980a). Although there is good evidence for similar precocial breeding in wild birds (Chapter 6) there is no conclusive evidence that birds in all populations constantly maintain gonads in an active or semi-activated state. Keast and Marshall (1954) discovered that Zebra Finches collected during a drought at two localities in inland Australia not only had minute gonads, but histological examination revealed them to be completely inactive. Recently, Dunn (1994) found that levels of androgens in the blood of wild Zebra Finches in northern Victoria, declined significantly over the breeding season. A long-term investigation of gonadal activity across different climatic regions would decide the issue.

Photoperiod

In most temperate species the increasing day length after the winter solstice is the main environmental trigger, or *Zeitgeber*, that stimulates gonadal development. Marshall and Serventy (1958) showed that captive Zebra Finches maintain active testes during total darkness, but increased spermatogenesis when subjected to both increasing and decreasing day lengths; nevertheless, they concluded that photoperiod was not an important *Zeitgeber* regulating breeding in the Zebra Finch. Sossinka (1970) found that wild-caught Zebra Finches from Katherine and Wittenoom (22°19′S, 118°21′E) layed more clutches with increasing photoperiod and fewer with decreasing photoperiods; there were no differences among domesticated females. In laboratory Zebra Finches long days increase body weight and readiness to breed (T. Meijer, pers. comm.).

In northern Victoria, there was no significant correlation over the eight to nine month breeding season between day-length and the number of eggs laid the following month (Zann and Straw 1984a). The pattern of spring and autumn peaks in Zebra Finch breeding, which are prevalent in more southern populations, suggests that day length may have some regulatory function. Lofts and Murton (1968) pointed out that the day lengths and temperatures of autumn mimic those of spring, a factor which both Davies (1977a) and Kikkawa (1980) believe may account for bimodal peaks of breeding. However, in some years my Victorian populations had an additional breeding peak in the middle of summer when day length was two hours longer than in autumn and spring and was neither increasing nor decreasing in length (Zann and Straw 1984a; see Figure 4.2).

Rainfall

In arid Australia, *Zeitgebers* that predict the impending availability of grass seeds, the prerequisite for breeding in Zebra Finches, have received much attention in the literature. Serventy (1971), Serventy and Marshall (1957), and Immelmann (1963a,b) in particular, were impressed by the strong association between the incidence of rainfall and aseasonal breeding in Zebra Finches and other arid species, and concluded that the rainfall itself must be the *Zeitgeber* in most populations and that photoperiod plays a minor role, if any. In no study however, was the link established between the occurrence of the rainfall and that of ripening grass seeds or any other essential requirement, nor was it specifically monitored.

I. C. Carnaby (1954), a professional egg-collector in Western Australia, provides the best anecdotal observations on the effects of rainfall on the breeding activities of arid zone species of birds. At Landor Station, a property on the Gascoyne River (25°10′S, 116°5′E), 200 km east of Carnarvon, Carnaby documented the breeding response of 58 species to good falls of aseasonal rain that fell 'at the end of March and in April 1934'. About 30 species nested soon after the rain event with the remaining species waiting until winter and spring, the normal time for breeding in the area. The first eggs of Zebra Finches were found on 24 April, about three to four weeks after the rain began. Twenty-two species, all insectivorous and/or nectarivorous ones, except for the Crested Pigeon, laid eggs earlier than the Zebra Finches, some as early as 7 April. Zebra Finches re-nested in May, July and August as did a few other species.

During his 12 months of fieldwork in Australia, Immelmann (1963a) was fortunate to witness the onset of two breeding episodes by Zebra Finches, one at Kununurra in the Kimberley region of northwest Australia, and the other at Alice Springs in central Australia. At Kununurra in November 1959 he describes how, on the first day of intermittent showers that heralded the approach of the impending wet season, Zebra Finches suddenly increased their rate of singing and courtship; 50 per cent of pairs began to refurbish nests the following day, and he even saw some copulations. When the showers ceased for a few days breeding activity waned, only to be rekindled with greater intensity across more pairs when showers resumed. Nesting in the population of about 200 pairs reached a peak in December. Unfortunately, the occurrence of the first seeds on the grass heads or in the crops of nestlings was not monitored so that the link between rain and seed production was not conclusively established.

After his stay in the Kimberley, Immelmann travelled to Alice Springs where he spent all of May 1960. No rain had fallen since February, but 22 mm of rain fell over a four-hour period in early May beginning around midday. He observed Black-faced Woodswallows *Artamus cinereus* starting to court within a few minutes of the downpour and the first copulation occurred about two hours later. Zebra Finches began courting just before the rain ceased and one pair was seen to copulate. The next day, Zebra Finches began to carry nesting material to refurbish roosting nests in which the first eggs were laid 13 days after the first rain. In addition to the woodswallows, which laid on the twelfth day after the rain, other species, including Budgerigars, Mulga Parrots *Psephotus varius*, Magpie-larks *Grallina cynoleuca* and Singing Honeyeaters were also observed to gather nesting material the day after the rain. The rainfall figures for the six months preceding Immelmann's visit to Alice Springs suggest that he may have witnessed a small resurgence in activity at the end of a breeding episode of four to five months duration rather than the onset of a new episode (Zann *et al.* 1995).

Immelmann believed the sight of the falling rain was the key stimulus that triggered a breeding response in these species: it was not the sound of falling rain, or its smell, or the changes in air pressure, or the increased humidity, or the wetting of the feathers. This conclusion was based on observations at Kununurra where neither 'dry' thunderstorms, nor heavy downpours at night, appeared to stimulate breeding behaviour. Furthermore, he concluded that the greening-up of the vegetation which followed rain could not possibly be a *Zeitgeber* in many cases because of the time-lag involved.

In anticipation of documenting similar events to those described so graphically by Immelmann 26 years earlier at Alice Springs we monitored nesting status weekly and breeding behaviour (singing and courtship) most days for one month before and six months after drought-breaking rains began in June and July 1986. However, courtship and nest building did not start until about six weeks after the first rains (Zann *et al.* 1995).

Our observations departed from those of Immelmann's (1963a,b) in two other respects. On his visit he found Zebra Finches nesting colonially, with up to nine active breeding nests in one bush, whereas we found nests widely dispersed in the corkwoods and two active nests in the one tree was the maximum found (Zann *et al.* 1995). Moreover, Immelmann found that in eight pairs he observed during nest construction in 1960 both sexes carried and built-in nest material; in contrast, we found, of 15 nests observed during construction, that the male carried all nest materials except in one pair when the female carried about seven %. Immelmann proposed that colonial nesting was one means of stimulating a rapid breeding response and that biparental participation in collecting nest material was an adaptation to hasten nest construction. According to Immelmann it reduced construction time from 13 days to 7–11 days. In contrast, we saw some nests built in five days despite the division of labour and some were built to the nest-chamber stage in two days. Rapid construction has also been observed in southeastern Australia (Baldwin 1973), and occurred without the urgency of drought-breaking rain.

The reasons for such diametrically different responses to heavy unseasonal rain in 1960 and 1986 are not obvious since the study sites were close (35 km apart) and timing of the rain was within a month, but, as mentioned above, the seeding of grasses in this region can be unpredictable despite good falls of rain. Immelmann's site was in mulga scrub and ours in corkwood woodland. In the intervening years there had been one major change in the composition of the dominant species of grasses in the district of Alice Springs that may partly explain the differences in urgency of breeding response. This was the introduction, establishment and spread of Buffel Grass, a drought-resistant perennial species that could withstand heavy grazing better than the native species it replaced. The original native species were short-lived annuals, Five-minute grass and species of Nineawn grass, that could germinate in both summer and winter and had the ability to produce seeds very rapidly (4–5 weeks; Lazarides 1970). Unfortunately, Immelmann could not monitor the diet and seeding phenology of the grasses during his short say. By contrast, Buffel Grass requires higher temperatures before it starts regrowing and tends to put on considerable vegetative growth before setting seed. This results in a longer lag between rain and the availability of seed. In both sets of observations, Zebra Finches may have been responding to the impending supply of ripening grass seeds, but the native grasses in 1960 probably produced seed more quickly than the Buffel Grass did in 1986.

Clearly, the response to rainfall by Zebra Finches at Alice Springs changed significantly since Immelmann's visit. The two month lag between the rainfall event and the first eggs in 1986 was not exceptional. In the six years after 1986 it happened at least four times when rain fell in autumn and winter, but the lag fell to about one month on several occasions when there were summer rains (Zann *et al.* 1995).

One could imagine how this change in breeding responsiveness came about. As the late-seeding exotic species of grass began to spread throughout the district and to dominate, then exclude, the early-seeding native grasses, Zebra Finches that continued to respond with breeding immediately on significant falls of rain would have had little or no seed for their nestlings. They may have weakened themselves in their efforts so that when seed did become available their renewed breeding attempts were less successful than those of other pairs that ignored the rain or responded more slowly. Thus, there would be increasingly strong selection against those individuals that responded immediately. This explanation requires that there exist a fair degree of genetic variability within a single population in the loci that control sensitivity to *Zeitgebers*. The ability to respond rapidly to the onset of rainfall should still be maintained by selection in those vast areas away from Alice Springs where the rapidly seeding native species of grasses still predominate. Efforts were made in 1986 to study populations at several of these sites but this proved impractical because trapping was too difficult and nests too dispersed, but observations of flocks suggested that breeding occurred no earlier than it did at Alice Springs.

However, breeding seasonality in other parts of Australia could not possibly be dependent on rainfall as a *Zeitgeber*. This is further evidence for Kikkawa's (1980) hypothesis that if proximate factors exist they must be different in different parts of the range and furthermore, subjected to strong selection.

Proximate and ultimate factors are one and the same

The availability of ripening grass seed could be the factor that triggers semi-activated gonads to full functionality. Immelmann (1963a) postulated that gonads of males in the non-breeding state are more reproductively advanced than those of females, and this enables males to respond to environmental cues first. Subsequently, courtship, singing and allopreening stimulate the females' gonadotropic system to a state where ovulation can occur. According to Sossinka (1980a) this takes a minimum of about two weeks.

Ultimate causes of breeding

'Ultimate factors' are those environmental conditions that control the efficiency of breeding in birds and are responsible for the evolution of species-specific breeding periodicities (Immelmann 1971). In most instances breeding is timed to coincide with that period of the year when maximum levels of the food required for the raising of young are available and accessible. Field studies suggest that Zebra Finches require a good supply of ripe grass seed for the production of eggs. Half-ripe seeds and some green plant material are considered essential for feeding growing young (Immelmann 1965a,b; Zann and Straw 1984a). Fledglings, require, in turn, a good supply of easily accessible, ripe seed in order to sustain them until they have become sufficiently competent foragers to compete successfully with conspecifics and other granivores (Chapter 9).

Domesticated and wild-caught Zebra Finches can breed in captivity on abundant dry seed alone without any special nestling food mixture to provide supplements of proteins, vitamins and minerals (Immelmann 1965a). Supplements, nevertheless, are strongly recommended by all authorities on captive breeding of Zebra Finches (e.g. Cayley 1932; Immelmann 1965a; Goodwin 1982; Martin 1985; Vriends 1980). Until recently there was no experimental evidence that these supplements enhanced breeding success, but there is now evidence that proteins are essential for egg production and for growth of nestlings.

Houston *et al.* (1995a) investigated the nutrients required for egg production in domesticated Zebra Finches. Birds were held on a pure panicum seed and cuttle-bone diet and it was estimated that a clutch of four eggs required about 233 mg of lipids, 540 mg of protein and 71 mg of calcium with a peak in demand on the day before the first egg of the clutch was laid. Calcium for the shells was obtained from cuttle-bone, and there was almost a fourfold increase in its consumption during laying. Surprisingly however, the lipid and protein requirements were not obtained from increased seed intake, since the female did not increase the amount of seed consumed during egg-laying; rather, these nutrients were taken from the female's body reserves—lipid from subcutaneous fat bodies and protein from stores in the pectoral muscles. Consequently, females suffered a serious decline in body condition over the laying period, losing about 14% in lean dry weight, mostly in protein from the pectoral muscles. Houston *et al.* (1995a) also detected a 65% reduction in locomotory activity by laying females and interpreted this to be a means of making substantial savings in metabolic requirements and protein turnover during a period of metabolic stress. In a follow-up study, it was discovered that only one particular protein was lost from the pectoral muscles and it was suggested that this protein, which was limited in the diet, provided the essential amino acids specifically needed for egg proteins, (Houston *et al.* 1995b). Most species of birds that have been investigated to date also show a decline in pectoral muscle weight during egg laying, but few species are adapted to breeding on a diet as low in protein as that of Zebra Finches. Despite this adaptation, laboratory Zebra Finches fed on a protein-enriched diet produce larger eggs (T. D.

Williams, pers. comm.) and larger clutches (Houston *et al.* 1995a). Therefore, Zebra Finches require proteins during egg production—some are available from the diet over the laying interval, but rare ones must be drawn from reserves stored in the pectoral muscles.

Whereas Boag (1987) found that diets differing in protein content affected growth rates of nestlings and final adult size of domesticated birds Haywood and Perrins (1992) found that only food quantity, not quality had a significant effect. It is difficult to reconcile these findings since both sets of investigators used similar sample sizes and the same analyses.

Assuming that additional protein enhances size and quality of off-spring it is surprising that wild Zebra Finches rarely supplement the diet of their young with additional protein in the form of insects (Chapter 4). Only half-ripe seeds and green leaf material distinguish the nestling diet from that of adults in the non-breeding season (Zann and Straw 1984a). Some other Australian estrildines normally eat many more insects than Zebra Finches and take additional quantities during the breeding season to feed their young (Immelmann 1962a). Interestingly, nestlings of those species that had a high insect component in their diet ate only ripe seeds whereas those that ate green material and half-ripe seeds ate no insects (Schöpfer 1989). Conceivably, half-ripe seeds and green leaf material provide vitamins and proteins essential for the efficient growth of nestlings and in some way compensate for the absence of insect protein. However, the nestling period of altricial birds is longer in granivorous than in insectivorous species (Lack 1968) so one would expect those estrildines that increase their intake of proteins via insects to have a shorter nestling period, but no differences have been reported to date (Immelmann 1962a). There is a need for more experimental work along the lines of Boag's (1987) study to examine the effects of insect and plant protein on growth rates of estrildines.

It is not simply the quality and quantity of food available that determines whether efficient breeding is possible in Zebra Finches, it is also the energetic cost of foraging relative to the energy gained from the food. Experiments by Lemon and Barth (1992) showed that lifetime reproductive success is greater in domesticated Zebra Finches where costs of foraging are low. Birds with higher foraging costs had smaller-sized broods, a greater interval between successive broods, and higher adult mortality than birds with low foraging costs. In the wild, not surprisingly, Zebra Finches prefer to forage at those sites where foraging is most efficient (Zann and Straw 1984a; Chapter 4).

Abundant and easily accessible dry seed is not a sufficient factor to stimulate nor maintain breeding in wild birds. During the 1981–1982 breeding season in northern Victoria the breeding performance of two neighbouring Zebra Finch colonies was compared. The Padgett colony consumed regular supplies of commercial bird seed at walk-in traps, while the Cloverlea colony had none. While there were no differences between the two colonies in the onset and termination of the breeding season, Padgett had a stronger breeding attempt in spring than Cloverlea (Zann and Straw 1984a). Early spring was a time of food shortage and it is plausible that the extra food provided energy for the breeding parents so that they might divert more wild seed to the young. However, timing of the major breeding peaks during the season correspond very closely to the flushes of ripening seeds produced by the wild grasses (Chapter 4).

In summary, the ultimate nutritional factor that controls the timing of breeding in wild Zebra Finches is most likely the time of maximum abundance, accessibility, and quality of half-ripe and ripe grass seed. The timing of optimal grass growth and reproduction across Australia must therefore determine when Zebra Finches will breed. This is the austral spring, summer and autumn across most of Australia (Nix 1976).

Factors that inhibit and terminate breeding

When the supply of half-ripe grass seeds is exhausted breeding should cease, but there is no good evidence that it does.

Low ambient temperatures

Low temperatures can stop spermatogenesis and ovulation in birds (Marshall 1949). Serventy and Marshall (1957) concluded that low temperatures across a range of latitudes in Western Australia inhibit or retard breeding in autumn and winter in most species of birds. Low temperatures also appear to inhibit and terminate breeding in Zebra Finches, thus forcing a non-breeding period during the austral winter months (June, July and August) throughout most of the range of distribution, including sub-tropical and tropical regions, although the effect is more pronounced at higher latitudes (Figure 7.1). The exceptions to this generalisation come from several arid inland sites with the best data coming from Alice Springs. For example, good autumn rains stimulated breeding throughout the winter months of 1987–1989 despite mean minimum monthly temperatures ranging from 2.2–5.5°C (median 4.1°C; Figure 7.2). In contrast, at the Cloverlea colony in northern Victoria, Zebra Finches did not breed when the mean minimum monthly temperatures fell below 6°C in the 1982–1983 breeding season (Zann and Straw 1984a).

Sensitivity of the breeding response to low ambient temperatures not only varies geographically, but appears to differ within a colony depending on whether breeding is being stimulated (or disinhibited) or inhibited. During detailed studies of breeding activity over four seasons (1985–1989) at the Danaher colony, spring breeding did not begin in three seasons until the mean minimum monthly temperatures exceeded

5°C; it did not begin in the fourth spring until it exceeded 6°C (Zann 1994a). However, as winter approached, the autumn temperatures that appeared to inhibit breeding were several degrees higher than those that stimulated it in spring. In three seasons breeding stopped at the end of April when the mean minimum monthly temperature for May fell below 8.2°C although one clutch was laid in May when the temperature was 8.6°C (Zann 1994a).

The inhibitory effects of low temperatures on breeding can be demonstrated when non-breeding wild Zebra Finches are captured in winter and placed in a warm laboratory, whereupon some will begin to lay within two weeks (R. Zann, unpublished observations). Experiments are needed to determine the temperature threshold for breeding disinhibition and to test for any interactions with day length.

Drought

Drought may inhibit gonad function directly through dehydration or indirectly through its general debilitating effects. Adult Zebra Finches shot during severe drought in central Australia had minute inactive testes devoid of any signs of spermatogenesis (Keast and Marshall 1954). In Davies' (1977a) three-year study at Mileura, he found that testis size was lowest during months of little or no rain. However, this does not necessarily mean that spermatogenesis had ceased, since Priedkalns *et al.* (1984) found no relationship between testis size and spermatogenesis. However, they did find that dehydration reduces testis size and rehydration increases it. When Sossinka (1972a, 1974) deprived domesticated and wild-caught Zebra Finches of drinking water for many months he found that while they could still survive, ovulation was completely inhibited, yet it commenced soon after rehydration. Finally, Menon *et al.* (1988) cite an unpublished study in which domesticated Zebra Finches bred successfully despite being completely deprived of drinking water but their diet contained apples as well as dry seed.

Captivity

In captivity aviculturalists find it difficult to prevent Zebra Finches from making breeding attempts. Dry seed, low temperatures and lack of nesting material do not prevent birds from laying eggs and attempting to raise young. If Zebra Finches are to be rested or prevented from breeding, nesting sites should be removed.

Nesting sites

In many areas of inland Australia adequate supplies of grass seeds and water make breeding possible but suitable nesting sites may be limiting. This is evident by the willingness of Zebra Finches in certain habitats to crowd nests together in the same small bush or to use structures made by man or other animals. In some instances this is an antipredator response but in others it a consequence of the limited nesting sites available (see Chapter 6).

Summary

Nothing is known about the breeding periodicity of the Lesser Sundas Zebra Finch, but it is expected to breed towards the end of the monsoon before the dry season sets in. The breeding period of Australian Zebra Finches is long and highly flexible. The minimum duration is eight months in southern latitudes, but may extend to fifteen months in central Australia in some seasons. The breeding period is longer than that of other Australian species of birds and increases in duration from high latitudes to low latitudes. Breeding can occur in any month of the year in some populations, but most show some seasonality with a winter lull and a spring peak, and some also show a second peak in autumn. Northern populations have a summer peak rather than a spring peak but more data are needed from this region. Breeding in central Australia is exceptional—although there is a spring peak, breeding is non-seasonal and a whole year may pass without any breeding attempts; when suitable conditions arise breeding will begin whatever the month. Availability of maximum supplies of ripening grass seeds is the critical factor determining when breeding will occur and this depends on the plant growth cycle for the particular climatic region. This is fairly seasonal in all locations except central Australia, where Zebra Finches are opportunistic breeders in the sense that they exploit any production of ripening grass seeds irrespective of the time of the year. Here the timing of grass growth and seed production is complex but mainly depends on when good falls of rain occur; however, this is highly erratic in space and time. A recent study in central Australia did not verify the renowned observations of Immelmann in which Zebra Finches made an immediate breeding response to rainfall, rather, there was a lag of two to three months between the first fall of rain and the first eggs. The proximate factors that stimulate breeding in Zebra Finches probably vary within and among climatic regions and are subject to strong selection pressures.

Except for the populations from central Australia, low temperatures inhibit breeding activity in Zebra Finches. Temperature thresholds for inhibition and disinhibition of breeding probably vary in different parts of the range. Dehydration inhibits spermatogenesis and ovulation.

8 Populations

'At the spring near Mt Ultim the Chestnut-eared Finches were present in countless myriads, rising in clouds with a noise like a rushing wind from every shrub and tree within a quarter-mile of the water, clothing the leafless branches of dead trees with their quivering forms looking in the distance like grey butterflies as they sought a resting place, and covering the rocks round the spring itself with their droppings till these at first sight and to the uncritical observation looked as if a light mantle of snowflakes partly hid their texture from view.'

J. B. Cleland in central Australia, 1931.

Fluctuations in size of populations

Many observers in the arid zone, especially around central Australia, have commented on the vast numbers of Zebra Finches that may assemble around water-holes during periods of drought. With so many birds coming and going observers find it difficult to make careful counts and terms such as 'abundant', 'numerous', 'countless', 'myriads', 'thousands', 'hundreds of thousands', and even 'millions', have been used. 'Literally thousands' have also been observed on dense patches of seeding grasses in semiarid areas (Rix 1943). The calls of large flocks around water can make such a din that some observers can find it a relief to leave the area. Careful counts of large flocks at water-holes have been made by Fisher *et al.* (1972) during their comprehensive survey of drinking patterns of desert birds. They found the Zebra Finch to be the second most common species observed after the Galah, and the largest number of Zebra Finches estimated drinking at a water-hole (70 km north of Alice Springs) during one day was in September 1967 when 17,750 visits were recorded. The same day 47,000 Budgerigar visits were also made. A year later Rix (1970) estimated that there must have been 'hundreds of thousands' of Zebra Finches drinking at the same water-hole.

When the last residues of water disappear from a drinking site Zebra Finches disappear, although this has rarely been documented. MacGillivray (1929) states that they die in numbers under surrounding shrubs and bushes but when good seasons return the few survivors breed rapidly and good numbers build up again. At Erldunda, 200 km southwest of Alice Springs, Corbett and Newsome (1987) monitored the abundance of Zebra Finches visiting watering points during their seven-year study of the diet of the dingo. Using direct counts of birds present around mid-day, numbers ranged from nearly 600 in 1969, a 'flush' period, to less than 100 during periods of drought (Figure 8.1a). The

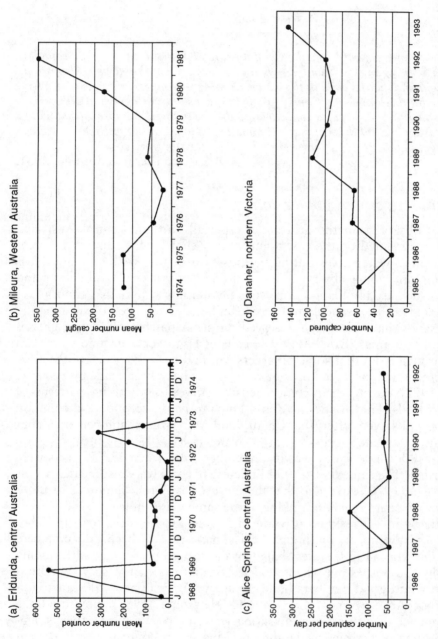

Fig. 8.1 Yearly changes in abundance of Australian Zebra Finches from four localities. (a) Mean number of Zebra Finches censused per watering point surveyed at Erldunda (25°S, 135°E), 200 km southwest of Alice Springs (modified from Corbett and Newsome 1987); (b) mean number of Zebra Finches mistnetted per day each October at Mileura station, Western Australia (from Table 4 in Davies (1986)); (c) number trapped by walk-in trap set for one day each August at Alice Springs (Zann *et al.* 1995); (d) number trapped by walk-in trap each August at the Danaher colony, northern Victoria (this study).

apparent crash in numbers after 1973 is an artefact of the method of census at water-holes—good rains fell at this time so birds no longer came to the permanent watering points surveyed. Large fluctuations in numbers of birds trapped at a watering point also occurred at Mileura (Figure 8.1b). This was a direct consequence of the amount of rainfall that had fallen in the previous 12 months (Pearson correlation between rainfall and numbers captured $r = 0.92$, $n = 8$, $P = 0.001$: S. J. J. F. Davies, pers. comm.). Fluctuations at Alice Springs were also a consequence of rainfall the previous year (Figure 8.1c; $r = 0.84$, $n = 7$, $P = 0.02$), but rain had no effect on abundance of wintering birds at the Danaher colony ($r = -0.18$, $n = 8$, $P = 0.67$), since this is an irrigated area (Figure 8.1d).

Two attempts have been made to estimate the size of Zebra Finch populations exploiting localised resources. At Mileura, Western Australia, Davies (1986) banded and released all Zebra Finches coming to a water point, 'Jindi Jindi', for a period of several days each October for eight years. He captured 2,867 individuals over the survey period, including 1,398 birds caught over a four-day period in 1981, but recaptures were quite low (total 120). He used the simple Petersen method to estimate the number of birds coming into drink by comparing the proportion of birds recaptured on the second day with that banded and released on the first day. After he banded at Jindi Jindi in 1981 he checked six adjacent water points for recaptures. He found none among 600 captured so assumed that there was no immigration or emigration over the few days trapping occurred at Jindi Jindi. His estimates of the population ranged from 686 in 1976 (137 banded in three days) to 22,258 in 1980 (351 banded in two days). While not much confidence can be placed in the accuracy of the Petersen method, since it does not produce standard errors, these values give some idea of the order of magnitude of fluctuations that can occur in natural populations.

In 1986 we also used capture-mark-recapture methods to estimate the number of Zebra Finches visiting a walk-in trap baited with seed at Alice Springs. Trapping data for four months satisfied the criteria for estimating population size using the Jolly–Seber method, which gives standard errors (Caughley 1977). Values ranged from 320 ± 130 in August to 198 ± 78 in November (Zann *et al.* 1995).

The size of Zebra Finch populations in arid Australia clearly undergo wild fluctuations in response to annual rainfall. Other aspects of the population biology of Zebra Finches were investigated using capture-mark-recapture methods at my study colonies in northern Victoria and at Alice Springs.

Survivorship

Banded nestlings

Survivorship of Zebra Finches banded at the Danaher colony was very low (Figure 8.2)—67% were lost between banding age and day 35, the age of nutritional independence, and this accounts for the initial steep decline in survivorship. Most losses probably occurred after young had fledged, hence the 15–20 day interval between fledging and nutritional independence appears the most vulnerable part of the life cycle at this colony. Unpublished observations of fledglings at Danaher are consistent with this level of losses: by day 35 only 6 out of 44 young that fledged were observed around the feeder at the walk-in trap and 9 out of 37 fledglings were observed at the roosting nests (Chapter 9). Most losses during this interval are probably mortalities rather than movements from the colony because extended flight of newly fledged young appears fairly limited, although this should not be underestimated. Predators are the likely cause of losses because food was supplied in abundance at the walk-in trap, but it is possible that parents may have neglected the young for some reason, perhaps when re-nesting. Losses after day 35 kept fairly constant and remained so until they suddenly dropped again after day 540. These losses arise from mortalities and emigration. By day 35 the improved flying ability enable young to avoid predators, and their competence in dehusking seed may approach levels of efficiency found in adults (Chapter 9).

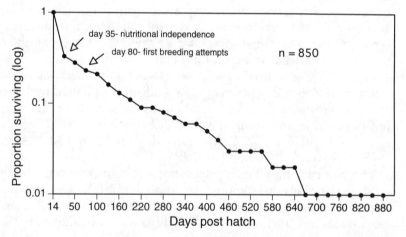

Fig. 8.2 Log-normal survivorship curve (proportion recaptured by days after banding) for 850 Zebra Finches banded in the nest at the Danaher colony 12–14 days after hatching, based on known age at disappearance. The study encompassed six breeding seasons 1985–1991. (After Zann and Runciman 1994.)

Only 23% of nestlings banded reached 90 days of age, the median age of first reproduction for birds breeding in their season of hatching. Just 65% of hatchlings reached banding age in the first place, consequently the probability that a hatchling will survive to breeding age was only 0.15 (0.23 × 0.65). Mean annual survivorship for the first 12 months of adult life was four per cent and the estimated life expectancy at hatching was 51 days.

Losses of Zebra Finches at Danaher between fledging and age of first breeding are extremely high, especially when one considers that such a short span of time is involved (about 70 days). Nevertheless, the value (77% losses) is within the range found for other species of birds (42–86%; Newton 1989), but few species reach sexual maturity as early as Zebra Finches. Data on survivorship of nestlings to age of first breeding are needed from Zebra Finch colonies in other parts of the range in order to place these findings from Danaher in perspective.

At Danaher, the proportion of males and females surviving from one month to the next (age-specific survival) was not significantly different. There was a trend for females at both Danaher and Padgett colonies to live slightly longer than males although this just failed to reach statistical significance (Table 8.1). The oldest known-age bird was a female that disappeared when four years and five months of age although a female at the Padgett colony disappeared five years and three months after first capture as a mature adult. The longest span between first and last capture at Alice Springs was 13 months for several birds caught as mature adults. At Mileura, Davies (1986) recaptured one bird three years after banding. Therefore, although the vast majority of individuals appear to have very short lives, some can live up to five years, or more, in the wild. The potential lifespan of domesticated Zebra Finches held under optimal breeding conditions is 5–7 years (Burley 1985a).

Table 8.1 Estimates of lifespan based on the interval in days between first and last capture at walk-in traps for Zebra Finches hatched in the Danaher and Padgett colonies that reached at least 35 days of age[a]

	Danaher		Padgett	
	Males	Females	Males	Females
Median	120	127.5	53	70
Inter-quartile range	48.5–211	70–275.2	35–156.5	35–185
Maximum	911	1612	1668	957
n	198	216	189	147

[a] Data include young banded in the nest and those caught before 35 days. There were no significant differences between the sexes (Danaher: Wilcoxon two-sample Test $Z = -1.88$, $P = 0.06$ and Padgett: $Z = 1.86$, $P = 0.06$). (From Zann and Runciman 1994.)

Zebra Finches first caught as adults

Survivorship of Zebra Finches first banded as mature adults, that is, birds of unknown age, was much lower for Zebra Finches at the Alice Spring colony than it was at the three Victorian colonies. Significantly fewer birds were recaptured at Alice Springs one month after first capture (Zann *et al.* 1995) and their estimated annual mortality was higher (Table 8.2).

Survivorship curves (Figure 8.3) were characterised by a sudden drop the first month after banding, and this was much steeper for the Alice Springs population than the Victorian population. During Period I at Alice Springs, this high loss of birds continued for six months. However, in Period II the high losses during the first month after banding were arrested and survivorship was fairly constant thereafter, although there was another sudden decline in survivorship at the end, between the eleventh and twelfth month after banding. It is conceivable that this last decline in survivorship may be an effect of senescence, a factor responsible for terminal decline in survivorship of other species of birds (Newton 1989), but such a decline was not evident for the three Victorian populations even two years after banding (Zann and Runciman 1994).

By comparison with other species of birds (reviewed by Newton 1989) survivorship at all four study colonies was low. Survivorship and maximum longevity were also much lower than those found in old Australian endemics inhabiting forests (reviewed by Rowley and Russell 1991). However, a recent study of three other species of estrildines (Gouldian Finch, Long-tailed Finch and Masked Finch) in the wet–dry tropics of northern Australia found survivorships that were not much higher than

Table 8.2 Population parameters for three colonies of Zebra Finches in northern Victoria and one colony in Alice Springs; estimates are derived from the capture, mark and recapture of mature adults (≥ 100 days of age at first capture) at baited walk-in traps set at monthly intervals

Colony	Number trapped	Per cent recaptured one month after first trapping	Annual mortality (%)	S_{10} (years)[c]
Danaher[a]	687	59	82	1.2
Padgett[a]	185	53	72	1.5
Cloverlea[a]	268	61	72	1.6
Alice Springs I[b]	613	18	96	0.1
Alice Springs II[b]	454	31	83	0.7

[a] Zann and Runciman (1994).
[b] Period I (June 1986 to September 1987, Period II (July 1988 to August 1989); from Zann *et al.* (1995).
[c] Interval after first banding when 10% of birds could still be recaptured.

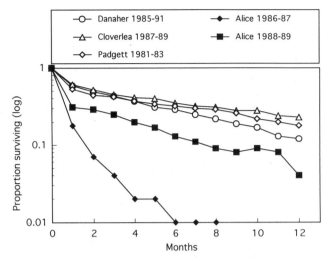

Fig. 8.3 Log-normal survivorship curves (proportions recaptured at baited walk-in traps by month after initial capture) for Zebra Finches classified as mature adults at first capture from three colonies in northern Victoria (Danaher *n* = 687; Padgett *n* = 185; Cloverlea *n* = 268) and one at Alice Springs. Survivorship is given for two sampling periods at Alice Springs: Period I (June 1986 to September 1987; *n* = 613) and Period II (July 1988 to August 1989; *n* = 454). (After Zann *et al.* 1995.)

those found here for the Zebra Finch (Woinarski and Tidemann 1992). Nevertheless, in a pilot study, Tidemann (1987) found that the Zebra Finch had the lowest recapture rate among seven species of estrildines trapped over a four day interval at drinking points in the Top End.

Low survivorships estimated from capture-mark-recapture studies do not necessarily mean that mortalities are extreme because 'dispersal' could also make birds unavailable for recapture. Trap shyness, which also reduces rates of recapture, is more significant with mistnetting than with walk-in traps (Zann and Runciman 1994). Unfortunately, in capture-mark-recapture studies, there is no easy way to distinguish effects of mortality from those of 'dispersal', but a number of lines of evidence suggest that 'dispersal' may be a significant factor accounting for the high losses of banded Zebra Finches from the trapping sites.

Sex ratios

Sex ratios of the Lesser Sundas Zebra Finches are unknown, however, secondary and tertiary sex ratios have been examined at a number of widely separated populations in the Australian subspecies (Burley *et al.* 1989). Secondary sex ratios are those found in young at the end of parental care and tertiary ratios are those found at adulthood.

Tertiary sex ratios were slightly male biased, but not significantly so, across six populations (Table 8.3). This trend was consistent (test of homogeneity: $G_1 = 0.53$, $P = 0.97$), and when pooled the sex ratio for 10,741 adults was 52.0% males which was significantly different from parity ($G_1 = 8.8$, $P = 0.003$). There was no significant between-year variation in the tertiary sex ratio at the northern Victorian colonies nor did method of trapping (mistnetting *vs.* walk-in trap) affect the sex ratio (Burley *et al.* 1989).

Reliable data for calculating secondary sex ratios were obtained from Padgett and Danaher where young were sexed between 35 and 40 days of age, just after they have become nutritionally independent of their parents. Whereas, Padgett birds ($n = 148$) were 52% male (Burley *et al.* 1989), Danaher birds ($n = 407$) were 44% male (Zann and Runciman 1994); however, the difference was not significant ($G_1 = 2.8$, $P = 0.7$) nor were either ratios significantly different from parity. Nevertheless, a female-biased secondary sex ratio (44–47% males) was found in each of the six years at Danaher, and because almost every banded nestling that reached 35 days of age was captured and sexed it suggests that a real, yet slight, bias towards females may exist.

The proportion of males and females natal to Danaher changed between independence and adulthood. Over an interval of about 60 days, significantly more females than males disappeared, so that the sex ratio shifted from one with a female bias to an unbiased one at 100 days of age. This tertiary sex ratio was 49% males and was not significantly different from that of adults arriving at Danaher from other colonies. It is not clear whether the differential loss of young females from their natal colony before adulthood represents higher rates of dispersal or mortality than those of their male counterparts.

Sex ratios of young Zebra Finches dispersing to Danaher from other colonies varied according to age, and to a lesser extent, year of study (Zann and Runciman 1994). There was a significant male-bias (60% males) among the younger class of dispersers (36–50 days of age) and a significant female-bias (37% males) in the older class (51–100 days of age). By the time these immigrants reached adulthood the sexes were almost equal (217 males and 214 females). These capture data suggest that males disperse to Danaher at a significantly younger age than females; alternately, males may simply become active at a younger age than females and so encounter the trap sooner, thus creating a false picture of a biased sex ratio (Burley *et al.* 1989). By the same argument, there should also be a male bias in the secondary sex ratio of birds hatched at the Danaher colony yet a female bias was found. Hence, the bias towards males in the younger age class is more likely to arise from either a male-biased secondary sex ratio, or sex differences in the age of dispersal from other colonies rather than from sex differences in activity

Table 8.3 Tertiary sex ratios (% males) of Australian Zebra Finches from six populations; modified from Burley et al. (1989)

	Shepparton		Padgett		Danaher[a]		Alice Springs[b]		Top End		Mileura[c]	
	m	f	m	f	m	f	m	f	m	f	m	f
Adults	223	209	402	381	372	315	1908	1783	308	290	1468	1334
Percentage male	52		51		54		51		51		54	
Young surviving to adulthood	29	18	110	96	121	125	11	6	–	–	–	–
Percentage male	62		53		49		65		–		–	
Total adults	252	227	512	477	493	440	2555	2385	308	290	1468	1334
Percentage male	53		52		53		52		51		52	

[a] Zann and Runciman (1994).
[b] Zann et al. (1995).
[c] S. J. J. F. Davies (pers. comm.).

levels. One can only speculate why young post-independent males at Danaher were more philopatric than their counterparts from neighbouring colonies. Secondary sex ratios may possibly vary with the quantity and quality of the food supply; a phenomenon found in some species of mammals (Krebs and Davies 1993). In northern Victorian colonies, Burley *et al.* (1989) found a significant negative correlation between sex ratio and the amount of precipitation that fell the preceding season and hypothesised that females may be selectively produced when food is abundant. The female bias in the secondary sex ratio at the Danaher colony is consistent with this idea because seed was provided *ad libitum* at the walk-in trap. However, wild food may be more critical for rearing young than dry commercial seed dispensed from a hopper, and flood irrigation ensures that its productivity is independent of local precipitation. Why abundant food should make males more philopatric yet not affect natal dispersal of females is a mystery. Conceivably, natal dispersal in males may be contingent on current environmental resources, such as food supply, whereas females may be primarily 'innate' dispersers which disperse irrespective of prevailing conditions. A further difficulty is that the secondary sex ratio appears female biased and the tertiary sex ratio appears slightly male biased yet no significant differences in survivorship between the two sexes could be detected, as one would predict if females suffer higher mortality during the course of dispersal. On the contrary, lifespan of females appears to be slightly longer than that of males (Table 8.1). These apparent contradictions in the secondary sex ratio data may dissipate when further long-term experimental studies accumulate sufficiently large data sets to detect, in a statistical sense, the small differences that appear to exist.

Dispersal and mobility

Natal dispersal

Between 1985 and 1989 only 23.5% (57/243) of individuals identified making breeding attempts at the Danaher colony were hatched in the colony, or nests nearby. Obviously, immigrants constitute the great majority of recruits to the breeding population here, and this may be representative of colonies elsewhere. More studies are needed to verify this.

At the Danaher colony there was no sex-biased philopatry among birds breeding in their natal colony nor was there any sex-biased dispersal among adult immigrants (Zann and Runciman 1994). Consequently, the costs and benefits of moving from, or remaining in, the natal colony fall equally on both sexes. This contrasts with the situation in most species of passerines where females are the dispersing sex (Greenwood 1980; Gowaty 1993). Most data on natal dispersal come from territorial species so it is conceivable that other examples of sexually unbiased patterns of dispersal may be found, especially in species that have a colonial breeding system similar to that of the Zebra Finch.

Dispersal to the Danaher colony

At all colonies where banding studies were undertaken, all-day trapping on a weekly, or even daily basis, yielded an unending supply of unbanded individuals that must have recently arrived from other areas. One consequence of this phenomenon was that only 22% (246/1,125) of all adults captured at Danaher between 1985 and 1991 were hatched in the colony itself or nearby. Thus, it was an almost a Sisyphean task to ensure all birds at the study colonies were banded. Newly arrived individuals were re-classified as 'transients' if they were not recaptured ≥ 28 days after first trapping and banding; if they were recaptured they were classified as 'residents'. Twenty-eight days was long enough for individuals to be trapped twice in the non-breeding season and three times in the breeding season and was sufficient time for them to make a detectable breeding attempt.

Both adults and young dispersed to Danaher from other colonies (Zann and Runciman 1994). Most adults arrived in the spring and fewest in winter, whereas young arrived almost exclusively in summer and autumn. While there was no significant variation in the proportion of males and females arriving from other colonies, their timing was slightly different, with male arrivals outnumbering females in spring and autumn. More adult males arrived from other colonies than young males.

The sudden fall in survivorship in the first month (Figure 8.3) after banding probably indicates that the transients had moved on, although an unknown proportion would have died after arrival at the colony. There was a significantly greater proportion of transients at Alice Springs than in the Victorian colonies (Table 8.2). Method of trapping also had a significant effect on the proportion of arrivals that stayed on to become residents. The recapture rate one month after banding improved at the Padgett colony when the walk-in trap replaced mist-netting (Zann and Runicman 1994). Presumably, the seed used as bait was partly responsible for inducing potential residents to stay longer, and may have reduced mortality rates. The seed bait did not affect recapture rates of adults beyond the first month, but it did significantly increase the survivorship of sub-adults.

Despite high rates of immigration into the study colonies and losses of banded birds from them there was remarkably little direct evidence of movement of individuals from one trapping site to another. Concurrent monthly trapping over a period of 33 months at Cloverlea and Danaher, neighbouring colonies only 6.7 km apart, yielded only four individuals that moved out of 596 birds banded. Leg bands of numerous Zebra Finches were also checked visually, via binoculars and telescopes, at all trapping sites in northern Victoria and in much intervening country, yet few birds were sighted more than 2 km from their site of banding. Movement of two birds from Padgett to Bunbartha, a distance of 9 km, was the maximum movement detected. Appeals to the public via the local media for information on banded birds were fruitless, nor were any recoveries of our banded birds reported to the Australian Bird and Bat Banding Scheme from other bird banders or members of the public. Similar results were found at Alice Springs in 1986. Only three birds from more than a thousand were retrapped away from their banding site and all were less than 4 km away. Again regular visual surveys failed to reveal any movements. Davies (1986) detected little movement of birds between sites of banding and recapture at Mileura although his recapture rate was very low (4%). The longest movement registered with the Australian Bird and Bat Banding Scheme is 20 km (Blakers *et al.* 1984).

Absences of 'residents' and excursions from colonies

Indirect evidence suggests that many residents of the study colonies of both natal and immigrant origins were temporarily absent for periods of variable duration, during which they presumably made excursions to other parts of the home range beyond the study area. This is inferred from the pattern of recaptures and resightings at Danaher over six study seasons in which many individuals were not seen or retrapped for long periods, after which they were regularly seen and trapped again. Twenty-nine per cent of the 960 residents at Danaher had periods between successive recaptures exceeding 100 days, and seven per cent of these had several such periods. In a few cases, birds were absent for periods up to 700 days. Birds that were hatched in the colony were absent for significantly shorter intervals than were immigrants; there was no significant difference between the sexes (Zann and Runciman 1994). Residents at the Alice Springs trapping site also went absent. In Period I only seven per cent of residents (n = 165) were absent for periods exceeding 100 days, but in Period II 38% (n = 375) were absent (Zann *et al.* 1995).

It is possible that some individuals classified as absent were actually present in the colonies, but were not detected because they did not attend the traps and were not observed at the known roosting and nesting sites. Trapping was intense at both colonies but observations were more thorough at Danaher where the colony was more discrete than at Alice Springs. No absentees were trapped or observed at nearby colonies or intervening areas nevertheless, the most likely explanation is that absentees had temporally vacated the study area. Many pairs were absent together, and the banding records often showed that a number of individuals with sequential band numbers also disappeared and reappeared together. This suggests that flocks of residents, many of which contained breeding pairs, went on extended excursions in company. Only once did I detect a returning flock of absentees: in January 1980 at the Padgett colony I mistnetted 11 birds that had been absent for intervals ranging from eight months to three years.

Synthesis

Patterns in recapture data of banded Zebra Finches in northern Victoria and Alice Springs not only indicate high levels of mortality, but more interestingly, suggest exceptional levels of mobility. This mobility produces continuous changes in the composition of birds at what can be more or less permanently occupied colonies. A casual observer of these colonies may falsely conclude that most individuals are permanent residents, and therefore sedentary, but the banding data refute this. The highly fluid membership arises from residents making frequent excursions of variable duration away from the colonies. This movement is set against a background of continuous arrival of birds from other colonies, only a small proportion of which stay long enough to make breeding attempts, that is, become 'residents'. The majority of new arrivals are short-term visitors in transit. Mobility, as reflected in turnover of colony membership and proportions of transients, is higher in the more arid Alice Springs population than in the less arid Victorian populations, but it appears to vary considerably within and among colonies and across seasons.

It is possible that in many parts of the range where Zebra Finches are described by local observers as permanent residents there are also high levels of individual mobility among members of colonies. One plausible interpretation of these movements is that a Zebra Finch population is composed of a number of nesting colonies of variable size and permanency among which the members continuously move. The population is based in a large home range in which is located at least one fairly permanent watering point that is the focus of movement of the members. The more isolated and widely dispersed the permanent watering points from one another, the more discrete neighbouring populations become. In the semiarid zone permanent watering points may not be the only essential resource that delimits neighbouring populations, but a fairly permanent source of seeding grasses, or nesting bushes, may serve the same function.

Consequently, the disappearance of a banded individual from its colony is not necessarily due to mortality, or dispersal—a one-way movement away from the natal colony—but may only be a case of 'quasi-dispersal' (Lidicker and Stenseth 1992), that is, an extended local excursion or exploration to other colonies in distant parts of the home range. The scale of these movements, in both the temporal and geographical sense, varies, and many individuals do not make the return journey to their natal colony, although they may still remain within the confines of the home range. Such movements may be termed 'extended local excursions'. They are not a type of nomadism since this term is used for movement where individuals are in a chronic state of dispersal and fail to establish a home range anywhere (Lidicker and Stenseth 1992).

The best evidence for such a structuring of a Zebra Finch population comes from observations in central Australia. When supplies of food and water occur in large patches, birds from different colonies in the same home range exploit the same localized resources. At permanent watering points during a long dry spell in the vicinity of Alice Springs, I observed that discrete flocks arrived from different directions and, after drinking, returned the same way, presumably to their nesting colonies and feeding grounds. Similarly, at Alice Springs, birds that attended the walk-in trap and irrigated areas of Buffel Grass came in flocks from a number of different directions, presumably from their nesting colonies. These observations suggest that flocks of colony members tend to remain fairly discrete during normal day-to-day searching for food and water. Nonetheless, regular encounters of different flocks at localised resources provide a simple means for individuals to travel from one nesting colony to another. Naturally, when the rains came in central Australia, and food and water became more abundant and uniformly distributed, attendance at these large and permanent resources dropped dramatically as supplies more local to the colonies were exploited. However as the country dried out again, locations with more enduring resources slowly re-asserted their attraction on the local population. Depending on local conditions, nesting colonies may be deserted in favour of others in the home range. Again the casual observer may gain the false impression that the members have dispersed to other regions or died, but on a longer time-scale the same individuals may return.

These extended local excursions within the home range appear to be at the lower end of a scale of movements that Zebra Finches may make. In an attempt to find evidence of more extended movements I surveyed contributors to the Australian Bird Count, an RAOU scheme designed to discover large-scale movements of birds based on patterns of seasonal abundance observed by amateur bird watchers scattered throughout the country. A majority of respondents described changes in abundance of Zebra Finches that were consistent with changing mortality rates and/or extended local excursions within the home range, as formulated above. Of more interest were replies from observers located on the margins of the distribution or in habitats where Zebra Finches were rarely seen. Here large flocks of Zebra Finches could suddenly appear for a few weeks, only to disappear again. There was no seasonal pattern and it suggests nomadic wanderings without any home range as a focus. Extra-home-range wanderings may also account for the very large concentrations of Zebra Finches found around the last reserves of water during periodic droughts in the arid zone. An extension of these drought-induced large-scale movements may result in long-distance dispersal to unoccupied regions where permanent water enables the establishment of a new and perhaps fairly permanent population. Similarly, new habitat, previously unsuitable because of dense vegetation, will be permanently invaded by Zebra Finches upon clearing by settlers; this is believed to account for the existence of many populations in northern Victoria. North (1909) reported that protracted drought in the inland of New South Wales and Queensland late last century forced large numbers of Zebra Finches to move to more coastal regions where they had never been previously seen. They appeared suddenly, and some bred before they eventually disappeared. Such reports have become less frequent this century, probably a consequence of the establishment of many permanent drinking points at bores and dams. A scheme showing the three classes of movements is found in Figure 8.4.

Rapid flight and the ability to withstand dehydration and high temperatures have enabled Zebra Finches to reach and exploit, even if only temporarily, many areas of the arid zone where man has provided surface water by digging wells and drilling artesian bores. When it comes to dispersal of this nature Zebra Finches have few peers; nevertheless, most authors agree that they are not among those species regarded as extreme nomads of the Australian arid zone.

Extreme or long distance nomads, such as the Budgerigar, Cockatiel *Nymphicus hollandicus*, or Flock Pigeon *Phaps histrionica*, appear suddenly in large numbers in areas where they have not been seen for a number of years and a long way from the nearest extant population (Blakers *et al.* 1984). Their appearance in a district normally occurs soon after good falls of rain, and after a period of breeding, they depart suddenly when conditions deteriorate again. In this way, they survive by avoiding the extremes of aridity. Extreme nomads typically have long pointed wings suitable for flying vast distances in the one haul (Schodde 1982) whereas Zebra Finches have shorter, more rounded wings, suitable for shorter hauls. Nevertheless, Zebra Finches are extremely mobile and efficient dispersers, and at any one time probably exploit a greater area of the arid habitat than the extreme nomads.

Although the situation is still far from clear, one can conclude that Zebra Finches demonstrate a range of movements from philopatry at one extreme, to long-distance dispersal at the other. In general, individuals tend not to be philopatric to a colony and its surrounds, but are site-faithful to a very large home range throughout which they frequent on extended local excursions as they move among colonies. Extra-home-range wanderings or nomadism occurs periodically and on infrequent occasions this extends to genuine dispersal over large distances to locations where new home ranges are established. Given this complex pattern of movements it is not surprising that the literature is inconclusive on the matter. Some authors believe that Zebra Finches are sedentary (Serventy 1971; Blakers *et al.* 1984; Davies 1986) while others believe they are nomadic (Cayley 1932; Immelmann 1962a; Rowley 1975).

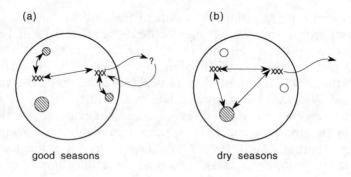

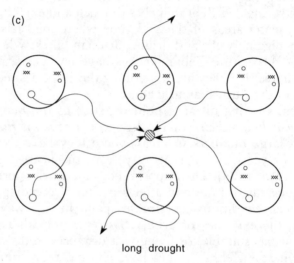

Fig. 8.4 Hypothetical schemata illustrating the structuring of populations and the types of movements made by Zebra Finches during favourable conditions (a), during dry seasons (b), and during prolonged drought (c). The large circles represent the home range of populations which are composed of separate colonies (XXX). The arrows show movements of individuals; arrows within the home ranges represent 'extended local excursions' among colonies; 'extra-home range wanderings' are shown by the arrows that leave the home ranges; 'large-scale movements' are represented by the long arrows in (c). The small hatched circles represent watering points and the blank ones represent watering points that have dried up.

The proximate factors that cause individuals to move from the nesting colonies are not known, but it is reasonable to assume that the quest for the key resources of seed and water, which are unpredictable in both time and space over most of the distribution, is the principal factor responsible. Increasing densities of predators and parasites may also cause departures from colonies. If movements are density dependent then the high levels of transients from study colonies where seed is provided *ad libitum* at walk-in traps require explanation. Why should birds leave sites with abundant, high-quality food, abundant nesting sites and abundant water? Conceivably, new arrivals at a study colony would follow locals to feed at the seed hopper in the trap and seeing only a small circle of seed surrounded by dozens of finches feeding shoulder to shoulder, may underestimate the sustainability of the resource and soon move off to another part of the home range. Also, the renewal capacity of the seed at the hopper can only be assessed from several visits and some newcomers may not stay long enough to make them. This hypothesis could be tested experimentally. It still does not explain why residents and birds hatched in the colony make excursions elsewhere when food, nesting sites and water are available *ad libitum*. There is no direct evidence of interference competition forcing some individuals to move away. Environmental factors may not be solely responsible for forcing birds to move to other colonies; some movement may be voluntary. There is some suggestion for this in recapture data at Danaher where there were sex differences in recapture rates between young males and females. Emigration by males appears to be affected by the local abundance of food whereas female emigration is not, and may be intrinsic, that is, under more direct genetic control. This is the subject of a recent review of dispersal in mammals by Lidicker and Stenseth (1992).

Summary

Size of populations fluctuates sharply from one year to the next, and there is a strong positive correlation between numbers and rainfall in the arid zone, but not in irrigated country. Banding studies reveal that Zebra Finches have very high rates of mortality (67%) between fledging and nutritional independence. The maximum life span is five years and females may live slightly longer than males. Survivorship rates after day 35 are extremely low, but losses due to mortality are confounded with losses due to movements away from the study sites. Immigration rates are very high; only 22% of all adults captured at the Danaher colony hatched there and only 23.5% of breeding adults were birds of the colony, the rest were from other colonies. Among breeding birds there was no sex-biased philopatry in those hatched at the Danaher colony nor was there any sex bias among immigrants. The sex ratio at adulthood (tertiary sex ratio) is slightly male-biased (52% males overall) and the sex ratio at 35–40 days of age (secondary sex ratio) is slightly female-biased. The secondary sex ratio may depend on the seasonal abundance of food. Emigration of young males may depend on availability of food at the colony but that of females is independent of food supply. The pattern of changes in the sex ratios are not clear and more data are required. Indirect evidence from a number of sources indicates that even when seasonal conditions are good, Zebra Finches are highly mobile. They travel continuously among the nesting colonies over a very large home range. More-extensive nomadic movements beyond the home range may occur under some conditions. A population consists of a number of different flocks that nest in discrete colonies, but rely on the same source of permanent water during unfavourable seasons. When this source dries up the surviving members of the population are forced to take more extensive movements beyond the home range in order to find new sources of water, and food, and birds appear in locations where Zebra Finches have never been seen previously. Although they are not included among the long-distance dispersers of inland Australia, Zebra Finches are extremely efficient dispersers, and have reached most of the artificial supplies of surface water created since European settlement.

9 Social and reproductive behaviour

'The courtship of the male Zebra Finch consists of auditory, static-visual, and dynamic-visual elements, whilst that of the female is primarily static-visual.'
D. Morris 1954.

Domesticated Australian Zebra Finches are the avian model of choice in many behavioural investigations. In this chapter I will first describe the relevant behaviour of free-living birds and then consider the more scientifically rigorous observations from the laboratory. Non-social behaviour of Zebra Finches, other than feeding and drinking (Chapters 4 and 5), is more or less the same as that found in other small passerines and is not dealt with here. I have listed non-social behaviours observed in wild birds in an 'ethogram' in Appendix 3. Finally, vocal behaviour and sexual preferences, areas of strong research activity, are treated separately in Chapters 10 and 11 respectively.

Social behaviour

Flying and flocking

There are several modes of flight. When travelling long distances, such as to and from food or water, flight is strong and determined. Birds quickly climb to an altitude of 10–20 m and fly rapidly in a straight direction with little undulation. They can be followed through binoculars for no more than about 60 seconds before the specks disappear into the distance, at approximately 0.5–1 km away. Pairs usually fly on the same level, about 1 m apart, one bird leading. When flying short distances, such as those between different parts of a colony, they fly more slowly, with greater undulations, at altitudes between 0.5 m and 2.5 m. Should they be surprised in flight by a predator, such as a falcon, they will descend to about 0.5 m and fly in a rapid zig-zag course to the nearest thick bush, where they shelter in its centre.

Wind tunnel studies show that Zebra Finch flight is geared for distance rather than speed. When the wings are folded between wing beats, maximum lift and gliding distance is obtained at the expense of speed (Csicsáky 1977).

Feeding and drinking flocks are reasonably tight and fairly well synchronised (Chapter 4). Coordination among flock members appears to depend mainly on vocalisations, and their cohesion and orientation in flight is probably helped by the conspicuous white tail rump. Zebra Finches flock throughout the year, but size of flocks depends on environmental conditions and whether birds are breeding or not (Chapters 4 and 8).

Clumping and allopreening

Zebra Finches are 'contact' species. During resting periods at any time of the year, sexual partners may be found perched or squatting in contact. Frequently, clumping is interrupted by auto- and allopreening. There are no special allopreening invitation postures other than erecting feathers on the head and throat and exposing them to the partner (Figure 9.1). In free-living flocks, members of pairs are usually identified as those that clump with, and allopreen, one another. Parents allopreen and clump with offspring, and siblings will also clump together and allopreen, but this ceases after 40 days of age when the male sexually dimorphic plumage begins to appear. Whereas unmated adult females may clump together, males do not and if their individual distance is infringed they move apart or bill-fence. Low temperatures do not increase the frequency of clumping, but sick Zebra Finches will clump more than healthy ones.

Domesticated adults are not as discriminating as wild birds and males may sit in contact and even allopreen. Males of white morphs, which are sexually monochromatic, clump together and wild-type males are not inhibited from clumping with white males (Immelmann 1959). This shows that the individual distance between males is a function of male plumage and that vocalisations and other behaviour have little effect. In a series of experiments, Immelmann (1959) showed that colour of plumage of a conspecific releases the clumping response, but the method and orientation of clumping depends on the red bill. Also, a model in a low squatting posture is preferred over one in a less squatting posture

Fig. 9.1 A female allopreening her partner's throat.

and explains why sick Zebra Finches, which sit low on the perch and fluff extensively, elicit strong clumping responses.

Aggression

In the first scientific study of domesticated Zebra Finches, Morris (1954) described the components of aggressive behaviour. This is qualitatively the same as that of free-living birds. Mildly aggressive rivals on the same perch settle disputes with bill-fencing where each bird jabs its closed bill at the head and bill of the opponent (Figure 9.2). Both sleek the plumage, especially that of the head, and the dominant bird adopts a more horisontal posture than the subordinate, which retreats or loses balance and flies off. The winner often follows up with a supplanting attack in which it flies directly at the perching opponent who flees a fraction of a second before it lands in its place. Fledglings and juveniles may not flee in time and the dominant may land on their back, often knocking them from the perch. The 'Wsst' attack call (Chapter 10) is often given by the dominant as it supplants. The supplanting dominant may snap at the opponent and very occasionally seize it by the wing or tail and suspend it mid-air eliciting distress cries from the subordinate. Sometimes furious brawls occur in mid-air and both birds may flutter to the ground as they pluck at one another's feathers. On the ground a supplanting bird will simply run at another who retreats. Threat and submissive postures are rare. Occasionally, a dominant may gape at an approaching opponent and force it to retreat and on rare occasions a defeated individual may food beg to its attacker, but it gets no quarter (Morris 1954; Immelmann 1962a). A fluffed out squatting posture does not inhibit attack (Kunkel 1959; cf. Morris 1954).

Zebra Finches fight over food, nest material, shade, roosting and

Fig. 9.2   A female uses bill-fencing to reject the unwanted advances of a male.

breeding nests, and the perches used to approach nest entrances. A pair fights as a team and can easily dominate superior numbers of single birds, for example, when fighting over a roosting nest. The dominance of a male can change depending on whether his female is present or not. The appearance of females will cause outbreaks of fighting among hitherto peaceful males (Immelmann 1959). Both males and females defend their partners against approaches by same sex-rivals, and I have seen females supplant single females courted by their partners. Immelmann (1962a) states that males defend food against every one, including the partner, but I have not seen this even among starving birds in central Australia. Nest owners, especially incubating females, will defend the nest strongly against strange pairs that might inspect the nest or come too close; long chases over distances up to 100 m may ensue. Neighbours are tolerated if they do not come too close. Copulating males also elicit supplanting attacks from nearby males (and sometimes females) who knock them off the females' backs. Females will use supplantings to reject the advances of courting males. If small passerines come too close to the nest, Zebra Finches will also supplant, and sometimes chase. These include competitors, such as House and Tree Sparrows (*Passer domesticus* and *P. montanus,* respectively), and nest predators, such as Brown Honeyeaters *Lichmera indistincta*; however, harmless species, such as thornbills (*Acanthisa* spp.), are ignored.

No dominance orders are evident in free-living colour banded birds, but in captivity some individuals are consistently more aggressive than others when held in unisex groups; they initiate more fights and win more, and a rough linear order may be formed, but this changes daily (Evans 1970; Ratcliffe and Boag 1987). The most dominant bird usually wins control of the highest nest box in the aviary. However, there are many retaliatory attacks from inferiors on dominants so that no linear orders form, furthermore inferiors may win over dominants when defending their own nest boxes. In domesticated Zebra Finches, aggression in all-female groups is not as frequent as that in all-male groups (Butterfield 1970). In both domesticated and wild-caught birds, homosexual bonds may form in unisex groups. Partners clump, allopreen and cooperate as a team in the defence of resources and may become dominant in an aviary. If reunited with their heterosexual partners there is an increase in the level of aggression even when density is held constant (Butterfield 1970).

In a series of carefully planned experiments, Caryl (1975) found that domesticated male Zebra Finches could be provoked to fight at the sight (but not the sound) of a female and the amount of fighting depended on how close the female was and whether the male was paired with her or not. Sexually provoked aggression cannot be entirely explained by competition among males for females, or defence of the sexual partner.

During pair formation Clayton (1990a) found that aviary-bred

Australian male Zebra Finches were more aggressive than their smaller counterparts from the Lesser Sundas.

Reproductive behaviour

Pair bond

Single birds of the Australian subspecies are found in most flocks throughout the year and here they meet potential partners. Single birds may be youngsters looking for their first mate or adults who have lost theirs. On a population basis, there was no significant departure from parity in the tertiary sex ratio (Chapter 8). However in 48 winter trapping sessions at the Danaher colony, adult males were in excess of females in 31 months, and females in excess of males in 15 months. Surpluses could be quite large, for example, in May 1986 there were 22 adult males and only 8 adult females, whereas in August 1989 there were 23 males and 36 females. Mortalities within the colony and high levels of immigration mean that pairing opportunities and levels of competition for partners are quite variable from one month to the next.

Bond formation

There are no special pair-formation displays. In the wild, single males, either mated or unmated, will approach any single female to initiate a sexual encounter. Some perform exaggerated greetings and begin the stage 1 courtship waltz (see later), others simply confront the female and direct a number of song phrases at her, and some mated males omit all courtship and simply jump straight on the female's back without preliminaries. If the female is already paired, but her mate is incubating or elsewhere, she may ignore him or supplant him. Uninterested females flee, but if interested they will respond with a Head Tail Twist greeting (see later) of their own and may even begin the stage 1 dance. Occasionally, single females will take the initiative and greet single males, but mated females have not been observed doing so. If both birds are receptive, courtship does not proceed very far, but gradually over the course of the day, or days, both birds spend more time together and courtship becomes more frequent, although it still does not lead to copulation. The first sign that a bond has formed is when the pair sit in contact and allopreen and this is the criterion chosen in most studies. Copulation tends to occur sometime after this when the pair begin to nest-build. Thereafter the partners are inseparable unless incubating or brooding and their activities are synchronised. Nest-building displays are not essential for pair formation (cf. Morris 1954) which can occur in both wild and domesticated birds in the absence of any nest site or nesting material.

In the wild some bonds form within a day, although Immelmann (1962a) found that some took many days. In unpublished experiments conducted at the Danaher colony during the non-breeding season we removed mates of mated pairs and found that replacement pairing usually occurred within several days. It appears that single birds in the wild will form bonds when and wherever possible, irrespective of season.

Pair formation also occurs quickly in captive Zebra Finches of both subspecies (Immelmann *et al.* 1978; Clayton 1990a). Using slightly different experimental designs, Caryl (1976) and Silcox and Evans (1982) focused on the changes in behaviour that occur over the course of pair formation in the Australian subspecies and arrived at almost identical conclusions. In each study, a pairing male was observed to fight more with females with whom he did not pair, but sang more directed song and copulated more with the female with whom he paired. He also stayed closer to her, clumped with her more and synchronised feeding and preening more. Most differences in behaviour between pairing and non-pairing individuals reached significance on the second day of the experiment, but clumping was not significant until day 4 or 5 in Caryl's study, and day 2 or 3 in Silcox and Evans' study. Preferences for a potential mate can form soon after meeting in both subspecies (Clayton 1990a); this was evident in most pairs in Silcox and Evans' study within 30 minutes of meeting. Dunn (1981) also found that most bonds were formed by the second day, with some forming as early as 20 minutes after introduction, whereas others took four or five days. Dunn concluded that clumping was a means of strengthening a bond that had already formed rather than part of the formation process itself.

Potential partners cannot pair unless they can make tactile contact. Visual and auditory contact and auditory contact alone were insufficient for a pair bond to form (Silcox and Evans 1982; Clayton 1990a).

In captivity, incompatibilities often exist among potential sexual partners and about 30% fail to establish bonds within about 10 days of contact even when the sex ratio is 50% (e.g. 7/26, Caryl 1975; 6/21, Dunn 1981; 3/10, Silcox and Evans 1982). Ratcliffe and Boag (1987) found that after 15 days, 10/36 females failed to pair despite a sex ratio of 66.6% males. By contrast, in two Bielefeld studies that used large aviaries to examine the effects of cross-fostering on assortative mating, the rate of pair formation was much higher: 95% (*n* = 64) of individuals of both grey and white morphs paired within three days (Immelmann *et al.* 1978) and 100% (*n* = 70) of available individuals of both subspecies of Zebra Finches paired within an unspecific time (Clayton 1990b). Possibly, large aviaries may be a significant factor in the rapid formation of pairs, since they allow the female space to escape overbearing males and opportunities to initiate their own preferences.

Who chooses?

On the basis of his observations in the wild, Immelmann (1962a) concluded that the ultimate decision to form a pair bond or not, rests with the female, and not the male—she may reject his proposal or accept, but evidence for this conclusion is not given. On certain occasions a surplus of single females will mean that they do not have a choice of males. It is difficult to determine who is selecting whom, since the process is a dynamic and reciprocal one. Male courtship is more active and conspicuous than female courtship so when Dunn (1981) found that males selectively courted the female with whom they eventually paired on first meeting, it was not possible to tell if the male had selected the female for her attributes or for the fact that she has selected him and encouraged him in subtle ways. However, by means of a multiple-choice arena (the 'Finkodrome') that allowed the test bird to make a simultaneous choice of 10 isolated stimulus birds, Clayton (1990a) found that both males and females quickly discriminated among potential partners. Moreover, within each sex there was agreement on which were the most desirable partners. If the preferred individual was available later in the aviary pairing occurred, but females were more successful in pairing with their initial preference than were males. This suggests that females have more of the final say in pair formation than males. In a second experiment Clayton (1990b) controlled the behaviour of the stimulus birds by placing them behind one-way glass so that they could not respond behaviourally to the test bird on approach and so influence its choice. When preferences tested this way were compared with those in subsequent tests in which test and stimulus birds could interact freely through a wire screen, there were significant changes in preference. Thus, for both males and females the initial choice is based on the appearance of the potential partner and the final choice is an outcome of a dynamic and reciprocal interaction between them.

Females prefer males of their own subspecies and discriminate against males that have been cross-fostered to the other subspecies (Clayton 1990b). Despite the tendency for the female to make the final decision at pair formation, Clayton (1990b) found that her initial preference could be overridden by that of the male if she was cross-fostered and he was not. Possibly, normally reared individuals are less ambiguous about their sexual preferences than cross-fostered ones, and their more persistent courtship prevails over other shortcomings.

Levels of sexual incompatibilities are impossible to establish in the wild, but given the tendency of males to court females indiscriminately, it is likely that the female is the more discriminating sex, although failure to form a pair bond during the breeding season could lead to lost breeding opportunities. In both free-living and domesticated birds, two males may compete for the same female. In domesticated Zebra Finches, both Butterfield (1970) and Silcox and Evans (1982) found that male–male aggression was very high at the start of pair formation, and Butterfield (1970) and Ratcliff and Boag (1987) found that only the dominant males formed pair bonds. Female–female aggression also occurs, but at a lower level than that of males and mostly after the bond is formed. Perhaps males compete for access to the females, and the females fight to protect the bonds that had started to form.

Pair formation in the Double-barred Finch is different from that of the Zebra Finch. According to Immelmann (1962a, 1965a), there are no special displays involved, but pairs form gradually at the onset of breeding by a 'negative process' in which the two slowly restrict their clumping and allopreening to the partner. In the non-breeding season, the pair bond weakens in this species and partners will clump and allopreen with other members of the flock once again.

Pair bond and pair formation in the three sexually monochromatic species of *Poephila* have been studied in controlled aviary experiments (Zann 1977). Their bond is as tight as that of the Zebra Finch, or tighter, and displays at pair formation are similar except that nest site displays are an integral part of the process. Although both sexes also show rapid preferences on first meeting, formation of the bond is slower than with the Zebra Finch. This is due to mainly to high levels of competitiveness in both sexes, and an 'unwillingness' to modify initial preferences so that reciprocal and mutual bonds can be established between partners. Consequently, depending on the species, between 35–100% of birds failed to form pair bonds over a 14-day period. By contrast, Zebra Finches appear more willing to pair with a non-preferred individual than not to pair at all. Conceivably, lives of Zebra Finches are shorter than those of the *Poephila* and a partner, even an inferior one, is better than none all, since a pair needs to breed at the earliest and slightest opportunity, whereas breeding in the *Poephila* is less urgent since its timing is seasonal. Possibly, extra-pair copulations (see below) are a more viable option for the Zebra Finch and could offset the disadvantages of an inferior partner.

Bond maintenance

Free-living Zebra Finches have a strong permanent pair bond which they service constantly. They are an inseparable dyad, with the male usually leading the female around the colony and in the flocks, except at breeding when the roles are reversed (Birkhead *et al.* 1988a). When active, the partners keep in contact and synchronise movements with Distance Calls, Tets and Stack Calls (Chapter 10). Most resting moments are spent in contact allopreening, and potential sexual rivals are driven away from the partner. At the start of nest building, nest ceremonies are performed, which Immelmann (1962a) believes serve to strengthen the bond before a breeding attempt.

The pair bond in domesticated Zebra Finches is also strong and permanent, but divorces can occur (Morris 1954). If partners are physically separated but allowed visual and auditory contact, or just auditory contact, they are able to maintain the bond through the exchange of Distance Calls even when caged with a member of the opposite sex. If held without any auditory and visual contact with their partners the bond is broken and there is almost immediate re-pairing if another individual is available (Silcox and Evans 1982). This new pair bond does not destroy the earlier one since birds will still clump with the old partner providing the new one is absent. Immelmann (1959) found that 35/42 pairs reformed after periods of visual and auditory separation ranging from two to 23 weeks, and pairs that had laid a clutch always re-paired. Similarly, Clayton (1990a) found that female Australian Zebra Finches forced to pair with Lesser Sundas males only maintained the old bond when Australian males became available if they had already laid eggs. A breeding attempt therefore seems to be the final confirmation of a permanent pair bond.

When members of a pair are visually separated, males are more active than females; they hop and fly more, and give more Distance Calls than females, but both sexes show an increase in defecation rates which suggests both are stressed by the separation (Butterfield 1970). Immelmann (1959) found that when previously separated partners were allowed to meet again in an aviary it was always the female that joined the male. He concluded that the pair bond was stronger in females than males. Males also sing more Undirected Songs (Chapter 10) when separated from the mate than in her presence. In an operant conditioning experiment involving an on–off perching response, Butterfield (1970) found that an isolated male Zebra Finch, found the sight of the female partner positively reinforcing. She was more reinforcing than other females, and most reinforcing when caged with another male. Presumably, the male is anxious to guard his female. Interestingly, females could not be similarly conditioned, perhaps because they are too inactive on separation from the partner to learn the operant response.

Auditory cues, not visual ones, are used for recognition of members of a pair. Immelmann (1959) discovered that pairs recognised one another irrespective of whether he painted their bills or plumage different colours.

After a period of separation from the female partner free-living and captive males court on reunion. This is particularly intense during the egg-laying period. Butterfield (1970) found a positive correlation between the amount of sexual behaviour and the duration of separation, and concluded that the pair bond has a sexual motivation. Her birds were not in the pre-laying and laying states when tested so the courtship on reunion cannot be regarded as a version of 'retaliatory copulation' (see below).

In order to alternate incubation and brooding duties, Zebra Finches must loosen the tight pair bond. This occurs gradually, and occurs in conjunction with bonding to the nest. Initially, the partner is always in sight, but when nest building commences the female remains at the nest site and when the male collects nest material nearby he is often out of sight. He now uses auditory communication to maintain continual contact: Undirected Songs are sung soon after leaving the nest and Distance Calls are given on the return trip, however, the female is silent. Should the female leave the nest the male will follow and attempt to lead her back inside with nest solicitation displays (below). By the start of incubation the bond has loosened to the extent that there is no need for auditory contact and the partners alternate between the nest and flock and no clumping or allopreening is observed. When the young are six to ten days old the bond begins to be strengthened again and the pair consort much of the time. Consorting begins earlier when high ambient temperatures make brooding unnecessary over much of the day; however, when nestlings reach ten days of age most day-time brooding has ceased and the pair bond is complete again. Initially, auditory contact via calls is resumed, then visual contact, and finally tactile contact where the parents once again allopreen and clump. Should the nest suffer predation the pair will begin clumping and allopreening within the hour. The first copulations in the wild were observed on the day young fledged, but males may initiate unsuccessful courtship with their partners even before the end of incubation.

Nest building

Searching for a nest site

When the male leads the female in search of a suitable nest site he gives Stack Calls (Chapter 10) much of the time. As he lands at a potential site, these switch to Kackle Calls, and if the female shows interest he hops to and fro from her to the site giving Ark Calls each time he lands on the site itself. If she approaches he performs the Head-down Tail-fan display in which he bows down and fans the tail and mandibulates at the female while slowly pivoting the body from side to side; Ark Calls are emitted continuously. At an old nest he may pause in the entrance and fan his tail at the female, his white rump highlighted like a beacon. Sometimes the male adopts the Head-down Tail-fan and nods in the direction of the nest entrance as he mandibulates. Females are discriminating about suitable sites and in most cases show no interest and leave, whereupon the male will lead the search elsewhere. Females reject many sites before finally settling on one. Strangely, females never lead the nest search, a strategy that would save much time. If she is interested in a site she may approach in a Head Tail Twist, mandibulate at the male and hop around the site, or in and out of an old nest; this will intensify the male's display and he will Nest Whine (Chapter 10) and start nibbling or pushing any nest material nearby. Males often give Undirected Songs (Chapter 10) while the female is at a potential nesting site.

Nest ceremony

If the female accepts a site she joins the male and both sit or sprawl in contact, side by side, on the exact location of the future nest, and intermittently stretch their necks out, point their bills at one another, mandibulate and give the Nest Whine (Figure 9.3). Later, the pair hop to-and-fro on the nest site greeting one another with Kackle calls followed by more Whines and Mandibulation Sprawls. Both nibble at nearby stems or twigs, and eventually the male flies off to get the first long grass stem for the new nest. According to Immelmann (1962a) the installation of the first stem of a new nest is an important symbolic moment in the nest ceremony. Amid much nest whining both male and female position and reposition the stem and the male may even take it away and bring it back again. More stems are fetched and the arrival and placement are accompanied by less and less ceremony so that the ritualised behaviour gives over to actual nest construction. Nest ceremonies may still given in the following days but are muted.

Nest solicitations and incipient nest ceremonies are more frequent in Zebra Finches than in other Australian estrildines (Immelmann 1962a). They can be heard giving Nest Whines throughout the year, even in roosting nests. Close relatives, namely the Double-barred and the three *Poephila* species, are almost identical in their performance of nest solicitations and ceremonies but they are more or less restricted to the final site chosen for the breeding nest.

Nest building—a division of labour

Nest construction follows estrildine conventions (Kunkel 1959), although Zebra Finches show a greater division of labour than that found in some species. In Zebra Finches both partners build the nest

Fig. 9.3 The nest ceremony is performed on the exact location of the future nest. The partners sit or sprawl, side by side, mandibulating and whining for an extended period before the onset of building.

with the male taking the initiative. In most free-living and domesticated pairs the male fetches the nest material and initially positions it, but the female puts it in its final position. In a few pairs, the female may fetch nest lining. The male will not carry stems to the nest nor do any building if the female is not present. Nest building occurs in bouts and most material is collected near the nest; at the Danaher colony 45% of items were collected within 5 m, the median distance travelled was 12 m and the maximum 87 m (Birkhead *et al.* 1988a). However, Immelmann (1962a) found that most pairs in his experience collected material away from the nest, usually flying up to 50 m away. The rate at which stems are added varies according to nest-building stage and individual pairs. The rate is highest about four to five days before the first egg is laid, and some Danaher pairs reached a maximum of 19 trips per hour (Birkhead *et al.* 1988a). At Alice Springs I saw one hyperactive male four days before the first egg make 42 deliveries to the nest in one hour, in addition to performing thirty-five bouts of Undirected Song, two bouts of courtship and several brief nest ceremonies. The maximum rate of stem deliveries to the nest observed by Immelmann (1962a) was 23 in 30 minutes. During laying the rate declines to around 3 trips per hour. Frequent performance of Undirected Song during the searching and collection of nest material is characteristic of Zebra Finches and the three species of *Poephila*, but Immelmann (1962a) reports that all remaining Australian estrildines collect material silently.

Although Immelmann (1962b) found that nest construction is poorly developed in domesticated Zebra Finches, frequently, excessive nest material is added on top of newly laid clutches making incubation impossible.

All nest material in the wild is collected from the ground, except feathers and other lining material which may be stolen from the nests of others, either conspecifics or heterospecifics. Previous experience with particular materials and nesting sites can determine what will be used (Sargent 1965). In the wild most stems are collected where they have fallen, but some are snapped off with the bill at ground level or pulled back until they break. Stems are tested by nibbling and several may be rejected before the right one is found. Like all estrildines, stems are carried by the thick end, and the flight may be laboured when holding a long stem—difficulties can arise if they get entangled in twigs. Nest lining is bundled together and carried in the middle; feathers are preferred and larger ones are carried by the shaft tip.

A foundation platform is built if none is present. Stems are crisscrossed horizontally with no special fastening to the supporting twigs. The nest is built from the bottom upwards with the material added from the inside and worked into the surrounding twigs. The nest chamber is built first, the floor, walls and roof are added in that order, then the nest tube and lining are completed together. In nest boxes or hollows the roof and entrance tube are omitted. Like most estrildines, there are three basic building movements, each with a quivering vibration that anchors the stem: push-away, lateral-pull and pull-in (Kunkel 1959). Push-away has several forms in the Zebra Finch; the male drags a stem into the nest, fastens the end at the rear with a push up movement by stretching the legs, neck and body and releasing the end of the stem. The head and breast are also used to push the framework of the chamber out and up in order to enlarge it, and the bill and head can be used to lever the lower framework up. With the lateral-pull the bird, usually the female, sits in the middle of the cup reaches forward, grasps the middle of a stem and draws it diagonally across the body and pushing it up behind so that the stem forms the lower walls and floor of the nest. The floor and walls of the entrance tube are formed by the pull-in movement where the sitting bird stretches forward, grasps the free end of a stem and draws it into the wall.

Zebra Finches nest build actively throughout the year. If they are not building breeding nests they are continually renovating and extending their roosting nests, sometimes building nests they never use. Immelmann (1962a) listed them among the three species of Australian estrildines that have a strong building drive throughout the year.

Courtship and copulation

Copulatory behaviour and sperm competition in birds is a new and rapidly expanding area of evolutionary biology. It provides a new facet to our understanding of competition and sexual selection in animals. Not surprisingly, research on Zebra Finches has been at the forefront of a number of important discoveries in the field, due principally to the work of T. R. Birkhead and his students at Sheffield University. They have fruitfully combined laboratory experiments with field studies and in the process have led to a better understanding of reproductive behaviour and physiology of birds in general.

In 1988 Birkhead *et al.* (1988a) investigated copulatory behaviour of at least 15 pairs of Zebra Finches at the Danaher colony, and followed four pairs in detail. They estimated that a pair performed at least 15 copulations for each clutch and began copulating up to 11 days (day −11) before the first egg (day 0) is laid after which the rate of copulation fell markedly, due mainly to lack of female participation. Copulation occurred at a median distance of 5 m from the nest, although on several occasions it occurred on the nest itself, while on other occasions pairs flew 35 m to the branches of a dead tree. Dead horizontal twigs are the preferred courting site because they allow birds to dance unhindered, although they are more exposed to interference from conspecifics and predators. Most copulations occur in the morning soon after pairs leave the roost (Birkhead *et al.* 1988a).

Domesticated Zebra Finches do not court and copulate as frequently as wild-caught Zebra Finches or their first-generation offspring. In systematic tests Sossinka (1970) found that domesticated wild-type males, both unpaired ones and paired ones with nestlings, courted and copulated less frequently than their wild-caught counterparts. Initially, Immelmann (1962b) believed erroneously that domesticated Zebra Finches were 'hypersexual' in comparison with wild birds. Domesticated Zebra Finch pairs copulate about 11–12 times per clutch (range 2 to 23) peaking on the day before the first egg is laid; only 67% of copulations result in sperm transfer and fertilisation of eggs (Birkhead *et al.* 1989).

In comparison with Zebra Finches, Immelmann (1962a) found that Double-barred Finches court very infrequently, and hypothesised that birds of more arid regions court and copulate more frequently than those from more mesic environments. In captivity all three species of wild-caught *Poephila* copulate about 15 times per nesting cycle (Zann 1976a), a rate comparable to that of the Zebra Finch.

Display elements of courtship

The components of courtship and copulation have been described by Morris (1954), Immelmann (1959, 1962a), and Kunkel (1959), and are shown schematically in Figure 9.4. Like most Australian estrildines that omit the primitive stem dance, Zebra Finch males normally have two stages of courtship which precede mounting and copulation. Stage 1 is an introductory waltz, which may constitute 50–70% of the whole courtship sequence. Stage 2 consists of the song and dance. Either sex may initiate a courtship sequence, but it is usually the male who lands next to the female in a stiff, upright posture and holds the body parallel to hers and greets with the Head Tail Twist and bows while emitting Distance Calls and Kackle Calls (Figure 9.5a). Immelmann (1959, 1962a) termed this the 'exaggerated greeting' ('überbetonten Begrüßung'). The male may fly to and from the female several times and greet before the waltz begins. Whereupon he hops to and fro between neighbouring branches and is joined by the female in a type of waltz; both try to keep their flanks parallel to one another. The plumage of the male is fluffed, especially the ear coverts, abdomen and flanks; however, the feathers of the forehead are flattened, so that an 'angular head' is formed with the feathers of back of the crown fluffed. Lesser Sundas males court in a more upright position and raise, rather than flatten the feathers of the forehead so do not form the angular head (Clayton 1990a). Feathers of the female are less fluffed than those of the male, and those of the forehead are not flattened. As the partners hop around one another, each maintains the head and tail twisted towards the partner and intermittently bill wipes on one or both sides of the perch, mostly on the partner's side. These wipes vary in completeness from just a slight nod of the

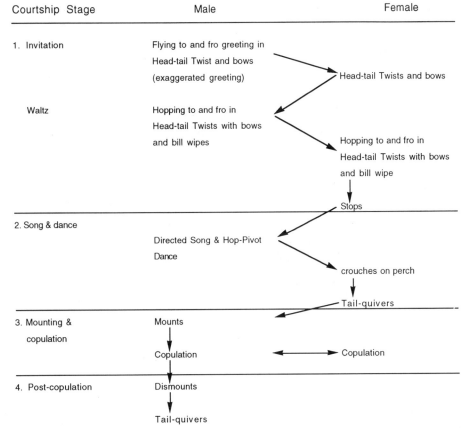

Courtship Stage	Male	Female

1. Invitation   Flying to and fro greeting in Head-tail Twist and bows (exaggerated greeting) → Head-tail Twists and bows

Waltz   Hopping to and fro in Head-tail Twists with bows and bill wipes → Hopping to and fro in Head-tail Twists with bows and bill wipe → Stops

2. Song & dance   Directed Song & Hop-Pivot Dance → crouches on perch → Tail-quivers

3. Mounting & copulation   Mounts → Copulation ← Copulation

4. Post-copulation   Dismounts → Tail-quivers

Fig. 9.4   Schemata of courtship behaviours leading to copulation in the Australian Zebra Finch. (Based on Immelmann 1959.)

head to a normal double wiping action in the air just above the perch; most appear as stiff bows (Figure 9.5b). Sometimes a genuine bill wipe after drinking appears to inadvertently initiate a courtship sequence by stimulating the partner to begin stage 1.

Workman and Andrew (1986) noticed that courting males keep the right flank orientated towards the female significantly more than chance expectation. They believe this occurs because the right eye is preferred for observing the female because of brain lateralisation processes. However, ten Cate *et al.* (1990) could not verify any eye bias in their tests although both sexes had a locomotory bias to move in a clockwise direction. Further experiments are needed for a conclusive finding on eye-bias in courtship (ten Cate 1991a).

When the female stops waltzing and crouches on the perch the male approaches within 20 cm and switches to his song and dance—stage 2. He adopts the Upright Fluffed singing posture, where the head is held

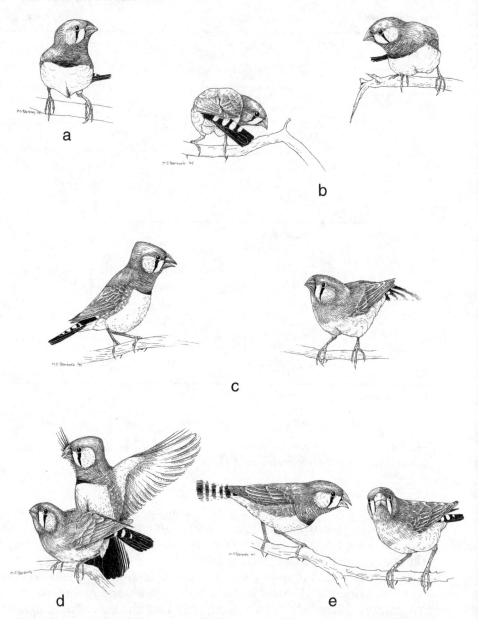

Fig. 9.5 Courtship behaviours and postures before, during, and after copulation. (a) exaggerated greeting; (b) waltz; (c) song–dance and Tail-quiver; (d) copulation; (e) male Tail-quiver.

erect, the bill and tail horizontal, and the legs bent so that the abdomen touches the perch (Figure 9.5c). Feathers of the abdomen, flanks and ear coverts are fluffed. As he sings he twists his head and tail towards the female and pivots the head from side to side as he 'beams' the song at her. On a straight perch he hop-pivots towards the female facing one way then the other as he swings the body from side-to-side through 180° as depicted by Morris (1954). Occasionally, when they court in dense twigs or on the ground the male hops in a semicircle around the female only pivoting at the end of a run. At all times the tail is twisted towards the female. Morris (1954) thought the hop-pivot was a ritualised movement, but Immelmann (1959) and Kunkel (1959) considered it the normal way Zebra Finches move along a straight perch, yet no evidence was given for either case. After a variable time the female crouches lower and performs the Tail-quiver in which the tail is quivered up and down while a horizontal or slightly head-down position is adopted (Figure 9.5c); this may last ten seconds or longer until the male stops singing, mounts and copulates. Both male and female give the Whining Copulation Call. Mounting and copulation only take one or two seconds. On dismounting, the male often crouches and performs the Tail-quiver himself for as long as 20 seconds (Figure 9.5e). Morris (1954) called this 'pseudo-female' behaviour and occasionally females respond by mounting the male, but Immelmann (1962a) thought that this was an artefact of domestication. However, Nancy Burley (pers. comm.) saw two reverse mountings by a pair at Alice Springs in 1986, but it has not been recorded elsewhere.

During stage 1, both birds give a series of calls which become more rapid and louder as courtship progresses, especially in the female. Males mainly give Kackles interspersed with a few Distance Calls and females mainly give Tets, but also Stacks (Chapter 10). The female calls reach a crescendo just before the Copulatory Whines (Chapter 10). These occur just as she begins to Tail-quiver. Song is interspersed with Distance Calls and the male stops immediately the female solicits copulation and begins to whine himself when his copulatory movements begin. Sonograms of copulation show the longer, higher-pitched whine of the female followed by the briefer, lower-pitched whine of the male on which is superimposed noises of rapid wing beats as he maintains balance. A bout of Undirected Song may follow copulation.

Most courtship in the wild does not result in Tail-quivering and copulation, but breaks off due to lack of female responsiveness. Courtships initiated by the female were more likely to proceed to copulation. Sometimes females Tail-quiver before the male has sung, and he may not mount for some reason, or she stops before he is ready. Males usually perform the song and dance, but not always, sometimes they omit the waltz after the initial greeting. Males may chase fleeing females, adopting the Head Tail Twists when they reach within about 40 cm; the singing posture is adopted when they get within about 20 cm. Courtship to non-mates may be very brief in both captivity and in the wild: simply a song phrase or two followed by an uninvited mounting attempt, and occasionally the song itself may be omitted. I have seen wild males interrupt incubation on hearing a female Distance Call, fly 50 m or more to her, court and try to mount. In contrast, Birkhead *et al.* (1989) describe a stage 2 sequence of extra-pair courtship lasting as long as two minutes.

Occasionally, free-living Zebra Finches perform a second type of pair courtship whose function is unknown. During the fertile period when both are sitting side by side in the nest the male will suddenly hop out, turn around in the entrance, face the sitting female and sing to her. She makes no overt response and the male usually rejoins her for nest ceremonies.

The displays and sequence of female Zebra Finch courtship closely resemble those described by Immelmann (1962a, 1965a) for the other Australian estrildines which conform, more or less, to the pattern found throughout the subfamily (Steiner 1955). Male courtship, in contrast, mostly resembles that of the three species of *Poephila*, especially the Masked Finch (Zann 1976a); stages 1 and 2 are similar in movements and posture, but feather erection is greater in the Zebra Finch. The Double-barred Finch, courts differently from the Zebra Finch in that the posture is more horizontal and the feathers more uniformly fluffed. Bill wiping is much more frequent in Double-barred Finches, but there is little dancing (Morris 1958). Domesticated Zebra Finches tend to abbreviate or omit the introductory stage of courtship (Morris 1954; Immelmann 1962b), but otherwise it is the same, except for white mutants (see below).

Pattern of pair copulation in domesticated Zebra Finches

As part of their investigations into the relationship between copulation and fertilisation in birds, Birkhead *et al.* (1989) continuously observed the courtship and copulatory behaviour of domesticated Zebra Finches throughout the hours of daylight by video-recording pairs (n = 10) in single cages. Males initiated 78% of copulations and females initiated the remainder with Tail-quivers. Cloacal contact was more likely in the latter. Males Tail-quivered after 62% of copulations. Female-initiated copulations occurred earlier in the breeding cycle relative to egg laying than those initiated by the male. All copulations occurred during the female fertile period and were most frequent in the first hour of daylight. Copulations were frequent between day −5 and day 2, peaking on day −1 and day 0 and dropping markedly after day 2 when the second egg was laid.

Similar patterns of copulation were found in aviary studies when birds were observed for the first six hours of daylight over the fertile period. However, there were two major differences:

(1) 26% of pair mountings were disrupted by other Zebra Finches (males disrupted mountings on 30 occasions and females on 8); and

(2) 47% of undisturbed copulations were followed by the female

soliciting a second time, and in 30% of these another copulation followed. Females in the cage situation, by contrast, did not invite successive copulations.

Significance of species-specific and secondary sexual characteristics

Immelmann (1959) conducted an exhaustive series of ingenious experiments on domesticated Zebra Finches in order to determine the most important visual releasers for the elicitation of species-, sex- and age-specific behaviours. Test birds of wild-type and white colour morphs were given a choice of two individuals in a small cage. The stimulus individuals were either painted models, live birds of different morphs, or white morphs painted with male or female marks or a combination of marks. The marks that elicited species recognition were the red bill, the eye stripe and the tail bands. These releasers stimulated initial approach and greeting and subsequent social interactions such as clumping and allopreening. Birds or models without head markings were ignored. White mutants, by contrast, have only the white plumage as a species character and use this exclusively.

Male secondary sexual marks had a strong releasing effect on female courtship, and marks tested on their own still produced a response. Stimulated females continually landed near the most effective model and greeted it with the Head Tail Twist, but further stage 1 courtship was prevented by the absence of male displays—only the male courtship dance elicited the female invitation to copulate. Presentation of painted models and white males showed that the most effective mark was the chestnut-coloured ear coverts, followed by the black breast band, and the white-spotted chestnut-coloured flank marks. Darkness of the bill was found to be of no importance (cf. Burley's experiments in Chapter 11).

When the releasing effect of female appearance on male courtship was tested, Immelmann (1959) found that the sign stimulus was a grey, cylinder-shaped object with a red front. Years later, in follow up experiments, Bischof (1980) showed that male courtship could even be elicited by two dimensional models of females. When Immelmann (1959) offered males a choice of females, live ones and/or models, those with pale red bills were preferred over those with dark red bills. Males directed courtship greetings and stage 1 courtship to the most effective models and some hypersexual individuals attempted to mount them. At the Danaher colony a continuous series of males courted (stage 2) and copulated with freeze-dried female specimens attached to branches of bushes in the colony (T. R. Birkhead and R. Zann, unpublished observations). Thus, courtship in male Zebra Finches can be elicited by two or three dimensional models as long as they contain at least one correct releaser.

In eliciting courtship, Immelmann (1959) found that males were more responsive to the appearance of the female, and while females initially chose males on the basis of their appearance, they quickly responded to the vigour of the courtship display (Chapter 11). Similarly males could overcome inhibitions based on incorrect bill colour if the female invited courtship vigorously. Morris (1954), for example, believed that female bowing was an important display in stimulating male singing and this was later confirmed by Garson *et al.* (1980) using models. Bischof (1985) used live birds to show that visual stimuli emanating from the female were sufficient to elicit all the male display elements in the song and dance, but acoustical stimuli were ineffective when isolated from the visual components of the display. The rate of male bill-wipes during the dance was not affected by the sight and/or sound of the female, and may simply reflect a state of heightened general arousal. In contrast, the sounds of the female heightened the male's sexual arousal, and in doing so, increased his rate of singing, but inexplicably, did not affect his rate of bowing and hop-pivoting, perhaps due to spatial limitations in the test situation.

White Zebra Finches are sexually monochromatic, consequently males respond to every stranger as if it was a female (Immelmann 1959). However, the bill of males is darker than that of females, and this is the morphological character used exclusively in this morph to identify the sex of strangers. It is a relative cue, that is, males will court strangers, even other males, if no other stranger has a paler bill. Vocalisations are not used to identify sex in face-to-face encounters in white Zebra Finches. Moreover, the absence of secondary sexual marks causes females to flee courting males, which in turn stimulates male aggression, so that courtship in white Zebra Finches is characterised by much fleeing by females and aggression by males. Eventually pairs habituate and copulation can occur, but the female courtship response remains muted.

Immelmann found that white males preferred to court white females and models, and vice versa for wild-type subjects. However, he also showed, through cross-fostering experiments, that this preference was a consequence of imprinting on the rearing parents in the first month or so of life (Chapter 11).

Further experiments by Immelmann showed that when the pale red bill of an adult female was painted black, the colour of fledglings, it inhibited male courtship for some time, and only strong female courtship invitations were effective in gradually overcoming his inhibitions. Similarly, an adult male given a black bill initially inhibited female courtship, but this was soon overcome by his song and dance. Adult males first learn to distinguish the sex of juveniles when they reach an age of 35–40 days. According to Immelmann (1959) this is not a consequence of patches of male-specific plumage which are beginning to appear in males at this age, but identification is based on sex differences in vocalisations. However, my analysis of sex changes in the Distance Call shows that sex differences do not develop until 40–50 days age (Chapter 10).

In summary, Immelmann (1959) concluded that the secondary sexual characteristics of male Zebra Finches

(1) maintain an individual distance between adult males and prevents them from courting one another,

(2) release the female greeting approach flight at the onset of courtship; and

(3) suppress the female fleeing response during his courtship dance.

While sexually dichromatic features are of primary importance in the initial choice of a partner, and in stimulating the first stage of courtship, any shortcomings in appearance can be eventually overridden by vigorous sex-specific behaviours.

Extra-pair copulations

DNA fingerprinting of nestlings and their putative parents at the Danaher colony showed that there were two cases of extra-pair paternity among 82 birds whose maternity was confirmed (Birkhead *et al.* 1990). So, for 2.4% of offspring (8% of broods and 12.5% of females) the mother copulated with a male other than the rearing father. This incidence of extra-pair paternity is lower than that found in an aviary study of domesticated Zebra Finches (Birkhead *et al.* 1989) in which genetic plumage markers indicated that extra-pair paternity occurred in 5.6% (4/71) of offspring and 11% of broods (2/18). Differences are probably due to effects of aviary confinement and domestication since the two methods of establishing paternity have been shown to be equally reliable (Møller and Birkhead 1992). Extra-pair paternity in wild birds of other species ranges from 0–60% (Birkhead and Møller 1992).

Theoretically, extra-pair paternity can happen in three ways—through forced and unforced extra-pair copulation, and by rapid mate switching. All three occur in Zebra Finches and have been intensively investigated by Birkhead and co-workers in domesticated Zebra Finches and in their wild-caught and free-living counterparts. Rapid mate switching happens when a female is inseminated by one male, usually the mate, but switches to a second male with whom she pairs before the eggs are laid. This can be manipulated experimentally in domesticated Zebra Finches so that the second male not only pairs with the female but can fertilise part of the clutch (Birkhead *et al.* 1988b). In the two broods where extra-pair paternity occurred at Danaher we observed the rearing parents together before and after the eggs were laid so no mate switching occurred here, but given the high rate of mate loss at Danaher rapid mate switching could be a possibility. Of course, there is a reproductive cost to the replacement male if he ends up rearing offspring of his predecessor, but if he gets a chance to copulate with the female before the first egg is laid he has a good chance of being the father (see below). If eggs are already laid he should postpone pair formation.

In the two cases of extra-pair paternity at Danaher we did not observe either female making extra-pair courtships, that is, mating with non-mates, although we did not focus specifically on copulatory behaviour when watching their nests. However, six weeks earlier when copulatory behaviour of four pairs was the subject of detailed study, two of 44 copulations observed were extra-pair copulations, the remainder were pair copulations (Birkhead *et al.* 1988a). This proportion (4.5%) was not significantly different from that of offspring extra-pair paternity (2.4%) and is consistent with the idea that extra-pair copulations are correlated with extra-pair paternities, but this is contrary to recent findings made in a number of other species (Dunn and Lifjeld, 1994). These two extra-pair copulations were not forced. The male performed the song and dance and one female invited mounting with Tail-quivers but the second did not. Cloacal contact occurred in both cases and ejaculate was possibly transferred.

Extra-pair courtship at the Danaher and Alice Springs colonies was commonly observed. Over a six week period at Danaher in January and February 1988, Birkhead *et al.* (1988a) observed 84 extra-pair courtships or copulation incidents between 10 males and 15 females; 67 involved male song and dance towards the female and 15 were forced extra-pair mountings, two of which achieved cloacal contact and presumably sperm transfer. Of the forced extra-pair mountings, in 14 instances the female was retreating when the male mounted uninvited, and in one instance the male disrupted pair copulation by knocking the pair male off the female's back and substituted himself—all without upsetting the female. This extraordinary method of forced extra-pair copulation is believed to be the most successful, since females will otherwise fly out from beneath uninvited males that pounce on their backs (Birkhead *et al.* 1989). Females may be more willing to participate in extra-pair matings than they appear to be since they may be testing males to ensure that they mate with the best or most persistent individual. Males can be very persistent and will still attempt to court a female even when they have been vigorously supplanted by her up to six or more times. Females can probably control the frequency and timing of copulations despite the constraints of mate guarding and the limited opportunities for extra-pair matings. Studies of other monogamous species have reached a similar conclusion (Birkhead and Møller 1993). Why a mated female should seek, or tolerate, copulations with extra-pair males, and avoid it with others, will be considered in Chapter 11.

Although a single copulation can fertilise an entire clutch (Birkhead *et al.* 1989), sperm competition experiments in domesticated Zebra Finches established that the timing of copulation has a profound effect on the number of offspring fertilised. Despite the fact that female Zebra Finches are fertile over a 14-day interval the last copulation she has fertilises a disproportionate (50–80%) number of eggs (Birkhead *et al.* 1988b;

1989; Birkhead and Møller 1992). Thus, if two males inseminate a female during a single reproductive cycle, the sperm from the last male have precedence over those of the first. This phenomenon is termed 'last male sperm precedence', and an interval of about four hours duration is needed for an effect. This accounts for the high incidence of extra-pair paternity since a single extra-pair copulation timed to occur last in a female's copulatory series will have a high success rate. Hence, pay-offs can be great for a male making extra-pair copulations at the right time, and the losses to the cuckolded male partner devastatingly high.

Males should not only carefully monitor the reproductive cycle of their own female partner, but also those of other females breeding in the colony at the same time. At the Danaher colony there may be up to four or five females out of a population of 20 or 30 breeding pairs having a fertile period at any one time, and somehow males need to detect them. High nesting density must facilitate the monitoring of female fertile periods (Møller and Birkhead 1993).

In birds, the male partner has two options to minimise cuckoldry,

(1) frequent mating with the female to dilute the effect of an extra-pair mating and to ensure that he is last; or

(2) to guard the female constantly during her fertile period to prevent any extra-pair mating.

Zebra Finches adopt a combination of the two strategies (Birkhead *et al.* 1988a).

Mate guarding

At the Danaher colony males timed their extra-pair copulation attempts for the female's fertile period and the rate peaked on day 0 and day 1, the day of the first and second eggs respectively. Nevertheless, males making breeding attempts did not seek out extra-pair copulations until their own partners had started laying and the fertile period was more or less complete (Birkhead *et al.* 1988a). The same pattern was found in domesticated Zebra Finches (Birkhead *et al.* 1989). Zebra Finches of both sexes normally guard their partner against sexual rivals at any time of the year, and are almost always inseparable, so that in contrast to many species there is no need for cues to initiate mate guarding. The problem during the fertile period is that the division of labour in nest building requires the male to leave the female at the very time when he cannot afford to do so. The increasingly longer periods the female spends in the nest each day after the first egg, and the increasing warmth of the eggs, may be the cues that indicate to the male that mate guarding is no longer necessary.

Males are constrained in their extra-pair copulation activity because they need to guard their own partner from other males during her fertile period. Females unattended by their male partner were the object of

86% (58/67) of extra-pair copulation attempts at Danaher so males that fail to guard the partner adequately during the fertile period were vulnerable to being cuckolded. A guarding male closely follows the partner every time she leaves the nest and attacks any males that attempt to court her. Consequently, during the fertile period the female initiates significantly more flights away from the nest site than the male, he follows a short distance behind (Figure 9.6).

Birkhead and Fletcher (1992) investigated the possibility that a male also refrains from extra-pair copulations during his partner's fertile period because his sperm may be in short supply. He needs to mate frequently with his own female to ensure his paternity of the young he helps rear, and this has priority over extra-pair matings. It was found that a male transferred less sperm in an extra-pair copulation during his partner's fertile period, than afterwards. This strongly suggests that once a male has ceased copulating with his own partner there are more sperm to spare. The amount of sperm the female partner received during each copulation in her fertile period was not assayed, so she too may receive less, simply because less sperm was available, or he may husband his supply to ensure it meets the demands of his copulating schedule. That

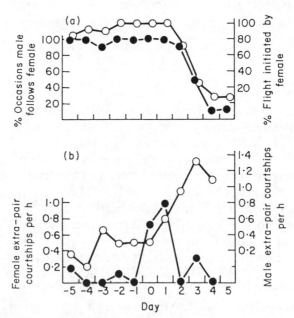

Fig. 9.6 (a) Mate guarding in Zebra Finches at the Danaher colony shown by the percentage of flights away from the nest initiated by the female (o) and by the male (•) towards the end of the fertile period. After day 2, the day the third egg is laid, the female fertile period ends, and the both birds spend more time in the nest. (b) Rate of extra-pair courtship performed by focal males (o) to any female, and towards focal females (•) by any male. (From Birkhead *et al.* 1988a.)

is, he may ration the number of sperm transferred to the female partner throughout her fertile period, even though only one copulation may be all that is necessary to fertilise the entire clutch. In order to maximise the likelihood of his sperm fertilising his mate's eggs he must copulate frequently with her and space the copulations to ensure that he has last-male sperm precedence, and moreover, he must save enough for an emergency retaliatory copulation if he suspects his partner of an extra-pair copulation. In the Bengalese Finch, sperm is limited to the extent that three copulations within three hours will deplete the supply, which takes 24 hours to replenish, therefore the male needs to allocate sperm carefully in order to maximise his fertilisations (Birkhead 1991). A new method of ejaculate collection recently devised by Pellat and Birkhead (1994) for Zebra Finches has confirmed that similar depletions occur in the Zebra Finch (T. R. Birkhead, pers. comm.). While Birkhead *et al.* (1993) found large differences (up to tenfold) in the number of sperm stored in the glomera of individual males there was no evidence that male Zebra Finches can adjust the size of their ejaculate in order to ration supply of spermatozoa (Birkhead and Fletcher 1995). These males were held under identical conditions, and presumably were in a similar physiological condition, but this aspect was not investigated.

There are constraints on the time a male has for guarding the female partner during her fertile period since he must leave her to collect nest material. His schedule is hectic at this time: he brings many stems to the nest, often performing an abbreviated Nest Ceremony with each stem added, and on leaving the nest he usually sings some phrases of Undirected Song nearby before dashing off for more stems. The male tends not to leave the nest if the female hops outside; he will entice her back inside with Nest Whines or lead her back into the nest. He leaves immediately she enters, gives more Undirected Songs and is off collecting more stems. Normally, the female is fairly cooperative in respect to guarding. She waits for the partner's return before flying off, and often gives prolonged pre-take-off calls as if to warn him of her impending flight. Nevertheless, in the wild I have seen some desperate males rush after their females, occasionally while still holding onto nest material. Despite male vigilance, and apparent female cooperation, some females still get off on their own during the fertile period; sometimes the male fails to notice her departure and sometimes he simply fails to follow. Males then normally leave the nest in search of the female, giving Distance Calls continuously. Once I saw a male in a nest at Alice Spring deliver a volley of loud Distance Calls into the face of his female when she silently returned to the nest after a long absence, and on another occasion he invited courtship and tried to copulate when she returned. In aviaries, Birkhead *et al.* (1989) observed on four occasions that males copulated forcefully with the female partner after having seen her involved in extra-pair copulations ($n = 20$). Unlike normal copulations between partners, which are unforced, the male holds the female by the nape feathers during a forced 'retaliatory copulation'. This is more a function of balance than force since wild males will occasionally hold the mate's nape in invited matings if he is prone to falling off. Experiments show that retaliatory copulations after extra-pair copulation reduce the likelihood of fertilisation by the non-mate by an average of 50%, and if pair copulations occur again after a four hour interval the benefits of last-male sperm precedence will further increase his fertilisations to around 80%, thus effectively neutralising the extra-pair copulation (Birkhead and Møller 1992). Males would be expected to attempt to copulate with their partners after any period of separation as a form of paternity 'insurance' during the fertile period; in domesticated birds this also occurs outside the fertile period.

Late in the fertile period, the nest appears to play an important supporting role in mate guarding in the Zebra Finch; the male acts as if he is safe from cuckoldry when his partner is inside the nest. Nevertheless, he keeps in contact with her acoustically, but will not leave if she is not inside the nest. During the early egg-laying period, when she is still fertile, the female spends between 63% to 90% of her time in the nest (Birkhead *et al.* 1988a), although this proportion is lower if she must forage further afield (Zann and Rossetto 1991). In the nest the female is fairly sheltered from other males and is likely to respond aggressively to any trespasser, of either sex; moreover, she has a developing bond with the nest which she must break in order to leave for a sexual encounter outside the pair bond. Her presence in the nest before laying has another advantage in that it may pre-empt egg dumping by conspecific brood parasites (Chapter 6).

In comparison with other species of birds studied to date (Møller and Birkhead 1991), the Zebra Finch is unusual in its method of preventing cuckoldry in that it uses both guarding and frequency of copulation (Birkhead *et al.* 1988a).

The extent of cuckoldry in the Zebra Finch will depend on the size and density of breeding colonies and the synchrony of breeding. For example, in the more arid parts of the range where breeding is opportunistic there may be little opportunity for extra-pair matings even when breeding densities are high because female cycles may be highly synchronised, at least for the first breeding attempt, but nest failures will quickly produce asynchrony. In contrast, in other parts of the range, in locations such as the Danaher colony where breeding extends for eight to nine months annually, female breeding synchrony is low throughout the colony so that the proportion of males available for extra-pair matings will be high and mate guarding correspondingly high as well. Birkhead and Biggins (1987) devised a simple simulation model to describe the relationship between breeding synchrony and the probability of extra-pair copulations; males are predicted to respond differently depending on changing circumstances that affect the costs and benefits of extra-pair matings.

Sperm storage and the fertile period

During their research into the timing of avian fertility Birkhead *et al.* (1989) discovered that female Zebra Finches could store sperm from copulations that occurred 10–13 days previously. This was determined by removing males from partners at different intervals during the pre-egg laying period and examining the fertility of the eggs subsequently laid. Similar storage durations have subsequently been found in a number of other species of small to medium-sized birds (Birkhead and Møller 1992).

The female's sperm storage tubules are located at the junction of the uterus and the vagina where they quickly take in the ejaculate from the cloaca soon after copulation (Birkhead 1987; Birkhead and Hunter 1990). However, there is an astonishing wastage of sperm. Only 0.1% (mean number 6,027) of spermatozoa in the ejaculate (mean number 5.8 × 10⁶) reach the sperm storage tubules, the rest being ejected with the faeces, and of those stored only 1.4% (82 spermatozoa) reach the infundibulum and the eggs, so that of the total spermatozoa estimated to be received by the female in one ejaculate, only 0.001% reach the eggs (Birkhead *et al.* 1993).

Since eggs are fertilised about 24 hours before they are laid, Birkhead *et al.* (1989) determined that the fertile period lasts about 14–15 days and starts 11 days before the first egg is laid and ends 24 hours before the last egg is laid, that is, the evening before the day the penultimate egg is laid. In wild birds where the modal clutch size is five, this will be the evening of the day the third egg is laid, and the evening of the day the fourth egg is laid in domesticated Zebra Finches where the modal clutch size is six. Birkhead *et al.* (1993) refer to a 'fertilisation window'—a space of about one hour after an egg is laid when sperm can travel up from the vagina or sperm tubules to the infundibulum to meet the ovum of the next day's egg. Since Zebra Finches lay most eggs in the first hour or so of daylight it is not surprising that most copulations occur in the early morning.

Parental care

Like most estrildines both parents help rear the offspring and there are no sex-specific caring roles. Incubation, brooding, allopreening of young, nest guarding and defence are performed by both sexes. Females take a greater part of the burden than males. Burley (1988a) estimated that domesticated males perform 43–45% of the daytime care which is consistent with other findings on domesticated birds (El-Wailly 1966; Delesalle 1986), but in wild birds there were no sex differences in incubation (Chapter 6) or in feeding rates (below). Females do all the nocturnal incubation.

Immelmann (1962a) noticed that most Zebra Finch pairs he observed practiced indirect relief of the incubating partner. Specifically, the sitting bird hears the Distance Calls, or Stack Calls of the relieving bird, and leaves before it arrives so that the two meet fleetingly some distance from the nest; in some cases they do not meet at all. Yet in 11 of 17 pairs I observed at Alice Springs and northern Victoria, the sitting bird did not leave until the relieving bird had entered the nest; in four pairs they met within 10 cm of the nest, and in two they met further away. Apparently, the method of relief may vary from one population to the next. Often the relieving bird will arrive with some nest lining material, usually a feather, and the relieved bird may return with nestling material before finally departing. Incubating birds sit so that the eyes are just above the rim of the nest cup enabling them to see out the entrance tunnel—a red bill just being visible to the observer. While incubating, birds often 'spruce' up the nest lining and tunnel. Most incubating birds are extremely wary of observers, including those in hides or sitting back 30–40 m away, and will repeatedly leave the nest at the slightest disturbance. The male is less wary than the female and will often lead her back to the nest, going inside first, giving Nest Whines and mandibulations, yet leaving as soon as she enters, although he usually sings several phrases of Undirected Song nearby before flying off. Sometimes, inexplicably, incubating birds sit so tightly that they can be touched during a nest inspection.

There is no change in the behaviour of the incubating parents to signal that young have hatched, unless one happens to observe the parent eating the egg shells. Nest attentiveness remains unaltered and incubating temperatures are maintained for at least five days after the first eggs hatch (Chapter 6). Gradually young are left uncovered during the day for increasingly longer periods so that by day 11 or 12 nestlings are no longer brooded, unless the weather turns cold. Only the female broods at night and this ceases several days before the young fledge.

Young are not usually fed on hatching day, but short feeding episodes occur during bouts of brooding for the next five or six days, after which feeding occurs as soon as the parents enter the nest, and a short bout of brooding may follow. In wild birds only one parent at a time enters the nest after day 6, the other remains on watch outside because the Begging Calls, which can be quite loud by this age, may attract potential predators. In domesticated Zebra Finches both parents may still feed the young together after day 6 (Immelmann 1962a). Contrary to Immelmann's (1962a) conclusion that females take the first feeding bout there was no trend in 24 wild pairs I observed (males first in eight pairs; females first in nine, and no precedence in seven). Moreover, there was no significant difference in duration of feeding of nestlings (based on Begging Calls) by mother and father (Wilcoxon signed-ranked matched pairs test: $z = -12.0$, $P = 0.5$, $n = 15$ pairs). In domesticated Zebra Finches held in small cages, ten Cate (1982) found that females gave more feeding bouts than males and the number of feeding bouts each day peaked on day 8.

Before a feeding bout the parent often gives regurgitation movements before it enters the nest: bill-gaping, lateral head shakes and pumping movements of the throat. The parent does not transfer food to the young from the entrance tunnel but hops to the back of the chamber, turns around so that it can see out the tunnel and straddles the young before it begins. Regurgitated seed mixed with water is transferred to the young in the standard estrildine manner. The gaping bills of parent and young are locked together and the parent's tongue is used to push the portions of food into the mouth of the nestling which in turn uses its tongue in a reflex action to pull the portions back into its oesophagus and into the crop. Crops of nestlings are often inflated with large pockets of air after feeding. Young are not fed continuously throughout the day; the most intensive bouts occur in the early morning and late evening and minor bouts in the later morning and early afternoon (Immelmann 1962a). A similar pattern is found in domesticated birds (ten Cate 1982).

Lemon (1993) investigated the energetic cost to domesticated parents of rearing nestlings. The cost includes foraging for extra food, carrying it to the nest and brooding. Costs increased steeply from hatching day to day 8, where it peaked, then it fell slowly until day 13 when it reached day 1 levels. This pattern coincides with the daily changes in feeding rate found by ten Cate (1982). Females showed a weight loss that mirrored these changes in energetic cost. Whereas Lemon (1993) could find no significant effect of reduced net daily energy gain on costs incurred by rearing females, Skagen (1988) found that limited food caused females to lose more weight than when food was not limited, moreover, females lost proportionally more weight than males, probably due to night-time brooding duties and heat loss through the brood patch.

Zebra Finches show rudimentary nest hygiene, the only Australian estrildine to do so. According to Immelmann (1962a) parents may occasionally swallow the droppings of nestlings until the latter are about seven days of age, after which they sporadically carry out dried droppings and drop them some distance from the nest.

Immelmann (1962a) found that free-living Zebra Finches fledge on day 22 post-hatch, while we found that they normally fledged around day 16–18. Offspring of wild-caught birds also fledged around 16–18 days. Most of Immelmann's observations were made in Western Australia and ours in northern Victoria and Alice Springs. Different weather conditions and access to rearing foods may be responsible for the discrepancy. The nestling period in domesticated Zebra Finches is around 18 days (Immelmann 1959; Bischof and Lassek 1985; Martin 1985; Corbett 1987). In the wild, young usually fledge sometime in the morning, and the parents may play an active role by hopping excitedly in and out of the nest giving Tet Calls and Distance Calls until one, or several young, leave the nest entrance and follow fluttering and jumping onto the branches of the nesting bush while giving Long Tonal Calls (Chapter 10). Should the parents fly away, the young will follow, but they are usually led back into the nest shortly after they first emerge. For the first two or three days young are usually led back to the nest before they are fed, but not always.

Fledging is a gradual affair; not all young fledge at once and each spends only a short time out of the nest on its first day. Should the parents re-nest, the male takes the more active role in caring for the fledglings, and if they venture too close to the new breeding nest they may be attacked by the female. Nest leading disappears about four or five days after fledging, but they are still led back to the sleeping nest each evening. At first the young are led in singly, but after a few days they enter as a group.

Parent–offspring aggression in free-living Zebra Finches is rarely observed. Females may drive away their fledglings from the new breeding nest, but aggression by the father towards the young was not seen. This contrasts with the situation in caged Zebra Finches where both parents frequently attack and drive away their young (Böhner 1983; ten Cate 1982, 1984; Clayton 1987a). This is probably an artefact of confinement.

Brood reduction

Even when predation is avoided, not all nestlings reach fledging age. About nine per cent of nestlings at the Danaher colony died before banding age. It was usually the smallest individuals that disappeared, although occasionally their bodies were found beneath larger nestlings. Whether parents purposely reduced the size of the brood through starvation of the smallest nestling is unknown. Parents were never seen to carry nestlings, alive or dead, out of the nest. Sometimes live nestlings a few days after hatching were found in the entrance tunnel or at the bottom of the nesting bush, but the causes were a mystery. Similarly, among domesticated Zebra Finches not all nestlings survive. For example, Skagen (1988) found that 13% of nestlings died, most before 10 days of age, but strangely, there was no significant difference between synchronous and asynchronous broods. Burley (1986a) observed parents removing struggling nestlings and dropping them outside in the aviary leaving them to die. She believes that this was one way to manipulate the sex ratio of offspring (Chapter 11). When Lemon (1993) experimentally manipulated the net energy gain of breeding Zebra Finches he found that the group with the least daily net energy reduced the brood size within three days of hatching by actively removing the youngest. He also states that 'broods were reduced by more subtle actions such as differential allocation of food between nestlings or sibling competition', but provides no details. However, Skagen (1988) found that when food was limited larger nestlings in a brood had faster growth rates than smaller nestlings and concluded that parents preferred to feed larger nestlings more than the smaller ones.

Behavioural development of young

Nestlings

Zebra Finches have two begging postures in which they gape upwards for food. For about the first five days after hatching the nestling begs in a typical passerine posture in which the neck is stretched upwards, the head tilted back so that the crown touches the crop and the gape is directed upwards; the body, head and neck are held still, but the tongue swings slowly from side to side. The wings are not moved. On the fifth day a new begging position, the head-down neck-twist, begins to replace the head-raised position. The neck is twisted and rotated almost 180° to one side so that the crown is almost level with the feet and the gape is directed upwards (Figure 9.7). The head is swayed slowly from side to side. The wings are not quivered as in most begging passerines, although the wing on the opposite side to that on which the head is twisted may be extended out during intense begging. This begging posture is unique to almost all estrildines. The period of transition in begging posture is quite variable within and between broods. The rate of tongue wagging increases with age from around one cycle per second on day 2 up to about four cycles at day 15, and it becomes faster during feeding bouts (Muller and Smith 1978). Both types of begging posture are almost identical to those of the three species of *Poephila* (Zann 1972), but Double-barred Finches move their heads, necks and lower mandibles more (Immelmann 1962a, 1965a).

Fig. 9.7 A mother feeds a fledgling that has adopted the head-down neck-twist posture.

Young nestlings will gape, and often call, when held in the hand. From hatching until about thirteen days of age tactile and auditory stimulation are sufficient to elicit begging in nestlings. This was shown experimentally by Bischof and Lassek (1985) who found that touching the head and bill with a stick and playing back begging sounds, stimulated gaping. Immelmann (1962a) noticed that vibrations of the nest caused by strong wind could also trigger long bouts of silent begging. Although young align themselves towards the nest entrance around day 7, gaping is still directed upwards until day 10, after which it is directed towards the parents. A red-coloured pencil held at eye level will also stimulate begging, but a grey-coloured one will not (Roper 1993). Begging in response to foreign stimuli begins to diminish around day 13 and ceases at day 16, because of an increase in fear (freezing, shuffling to the back of the nest), as young begin to learn the features of their parents and distinguish them from other objects (Bischof and Lassek 1985).

Nestlings do not give Begging Calls (Chapter 10) until the third day after hatching, and only the individual being fed vocalises; from day four to 12 other nestlings may vocalise during a feeding bout although they are silent before and afterwards, but after day 13 they detect the approach of the parent and begin to call before the bout commences (Muller and Smith 1978). Visual cues, initially provided by the mouth markings and tongue movements, are sufficient to elicit parental feeding, but acoustic stimuli soon become paramount. This was demonstrated by playback experiments of begging sounds which not only triggered begging behaviour in fledglings, but also stimulated the parents to eat seeds and to provision them (Muller and Smith 1978). Also, the rate of Begging Calls emitted is affected by the level of hunger (Roper 1993).

Nestlings defecate as high up as possible around the wall of the nest chamber so that a hard dry ring of concreted faeces forms out of normal reach of the nestlings, which manage to stay fairly clean. By day 10 most young back up to the entrance tunnel and defecate on the rim of the nest cup which becomes solid with droppings. Larvae and pupae of a species of tenebrionid beetle *Platydema pascoei* have been found feeding on old faeces in nests of Zebra Finches and other species of estrildines and can eventually clean up old nests in some cases (Hindwood 1951). When inspecting nests of wild birds at night I have seen parents and brood surrounded by mosquitoes with unidentified species of beetles and mites swarming over the droppings.

By 13 days post-hatch young back up against the rear wall of the nesting chamber when the nest is inspected and resist removal by hanging on to the nest floor with their toes. After this age they force fledge, exploding out of the nest with Long Tonal Calls and fluttering and flying as best they can, followed by the parents giving Distance Calls. On the last few days before fledging, young become active in the nest, flapping their wings, self-preening, ruffling feathers and moving around and calling back and forth to their parents outside.

Post-fledging behaviour

In the first few days after fledging young are inconspicuous, spending much of the time sitting silently and motionlessly clumped together in a row among dense bushes in the nesting colony. Parents return together at regular intervals to feed them. At first the young reply with Long Tonal Calls to most adult Distance Calls (Chapter 10), but after a few days only reply to their parents and siblings. They hop to their parents while giving Long Tonal Calls that flow over into Begging Calls then gather either side and begin to beg in the head-down neck-twist posture. If fed on the ground young beg in a semicircle around the parents. They thrust themselves in front of the parents then shuffle back in the head-down neck-twist amid shrill, irresistible begging cries. Feeding parents are most intolerant of other young nearby that might attempt to join in and beg. Fathers often sing Undirected Songs before and after they feed the young. After feeding, parents seem to avoid perching near their young at this vulnerable age, possibly because their brightly-coloured red bills might attract predators; nevertheless, this behaviour persists in domesticated Zebra Finches (Immelmann 1959). Several days after fledging, young follow their parents to other parts of the colony, especially the social area and local feeding areas. Here they encounter other juveniles, and soon spend a good part of the day sharing the same bush where up to 16 have been observed at Danaher waiting for parents to return to feed them. These juvenile groups may wander some distance from the colony and join the main foraging flock if nearby.

Elements of male and female courtship behaviour first begin to occur several weeks before the onset of juvenal moult. Subsong is evident in young males as early as 25–30 days after hatching when they sing quietly to themselves while perched alone in the bushes of the colony. By day 40, when patches of sexual plumage begin to appear in most males, this undirected type of singing is frequent, and occasionally the directed type sung to the female during stage 2 courtship can be seen, but without the accompanying dance, which does not occur until about day 60 or later. Exaggerated greetings and the stage 1 waltz with its hops, bows and bill-wipes first appear in young less than 30 days old, but again are not regularly observed until after day 50 or 60. At first, courtship stages 1 and 2 may be directed indiscriminately to males and females, siblings and peers alike, but by about seven weeks of age courtship is directed exclusively to members of the opposite sex. In the wild, the first full courtship of a female including Tail-quivering, copulation call and copulation was observed in a 30–35-day-old individual when courted by an adult male, but full courtship is generally rare until at least day 60. Mounting and copulation by young males has not been observed before day 90, however, it must first occur around 60 days, since the youngest male detected breeding at Danaher was only 67 days old when his mate laid her first egg. The youngest male seen carrying a grass stem was 30 days old, and although they begin to test stems and make the three nest building movements around this age, the earliest functional nest building is not seen until pairs are at least 50–60 days old. The first nest ceremonies and defence of the roosting and breeding nest against conspecifics and sparrows also occurs around 50–60 days. The time of first appearance of behaviours described here are earlier than those described for domesticated (Immelmann 1959; Sossinka 1970) and free-living (Immelmann 1962a) Zebra Finches except for subsong which was seen in captive males as early as 24 days after hatching. Immelmann (1962a) believes that the components of each stage of courtship appear first and are later organised into functional sequences directed at members of the opposite sex. He never saw young females courting young males with incomplete sexual plumage, whereas two young males may court one another and even form a pair bond, but this ruptures once most of the sexual plumage comes out.

Development of seed-dehusking

Nelson (1993) investigated experimentally how dehusking behaviour develops in domesticated fledglings. As expected, both maturation and experience are important. Young birds drop more seeds accidentally before they can be dehusked and swallowed than do older, more mature individuals, of equal experience. The larger the seed the more clumsy they are. Differences in length of bill and neural coordination are believed to be partly responsible. Nevertheless, the time it took to learn to dehusk a seed as fast as an adult varied from 10 to 13 days, and depended on the type of seed rather than the age of the individual. There was a tenfold reduction in dehusking times over this interval. In these learning trials, youngsters had 30 minutes dehusking practice each day, so that in the normal laboratory situation where seed is available *ad libitum*, about five to six hours of dehusking practice is the minimum needed to achieve adult levels of competency. This would take from two to four days to acquire in the normal captive situation and much longer in the wild where seed is more difficult to locate and access and more diverse in type and condition. Additional experiments showed that skill acquired in dehusking one type of seed could be carried over to different types of seeds, and that dehusking skills could diminish without regular practice.

Parent–offspring behaviour

Transition to independence

To investigate when young Zebra Finches forage and roost independently of their parents we focused observations on young at the Danaher

Table 9.1   Stages leading to foraging independence in fledglings at the Danaher colony based on data from two breeding seasons

Foraging stage	Age after hatching (days)		Number of broods
	Earliest	Latest	
Begging to parents and being fed	17	40	22
Begging but not fed by parents	25	40	19
Foraging with parents and fed by them	24	40	11
Foraging with parents, but not fed by them	26	40	4
Foraging without parents	30	56	15

colony for two breeding seasons, following them as far as possible from fledging to 50 days of age and beyond (R. Zann and B. Quin, unpublished observations). When banded at 12–14 days of age, all the brood were marked with non-toxic dye to aid identification around the feeder in the walk-in trap and at the roosting nests. Unfortunately, the early post-fledging stage is characterised by heavy losses so sample sizes diminished dramatically after fledging (Chapter 8).

The transition from complete dependence on the parents for food to complete independence began on the day of fledging (day 17–18) and was completed after day 40. At first, young nibble anything: grass stems, feathers, twigs, etc., but soon concentrate on seeds, which they pick up and drop. By day 28, half of all fledglings observed (44 individuals from 21 broods) were foraging independently at the feeder and half were still being fed by the parents, although not exclusively so. By day 35, only 10% were fed on some occasions by their parents. The timing of the steps to feeding independence are shown in Table 9.1. Immelmann (1962a) observed one free-living pair continue to feed their young until they were seven or eight weeks old. With domesticated Zebra Finches, Immelmann, (1959) found that parents normally feed their young for two weeks after fledging; however, in exceptional circumstances it can be extended for an additional week, but by the father only.

Within a day or two of fledging, young are led, usually by the father, to a roosting nest. The breeding nest may be refurbished for a new breeding attempt or it may be too fouled with droppings for further use. The roosting nest may be in the same bush as the breeding nest or up to 100 m away if the parents relocate for another breeding attempt. The father spends more time with the fledglings than the mother. Fledglings are not permitted to roost in the new breeding nest unless there is an emergency. From fledging day to day 22 they are led to the roosting nest during the day for feeding purposes, principally by the father, and at night for roosting up to the age of 40 days. One 40-day-old youngster was even fed by the parents after it followed them to the nest. After 30 days of age, young begin to enter the roost independently and continue to do so until day 48, after which they no longer roost in the family nest. Day 42 is the age when half the young observed (n = 14 young from 4 broods) roost in the family nest and half roost on their own in a different nest that may or may not be in the same bush as that of their parents. Young Zebra Finches normally go to roost about an hour before adults, but up to 30 days of age will enter the roosting nest during the day to shelter from heavy rain, something adults never do. Of 16 clutches observed, 11 roosted with both parents, three with the male only and two with the female only. On attaining roosting independence, young from several different clutches may share the same roosting nest.

A cautious interpretation of these observations suggests that most Zebra Finches reach nutritional independence from their parents by 35 days of age, but can maintain contact with the parents, as demonstrated by their roosting association, until 48 days of age at least. We also saw parents and young together around the feeder after this age but did not observe any interactions or spacing that might suggest that the family bonds still exist—their behaviour was not distinguishable from that with other members of the flock. Nevertheless, it is possible that parent–offspring bonds are maintained in some subtle way beyond this age. The bond between parents and offspring endures longer than Immelmann (1962a, 1965a) realised and extends well into the sensitive period for song learning (day 35 to day 65; Chapter 10).

Nutritional independence is achieved in captive-reared Australian Zebra Finches by day 35, but Lesser Sundas Zebra Finches from Timor took 40 days (Clayton *et al.* 1991).

Releasers of parental and juvenile behaviour

By painting the black bill bright red in newly fledged members of a brood, Immelmann (1959) established that the black bill alone is the releaser that distinguishes young from adults. The father immediately courts his red-billed fledgling on its release, but stops instantly the youngster begins to beg to him. He courts again when begging ceases. If the red-billed fledgling goes near the nest it is chased away. The mother ignores it at first, but also chases it from the vicinity of the nest. Thus, for about the first hour the mother and father react as if the red-billed fledgling is a strange female, accordingly, they clump with it and allopreen. Eventually, desperate begging by the youngster will overcome the inhibitory effect of the red bill and it is fed.

When Immelmann (1959) painted the father's bill black his fledglings would not beg to him, but they quickly did so when he made the regurgitation movements that precede food transfer. When given a choice of models, one with a black bill and the other with a red bill, the response of young depends on their age. Young less than 35 days of age clumped with the black-billed model and older young clumped with the red-billed model, even if their own bills still have lots of black pigmentation. Newly fledged Zebra Finches however, do not respond immediately to models, rather the movements and calls of living birds are the stimulus for an approach from a distance. To conclude, Immelmann's (1959) experiments showed that the black bill is used by both young and adults to distinguish fledglings from adults. This prevents them from being attacked and courted at a vulnerable time since they are not old enough to defend themselves or escape.

Kin recognition

Laboratory experiments in which individuals were tested for their approach preference showed that visual and vocal characteristics of kin are learned and later used as cues for recognition.

When an unmated test bird was placed in a multiple-choice apparatus and given the choice of four unfamiliar stimulus birds with whom it could choose to perch opposite, Burley et al. (1990) showed, with rigorous methodology, that domesticated adult male Zebra Finches significantly preferred brothers over non-kin males. Personal recognition was not a factor. Preferences between sisters were not tested, but when males and females were tested with matched stimulus sets of unfamiliar birds of the opposite sex there was no bias towards siblings over non-kin. In a follow-up experiment, females were shown to prefer the company of male first cousins irrespective of whether they had met one another early in the rearing aviary or not. In the reciprocal experiment, males showed no significant preference for female first cousins.

Burley et al. (1990) concluded that the male preference for brothers is an outcome of a non-sexual affiliation for same sex kin, whereas female preference for male cousins is an expression of mate choice. Why males do not prefer their female first cousins is unknown. Perhaps females are better at recognising male kin than males are at recognising female kin. Burley and Bartels (1990) showed that it is possible for humans to distinguish sets of same-brood brothers from matched sets of non-brothers by similarities in the detailed markings of the eye stripe and throat stripes. That is, there were strong family resemblances in these features. Assuming that female Zebra Finches can discriminate at least as well as humans, it is possible that they use these features to discriminate between strangers and kin. Females also have eye stripes and Burley and Bartels (1990) showed a photograph of two sisters with a family resemblance, therefore, one might expect female eye stripes to be used by males to identify their female cousins.

Vocalisations might also provide cues for kin recognition since the experiments of Burley et al. (1990) did not exclude vocal communication between test and stimulus birds. Family resemblances exist in Distance Calls between mothers and daughters and in Distance Calls (Zann 1985b) and songs of fathers, sons and brothers in the wild (Zann 1990) and in captivity (e.g. Immelmann 1969; Arnold 1975a). Miller (1979a,b) showed that daughters can recognise the songs of their fathers and females can recognise their partner's song. Both males and females learn to recognise their father's song before 35 days of age (Clayton 1988a; see Chapter 10). In nature it is likely that visual and auditory cues work together to aid recognition of kin. Vocalisations, especially Distance Calls, would be the long-range signals that lead to the initial approach of potential kin, subsequently, song and close inspection of the visual features of the head could verify first impressions.

Where learning is involved, errors can occur in the formation of kin-specific cues and this happens with male Distance Calls and songs. However, no learning is evident in the development of female Distance Calls (Zann 1985b) so these would be reliable indicators of relatedness. Facial markings of parents and siblings are probably learned and may also be subject to error, but this has not been investigated. The incidence of brood parasitism and extra-pair copulations would devalue genetic kinship among siblings such that templates developed for phenotypic matching would be based on rearing parents and same clutch siblings to whom an individual might not be related.

Why brothers should prefer each other's company during choice experiments is not clear. Mobility within and among colonies is so high that it is difficult to track subgroups of individuals that might be kin. There are definite roosting groups within colonies; in addition, the retrapping data suggest that small groups move about the home range together (Chapter 8). Conceivably, these groups may consist of brothers or other kin, but what advantages are gained over those that do not move in kin groups are unknown, and would be difficult to test in the wild. Reproductive success might be higher among individuals that breed in kin groups. In an aviary experiment, Williams *et al.* (1993) found that males from the same aviary population as their females had a higher reproductive success than those mated to females from different aviaries.

Summary

Wild Zebra Finches are highly social and flock throughout the year. Aggression is common, and mostly occurs over mates and roosting sites; no dominance orders are formed and agonistic behaviour is weakly ritualised.

Flocks consist mostly of adult pairs and immatures. Any unpaired adults meet and pair up in the flocks but have no special pair-formation displays: they simply use the early stages of courtship. Pair bonds are formed rapidly, and clumping and allopreening are the signs that a bond is complete. Both sexes have preferences and these may form on first meeting, but those of females normally prevail. Partners are initially chosen on the basis of appearance, but subsequent behavioural interaction determines whether a mutual bond will form. Pair bonds, which are primarily sexual in motivation, are strong in breeding and non-breeding seasons, and are constantly serviced and defended. Partners may be stressed when forcefully separated and the male is more active than the female in seeking reunion.

The male proposes suitable nesting sites to the female and she makes the final decision. This is confirmed by a Nest Ceremony. The male fetches the building material for the nest and the female works it in. Courtship is initiated with an 'exaggerated greeting' followed by a 'waltz', after which the male sings and dances to the female who eventually invites copulation. The male dichromatic plumage elicits the female sexual response and prevents her fleeing, but male courtship is necessary for the female to invite copulation. Copulation is most frequent around the day the first egg is laid and about 15 copulations occur for each clutch of eggs. Extra-pair copulations can result in extra-pair paternity (2.4% of offspring), and although forced copulation is possible, females appear to control the number and timing of copulations within and outside the pair bond. When two males copulate with the same female, sperm of the last male have precedence over those of the first. Females are fertile for a 15-day period ending before the laying of the second-last egg; males guard females over this period in order to prevent copulation with others. Enticing the female to stay in the nest is a form of mate guarding. Females can store sperm for 10–13 days.

Zebra Finches have biparental care, but the female does the most and loses the most weight. Under some conditions the size of the brood will be reduced, but the mechanisms are largely unknown. Young beg for food in the typical estrildine neck-twist posture and leave the nest 17–18 days after hatching. Nutritional independence is reached around 35 days, but roosting associations can extend another 15 days or so. The black bill is used as a releaser to distinguish fledglings from adults. Zebra Finches can recognise strange kin: brothers prefer each other's company and females prefer to mate with first cousins.

10 Vocalisations

'Two species of grassfinches, the Australian Zebra Finch and the Bengalese Finch, possess an instinctive song model that consists of only a few elements. Song develops in the young through listening to the species-specific song: however, if reared by a heterospecific it will give the foster-father's song and completely mask the instinctive component.'

Translated from K. Immelmann 1967.

The Australian subspecies of Zebra Finch is highly vocal, and in the wild one usually hears them before seeing them. They have ten or eleven distinct calls in addition to the song. The Lesser Sundas subspecies may have a similar number, but to date, only the song and three calls have been described. In general, aviary-bred Lesser Sundas Zebra Finches are less vocal (both songs and calls) than their Australian counterparts in both aviaries and cages. The sounds of each subspecies are qualitatively distinct: Lesser Sundas birds have a high-pitched whistle while Australian birds have a heavy nasal twang.

Immelmann (1962a, 1965a) was the first to describe the vocalisations of Zebra Finches, and those of the other Australian estrildines, which he did without the benefit of tape-recorders and sound spectrograms. When analysed on a sound spectrograph, the vocalisations of Zebra Finches, like most estrildines, display a complex structure in which the sound energy is concentrated into multiple bands that occur at harmonic intervals, and are referred to as 'harmonics' by most workers. The lowest harmonic is the fundamental. The Australian subspecies has more harmonics stacked into its sounds than does the Lesser Sundas subspecies and this accounts for the nasal-sounding timbre. The Double-barred Finch and the Masked Finch also have many harmonics in their sounds and in the field their calls can be mistaken momentarily for those of Zebra Finches.

Call repertoire

When active, Australian Zebra Finches call much of the time, except when feeding. The three most frequent calls are the 'Distance' Call, the 'Tet' and the 'Stack' (Figure 10.1), which are given on the move anywhere at anytime. Around nesting colonies less-frequently uttered sounds are also heard. During the breeding season members of a pair give special calls at the outset of nest-building ('Kackles', 'Arks' and 'Whines'); calls are given before ('Wssts') and during combat ('Distress' Cries).

Young give a range of Begging Calls and have their own distance call (Long Tonal Call). Parents have a call (the 'Thuk') that warns of the approach of potential predators of nests and young. The 'Undirected Song' is given at anytime of the year in both breeding and non-breeding seasons while 'Directed Song' is only given during pre-coital courtship and during pair formation.

The contexts, and putative messages and functions of the various vocalisations emitted by the Australian Zebra Finch are summarised in Table 10.1.

Tet (Figure 10.1a)

A series of soft 'tet-tet' sounds are emitted during most hopping movements, either between perches or when on the ground. These are probably the sounds most frequently uttered by Zebra Finches, and in many situations form a soft background hum in which other calls are embedded. That is, they are not directed at specific individuals and do not stimulate specific replies. Tets do not carry beyond about 5 m and are mostly heard by the mate or family members nearby although in captivity the calls are given automatically by visually and acoustically isolated individuals. Tets are given frequently when birds are excited and the series becomes most rapid just before flight. In this way, take-off may be coordinated among mates, family and flock members. The Double-barred Finch and the three species of *Poephila* also use pre-take-off warning calls (Immelmann 1962a; Zann 1975).

Tets are stereotyped in physical structure. Sonograms show a pile of crowded chevron-shaped harmonics with a duration ranging from 46–52 ms with a mean of 50 ms; the fundamental frequency ranges from 0.40 to 0.48 kHz. There are no sex differences, but birds from central Australia have significantly longer calls of lower fundamental frequency than birds from northern Victoria (Zann 1993a). Tets from Lesser Sundas birds are longer and have a higher fundamental frequency than those of Australian birds (Figure 10.1a *vs.* Figure 10.2a).

Distance Call (Figure 10.1b)

The short, ringing 'tia' of the males and the longer, more nasal and flatter, 'tiaah' of the females are the loudest and most penetrating sounds given by Australian Zebra Finches. The bill is barely opened when softer versions of the call are emitted, but opened more for louder ones. The Distance Call in the Lesser Sundas birds is a loud high-pitched whistle in both sexes. In the field the Distance Call of the Australian subspecies may be heard from 80–100 m away. It is the most characteristic sound heard from wild Zebra Finches and is most persistent in scattered and isolated birds. It is emitted in a variety of contexts: on take-off, in flight, during mild alarm, greeting newcomers, stage 1 of courtship, between bouts of singing, and between bouts of feeding of the young. Partners

Table 10.1 Commonly uttered vocalisations of Australian Zebra Finches and their putative messages and functions.

Name (synonyms)	Figure	Contexts	Messages	Functions
Tet Call (Communication Call)	10.1a	(1) Most hopping movements (2) Pre-take-off (3) Between songs and song phrases	'Partner/family: keep close!' ?	Close contact.
Distance Call (Long Call Identity Call)	10.1b	(1) Isolated from partner and/or flock (2) Take-off (3) Stage 1 courtship (4) Sudden danger/excitement (5) Greeting	'Join me!' 'I am here!' 'It's me!' 'Flee!'	(1) Distant contact and localisation (2) Identity (3) Alarm
Stack Call	10.1c	About to take-off and to land.	'Watch/join me?'	Intention
Wsst Call	10.3a	About to supplant.	'Move!'	Intimidation
Thuk Call	10.3a	Approach of potential danger from birds reluctant to flee	'Prepare to flee!'	Warning to young
Distress Cry	10.3a	(1) Intense combat (2) Contact with potential predator.	Pain/fear	?
Kackle Call	10.3b	Landing next to partner near potential nesting site	'Greeting; follow me to a nesting site!'	Nest-site attraction?
Ark Call	10.3b	Landing at potential nesting site.	'Join me here at this nesting site!'	Nest-site solicitation
Whine Call	10.3b	(1) Nest ceremony: partners at nesting site or in a nest (2) In nest mate/ young outside	'Nest here!' 'Come to me in the nest!'	(1) Nest-site bonding. (2) Nest-attraction.
Copulation Call		During copulation	'Do not flee!'	Sexual appeasement
Begging Calls	10.3c	Begging for food by young	'Feed me!'	Food solicitation.
Long Tonal Call	10.3c	Isolated from parents and/or sibs	'Join me!'	Localisation.
Directed Song		Directed at female during pre-coital courtship	'Let's copulate!'	Sexual
Undirected Song	10.6	(1) Near nest, female inside (2) In/near flock/nesting colony (3) Unmated males in flock (4) Visual isolation from conspecifics	'Stay in the nest!' 'I could be available' 'I'm unmated!' ?	(1) Nest bonding/guarding (2) Sexual advertising? (3) Weak sexual? ?

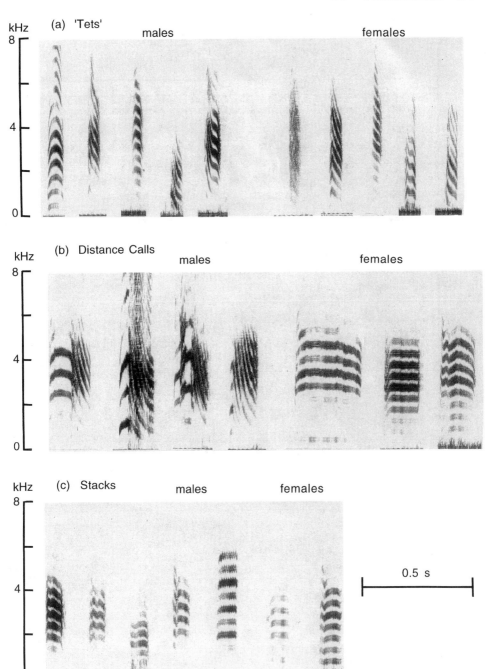

Fig. 10.1 Sonograms (wide-band filter) of the three most frequently uttered vocalisations emitted by Australian Zebra Finches. Recordings of wild birds from northern Victoria and Alice Springs are shown. Each sonogram is from a different individual.

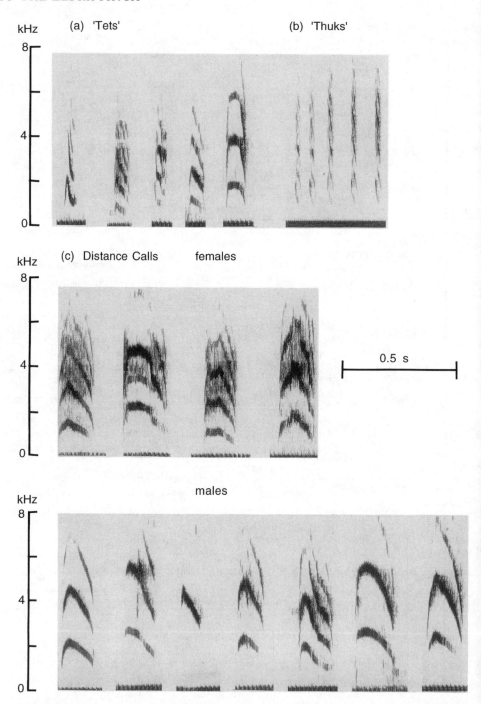

Fig. 10.2 Sonograms of three of the most frequently uttered calls made by the Lesser Sundas Zebra Finches. Recordings were made from captive-bred Timor birds, and each call shown, except for 'Thuks' is from a different individual.

may exchange Distance Calls with one another some distance away when one is in the nest and the other is flying to, or from, the nest. Males call more frequently than females (Zann 1984). In captivity the Distance Call is elicited by visually isolating a bird from its mate or companions. It has a high reply rate especially between mates and members of the immediate family and flock, but newcomers are also greeted from afar with a loud exchange of Distance Calls. When flocks of several hundred or more assemble, prolonged volleys of Distance Calls can be irritating to people. The call appears to function as an identity call, a lost call and a flight call and is the most important one given by the species. Details of sexual and geographic variation in structure, ontogeny and control of the Distance Call are given later.

Stack (Figure 10.1c)

Louder, longer and higher pitched than Tets, but softer, shorter and lower pitched than Distance Calls, Stacks are emitted at the moment of take-off. Thus, they are given after the series of pre-take-off Tets and usually precede the Distance Call which is given a second or so after take-off. Stacks are also given in flight, especially when hovering, or when birds are hesitant about flying down alone to feed or bathe, or to a nest at incubation change-over. Males leading females from one bush to the next in search of a favourable site for nesting give Stacks. There is little frequency modulation so that their harmonics are flat, horizontal bands, although there may be some modulation at the onset. There are no obvious sex differences in structure. Birds isolated visually and auditorily in a cage give this call more often than Tets and Distance Calls and hundreds can be emitted in an hour. The call is structurally stereotyped in males, but female calls are variable in duration, and this may be a reflection of their level of motivation. Female Stacks have a bimodal distribution in duration with a set of brief calls 0.05–0.1 s, and another set around 0.15 s duration; both are identical in harmonic structure to Distance Calls (0.2–0.3 s) but with a slightly lower pitch.

Wsst (Figure 10.3a)

A hissing 'wsst' sound 'like the sudden ripping of a piece of cotton cloth' (Immelmann 1965a) is given by Zebra Finches a fraction of a second before they supplant (Chapter 9) an enemy from a perch. The calls are given singly and the sonogram shows a burst of noise of low frequency. This is the noisiest call given by Zebra Finches and the last sounds heard at night as birds go to roost.

Thuks (Figure 10.3a)

A short, thick-sounding 'thuk' is punched out by parents at the approach of potential predators of nests or newly fledged young. It is an alert to get ready to flee when danger is not immediate but suspicion is aroused.

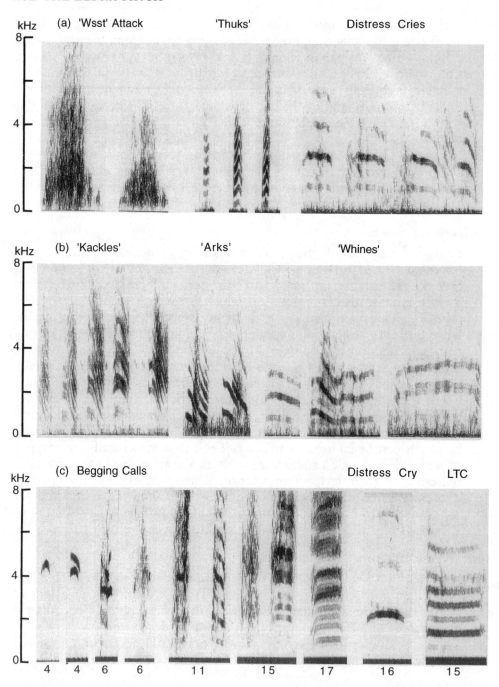

Fig. 10.3 Sonograms of nine less-commonly uttered vocalisations from wild Australian Zebra Finches. Numbers below each sonogram in (c) give the age in days after hatching when the recordings were made.

The threshold for calling is lowest in parents on the day their young fledge. Immelmann (1962a) observed that the frequency of calling begins to increase when young are about seven days old. By this age the Begging Calls are loud and penetrating and can attract the attention of predators. Typically, the parent waiting outside the nest gives the call as its partner feeds the young inside. The frequency of calling returns to normal levels about 10 days after fledging.

Kackles, Arks and Whines (Figure 10.3b)

These three calls are given at a potential nesting site by pairs searching for a suitable place to breed. Males, who take the initiative in nest site searching, normally call the most, but females also make identical calls. As the pair fly from bush to bush checking out old nests, the male gives Stack calls as the female follows. When he finds a site he attempts to lead the female to it; first, he lands next to her and gives a series of loud, raspy Kackle calls, then he hops to the site still giving Kackles, and once he lands on it he switches to Arks, which are low, long harsh, 'ark-ark' sounds. These calls are directed at the female and as he calls he will fan the tail slightly, bow down and mandibulate (nibble) towards her. The intensity of the display increases if the female looks at him or moves towards him. If she comes close he may switch over to the Whine. This is a long drawn out 'pleading' sound, like a small child whining for something and should the female find the site acceptable both partners will whine together as they sprawl side by side in contact and mandibulate at each other (Chapter 9). The Whine is the longest call given by Zebra Finches. The three calls may intergrade as the male moves from the partner to the nest site. The calling birds continually monitors its effect on the partner and modifies them accordingly. Kackles and Arks are also given by the male as he brings stems to the nest during building and may also be heard in the evening when birds are going to roost.

Copulation Call

During copulation both male and female emit a short series of whine-like calls. The female call begins as soon as she solicits mounting with her Tail-quivering display (Chapter 9) and the male begins when his copulatory movements commence. Clear sound recordings are difficult to make with both calling simultaneously and noisy wing flapping by the male, consequently no sonograms have been presented. My imprinted Zebra Finch, Fred (see Preface), gave short whines when he copulated on my finger.

Distress Cries (Figures 10.3a,c)

When distressed or in pain Zebra Finches may utter a high-pitched shriek with a wide-opened bill. It is most commonly heard when young birds are handled close to fledging. Adults rarely emit the call in these circumstances, and restrict the call to when in pain during vicious fights. Parents respond to the distress cries of their young with Distance Calls and by flying to them. Opponents, however, seem to ignore the call.

Begging Calls (Figure 10.3c)

For the first three days after hatching begging occurs silently in wild (Immelmann 1962a) and domesticated birds (Muller and Smith 1978). The only sounds are soft clicks that seem to emanate from full crops. After day three, soft cheeping sounds are given when gaping and these become progressive louder and more rasping as the bird gets older. In the wild, the penetrating staccato Begging Calls of older broods can be heard 100 m away; they are easy for humans to localise, and presumably for predators as well. The physical structure of Begging Calls change with age and the general pattern of changes are shown in Figure 10.3c. The soft cheeps first uttered are high-pitched tones (e.g. at 4 days in Figure 10.3c), but noise is added in subsequent days (e.g. day 6) and additional harmonics after that (e.g. day 11); by day 15 multiple harmonic bands are embedded in a column of noise. There are at least four age-related changes in call structure:

(1) an increase in unstructured sound energy so that the calls become progressively noisier;
(2) an increase in the number of harmonics and a lowering of the pitch;
(3) an increase in the duration of the call; and
(4) an increase in the loudness of the call.

In addition, the rate of calling per second increases, reaching a maximum around the tenth day after hatching. While this is the crude pattern of development, there is enormous variation in timing of the developmental sequence and the types of calls emitted and frequency of utterance within and between individuals of the same age and even within the same begging sequence. Roper (1993) concluded that much of the variation is probably due to motivational differences (mainly hunger and fear), changes in the physical dimensions of the syrinx and imperfect neural control of the synringeal muscles. Hungry birds call more rapidly and have harsher calls than those only moderately hungry.

The rate at which Begging Calls are given by nestlings of different ages does not differ between the sexes, nor are there differences in the types of calls produced, but Roper (1993) found that females have a greater diversity of calls than males after six days of age (see Chapter 11).

Long Tonal Call (Figure 10.3 c)

By day 15 young Zebra Finches have developed calls that are structurally and functionally different from the Begging Calls. These pure sounding 'dahr' calls are given singly and seem to 'float' in the air. They are

0.20–0.40 s long and have around 10 horizontal harmonics. At first they are emitted before, during and after a bout of Begging Calls, but are later given between feeding bouts in reply to the Distance Calls of the parents who in turn, reply once more (Zann 1984). On fledging day the calls are given spontaneously by the excited bird; they are also emitted when the fledgling is visually isolated, experimentally, from the parents.

Development of calls

Begging Calls develop first and the Distress Cries appear to arise from the more mature versions of these at around 15 days after hatching by eliminating most of the harmonics and noisy background. The Long Tonal Calls, which appear around the same time, may have a precursor in one of the more aberrant versions of Begging Calls that appear between 12 and 15 days. Tets, which suddenly appear on fledging in active excited young, have no identifiable precursor among the 75 variants of the Begging Calls identified by Roper (1993). Stacks appear to be abbreviated versions of the Long Tonal Call and also appear in the first few days after fledging; Distance Calls of both sexes also develop from the Long Tonal Call (see below). Arks are slowed down versions of Kackles and Whines are draw out versions of Arks. The Copulation Call is a type of Whine. The multiple chevron-shaped harmonics of Thuks and Kackles have similarities with Tets and may have arisen from them. The origin of the noisy Wssts is obscure.

Distance Call

This is the only long-distance signal given by Zebra Finches and other species of estrildines since almost all members of the subfamily lack a loud territorial song. The Zebra Finch Distance Call has a number of long-distance signalling functions. Moreover, experimental studies have shown it to be unique in its ontogeny and physiological control.

Structural Variation

When analysed on a sonograph physical variation is found in the structure of Distance Calls given by Australian Zebra Finches between the sexes and among individuals, colonies, and geographic localities; furthermore, the calls are clearly distinguishable from those of the Lesser Sundas subspecies (Zann 1984).

Individual variation

Each individual sampled from the same colony reproduces its Distance Call with great consistency such that a naïve independent observer can quickly classify sonograms by eye according to features unique to each individual with 100% success. Univariate and multivariate statistical analysis of various morphological characters of sonograms of Distance

Calls can also distinguish individuals. Thus, the Distance Call is structurally unique to each individual. Variation in the Distance Call is greater among males than females (Zann 1984).

Sexual variation

Sonograms of male calls of the Australian subspecies are immediately distinguishable from those of females because they are more complex. While the female call is a long complex tone in which the multiple harmonics show little frequency modulation and so appear as flattened or slightly arching bands on the sonograms, the male Distance Call has harmonics with two sharply contrasting, temporally distinct features. The first part is a complex tone in which the harmonics have a slight upward inflection in frequency, followed by a second part where the harmonics sweep dramatically downwards (Figure 10.1b). The first part of the call is termed the tonal component (T), because of its structure, and the second, the noise component (N), because it gives the call a harsh, grating quality (Zann 1984). Simpson and Vicario (1990) have termed the second component the 'rapid frequency modulated segment'. Ninety-nine per cent of free-living males (n = 258) have Distance Calls of this type (TN) with the noise component ranging from 50% to 90% of the total duration of the call; the exceptional individuals may omit the tonal component (all N), or have the noise component first (NT), or some have two noise components (NTN), but they never omit the noise component (i.e. all T). Gross sex differences in call structure are not evident in the Distance Calls of the Lesser Sundas Zebra Finches because the males do not have a two-component call. The whistle-like call has harmonics with an upslur–downslur configuration which is more or less identical in both sexes.

Additional sex differences in call structure become evident when the sonograms are measured (Zann 1984). Male calls are significantly briefer than female calls in both subspecies: Lesser Sundas subspecies from Timor: 0.13 ± 0.03 s ($\bar{X}$ ± s.d.) (males) vs. 0.17 ± 0.01 s (females); Australian subspecies from northern Victoria: 0.14 ± 0.02 s vs. 0.20 ± 0.03 s. The frequency of the fundamental, the lowest harmonic band, also differed between the sexes in the Australian subspecies (1.15 ± 0.23 kHz (males) vs. 0.62 ± 0.06 kHz (females)), but not in the Lesser Sundas subspecies (Lesser Sundas subspecies: 2.59 ± 0.65 kHz (males) vs. 2.08 ± 0.42 kHz (females)). Thus, more harmonics can be stacked into Australian female calls than male calls and this accounts for the more nasal timbre. The frequency of the harmonic with the greatest amount of sound energy, the emphasised harmonic, did not differ significantly between the sexes in either subspecies (Lesser Sundas subspecies: 4.47 ± 0.75 kHz; Australian subspecies: 3.6 ± 0.44 kHz).

Sexual differentiation in the Distance Calls of estrildines equivalent to that found in the Australian Zebra Finch have, to date, only been described for the Bengalese Finch (Zann 1985b; Yoneda and Okanoya

1991). Sexual differentiation in Distance Calls of Double-barred Finches have not been described, and only slight quantitative differences in frequency and duration exist in the three *Poephila* species (Zann 1975). Güttinger and Nicolai (1973) found no sex differences in the Distance Calls of 76 species of African and Asian estrildines.

Geographic variation

Female Distance Calls from individuals belonging to 10 colonies over southeastern Australia differed significantly in duration of the call, but they did not differ in the frequency of the fundamental, nor in the frequency of the most-strongly emphasised harmonic. Males differed in duration of the call and the duration of the tonal component, the fundamental frequency and the emphasised frequency (Zann 1984). Gross geographic differences in the sonograms of Distance Calls are evident, but more so in males than females.

Surprisingly, Clayton (Clayton *et al.* 1991) could find no geographic variation in the Distance Calls of the Lesser Sundas subspecies. The absence of call differentiation may indicate that populations are not sufficiently isolated or possibly that calls in this subspecies have rigid developmental programs with little genetic variation.

Subspecific variation (Figure 10.1b and 10.2c,d)

Inspection of sonograms of the two subspecies reveals gross differences in call morphology. There are more harmonics in the Australian subspecies and only males of this subspecies have the diagnostic noise component in the second part of the call. Measurements show that males of the two subspecies do not differ in total duration of the sound, but the fundamental and the emphasised harmonic are higher in the Lesser Sundas males. Calls of Australian females are longer, but the fundamental frequency and that of the emphasised harmonic are significantly lower (Zann 1984; Clayton *et al.* 1991).

Development

The adult Distance Call of Australian Zebra Finches is not fully developed in males until 60–80 days after hatching, while the female call matures much earlier (Zann 1985b). Calls of both sexes originate from the Long Tonal Call of fledglings. Between day 22 and day 30 the Long Tonal call becomes briefer, and the harmonics more modulated and amplified, so that by day 35–40 the adult version of the female Distance Call is complete. Around day 40–50 the flat harmonics of the male call begin to display rapid changes in frequency modulation where there is a steep upslur at the onset, a steep downslur at the offset, and a sustained unmodulated section of variable duration in between. By day 60, and frequently much earlier, amplitude modulation produces the essential elements of the adult Distance Call: the upslur at the onset disappears, so that the middle unmodulated section begins the call and forms the tonal component and the rapid downward sweep of the offset forms the noise component. Next, the amplitude of the higher frequencies of the call is greatly reduced so that the downward sweeping harmonics descend from non-existent high tonal components as if cut off by a filter. Concomitant with the changes in the configuration of the harmonics there is a reduction in number from six or seven harmonics at day 35 to three or four at day 60 with an associated increase in frequency of the fundamental from 0.6 kHz to around 1.2 kHz. By day 60–70 the male Distance Call has developed two components and has become higher-pitched than that of the female.

Cross-fostering

When foster-reared by Bengalese Finches from hatching to day 40 or day 60 the Distance Calls of female Australian Zebra Finches at 100 days of age were identical to those of their normally reared sisters except that the frequency of the fundamental was slightly, but significantly, higher (Zann 1985b). In contrast, every foster-reared male completely lacked the noise component at the end of the call, and in 84% of cases the tonal call superficially resembled the pure tonal calls made by females, but sounded less nasal (e.g. Figure 10.4 foster-reared son number 2). In other foster-reared males, the Distance Call was identical to that of their foster-mother (e.g. Figure 10.4 foster-reared son number 1) or foster-father, or resembled the harmonic configuration of their foster-parents' calls to some degree. The pure tonal call of foster-reared males was normal in all respects but for the absence of the noise component.

The length of foster-rearing (40 *vs.* 60 days) had no significant effect on the development of the Distance Calls in the experimental and control groups. Thus young males can learn the features of the noise component of the Distance Call before 40 days of age, and evidence suggests that the father serves as the model for this information (Zann 1985b). Foster-reared males exposed to the normal male Distance Call after day 40 did not develop a noise component into their call. This suggests that after the first 40 days of life the sensitive phase for learning the noise component of the Distance Call in male Zebra Finches ends. However, Slater and Jones (1995) found it was possible for a Zebra Finch kept with his father until 35 days of age to learn the Distance Call from a second male if subsequently confined with him until 70 days of age (5 males learned from the father and 12 males learned from the tutor). Therefore, it is likely that the sensitive phase for call learning extends beyond 40 days of age in some circumstances.

When Price (1979) conducted sound-isolation and deafening experiments on nestling Zebra Finches, he found no differences in structure of Tet, Stack and Distance Calls between treatments and controls and concluded that auditory experience was unnecessary for normal develop-

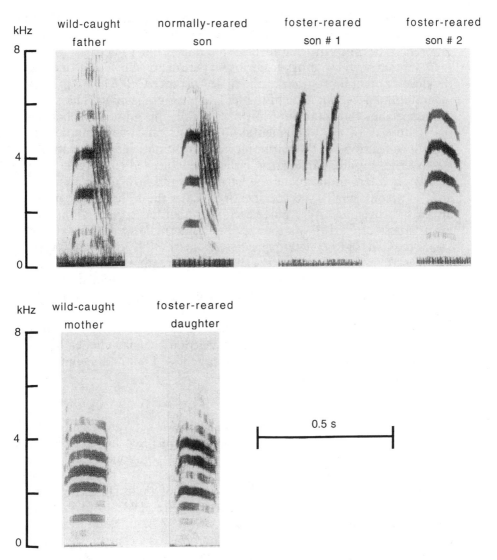

Fig. 10.4 Effects of foster-rearing on the structure of the adult Distance Call in Australian Zebra Finches. Distance Calls of a family of adult Zebra Finches are shown. Both mother and father were wild-caught birds and the four offspring were reared in captivity. The three foster-reared offspring were reared from the egg stage by Bengalese Finches for the first 40 after hatching and the normally-reared son stayed with his natural Zebra Finch parents until 40 days. Foster-reared son # 1 imitated the Distance Call of his foster-mother, while foster-reared son # 2 did not learn from his foster-father or mother but gave an aberrant Distance Call more like Zebra Finches than Bengalese Finches. The foster-reared daughter's call strongly resembles her natural Zebra Finch mother, not her foster-mother.

ment of these calls. Unfortunately, all his controls had the typical domesticated version of the Distance Call, which has no noise component, so
differences between the treatments were impossible to detect.

In summary, these cross-fostering experiments show that one part of
the developmental program of the male Distance Call is open to environmental influences, but little of the female program is open. The male program specifies the total duration of the call, the number of harmonics,
the frequency of the fundamental and the emphasised harmonic. However, specifications of the harmonic configuration need information from
the external model, namely, the father; he is needed to provide program
details on where to modulate the harmonics and how much amplitude to
apply. In some males the specification 'learn the father's Distance Call'
appears to override everything, even the inherited specifications, so that
when cross-fostered to another species some males will learn the whole
Distance Call of the heterospecific. At least 30% of normally reared
males gave identical Distance Calls to their fathers. In one case, two
brothers were foster-reared in the same nest, one copied the foster-
father's Distance Call and the other gave the deprived Zebra Finch version. It is not known whether these two developmental programs are
alternative strategies in the acquisition of the male Distance Call or if
they are just different responses to post-fledging environments, for example, the quality of parental care, the frequency of parental calling and the
number and sex of siblings.

Effects of domestication

Whereas only 1% of wild Zebra Finches have aberrant Distance Calls,
the incidence increases to about 3% (2/61) in offspring of wild-caught
birds. These birds may have either no noise component (all T) or no
tonal component (all N) (Zann 1984).

Effects of domestication on the structure of male Distance Calls has
been investigated in Australian stocks by Carr (1982) and in Japanese
and American stocks by Okanoya *et al.* (1993). In domesticated Australian stocks, the incidence of the components occurring in the same
order as that found in wild birds (TN) varied from 17% to 23%, and in
the remaining birds the N component was always present. In contrast,
20% of the American and Japanese stocks had no N component at all;
50% of American males had the components in reverse order, NT, and
50% of the Japanese had the species-typical TN.

Carr (1982) also found that sons of domesticated birds copied their
fathers' Distance Calls precisely, a finding confirmed by Simpson and
Vicario (1990). Moreover, the effects of cross-fostering of domesticated
Zebra Finches are identical to those found with offspring of wild-caught
birds.

The aviary environment, rather than regimes of artificial selection, is
responsible for the production of the aberrations in the domesticated

male Zebra Finch Distance Call. When male offspring of wild-caught and domesticated Zebra Finches are cross-fostered to one another, genetic sons of wild-caughts learn the aberrant Distance Call of their domesticated foster-father and his genetic sons learn the normal Distance Call from their wild-caught foster-father (Figure 10.5). Precisely what features of the captive environment are responsible for the increasing error rate in learning the noise component are unknown, nor is it known what selective forces in the wild make the structure of the Distance Call so stereotyped.

Conclusion

In most species of passerines the development of calls, as distinct from songs, has a closed developmental program, in the sense that specific social experiences are unnecessary for the development of species-typical calls. The Distance Call of male Australian Zebra Finches is one of the exceptions to this generalisation, since foster-rearing of a number of

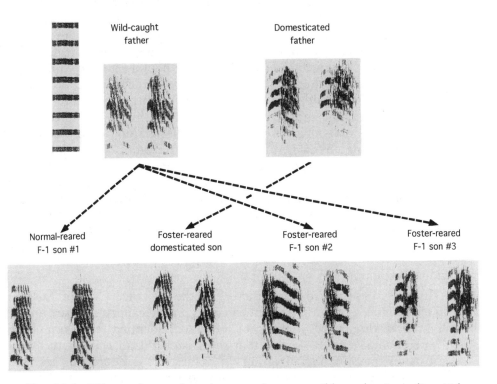

Fig. 10.5 Effects of cross-fostering eggs between wild-caught Australian Zebra Finches and domesticated white colour morphs on the structure of the male Distance Call. Eggs were swapped during incubation and young were removed from their natural and foster-parents 40 days after hatching and held in sibling groups until day 100. The calibration tone is set at 1 kHz intervals.

other species of estrildines has not shown any evidence of call learning (Güttinger and Nicolai 1973). Recently, however Yoneda and Okanoya (1991) found that male Bengalese Finches must also learn their Distance Call in a similar way to that of Zebra Finches.

Neural control

Neural control of Distance Calls differs between the sexes in the Australian Zebra Finch (Simpson and Vicario 1990). The simpler female Distance Call has a less complex, more basic, neural pathway, which is possibly located in the brain stem where it is associated with the centre that controls respiratory patterning. The more complex male Distance Call, in contrast, is controlled by the same complex neural pathway in the telencephalon that controls song production (see below). Simpson and Vicario (1990) established the function of this pathway in the production of the male Distance Call by destroying or cutting the different components. The noise segment, the short duration and the high fundamental frequency were all lost when the vital neural centres were lesioned, resulting in a Distance Call almost identical in physical structure to that of the female. Destruction of the pathway in females had no effect on the structure of the female Distance Call, proving that this pathway had no role in its control. Simpson and Vicario (1990) postulate that the male has two co-existing neural pathways for the production of the Distance Call: the song control circuit, which produces the learned Distance Call with its male-specific duration and fundamental frequency, and the more basic brain stem pathway that produces the female call. Young males produce the female version of the Distance Call before the song circuit is fully functional at around 60 days of age, and older males fall back on the female version of the Distance Call and its basic neural control if experimentally induced lesions damage the functions of the song circuit. If oestrogen is implanted in female nestlings soon after hatching their brains will also develop the song control circuit and they are then capable of producing the learned male-like Distance Calls (Simpson and Vicario 1991a,b).

Individual recognition

Theoretically, the acoustic structure of the Distance Calls of Australian Zebra Finches could impart information on the caller's age, sex and geographical origins as well as its individual identity. Two unpublished experimental studies suggest that the Distance Call enables recognition of mates, parents and offspring. In domesticated Zebra Finches, Silcox (1979) found that females discriminated between the Distance Calls of their partners and those of other males whereas males responded with equal fervour to all females. Using second and third generation offspring of wild-caught Zebra Finches, McIntosh (1983) found that members of a pair recognised the Distance Calls of their partners in simultaneous playback experiments where calls of birds of the same sex with whom they were familiar served as controls. Under the artificial conditions of this experiment, where the dynamic exchange of Distance Calls between stimulus and test subjects was not possible, females showed a stronger preference towards their mate's Distance Call than did the males. However, males called more than females during playback, and were more active, but females altered their rate of calling more than that of males. Butterfield (1970) made the same observation when members of a pair were visually, but not acoustically isolated from one another. McIntosh hypothesised that members of a pair relocate one another in the following way: the male calls more than the female when separated and moves around actively searching for her; when he comes within earshot she recognises his call and responds with her own Distance Call but remains stationary; the male homes in on the female that alters her call rate in response to his Distance Calls. This explanation is consistent with the finding that free-living males call more than females and that the physical structure of the male call is more individually distinct and thus easier to recognise than that of the female.

In other playback experiments McIntosh showed that parents discriminated the Long Tonal/Distance Calls of their own fledglings from those of others. Mother and father responded equally and responded to the calls of birds as young as 20 days of age, but the response waned when the young reached 32–35 days of age, and ceased completely soon afterwards.

Fledglings also recognise the Distance Calls of their parents: they suddenly cease moving and reduce their rate of Tet Calling when their parents call but they ignore the Distance Calls of others. McIntosh believes that the parental Distance Calls have a tranquillising effect on the newly fledged young under these experimental conditions. Naturalistic observations by Immelmann (1962a) suggest the opposite response; young call in reply to the Distance Call of all adults for the first five to seven days after fledging, but soon learn to recognise calls of both parents and restrict their replies to them. According to Immelmann, young can pick out their parents from a multitude of callers at distances of up to 100 m, or so, away.

Song

Of all Zebra Finch behaviours we know most about song, namely that series of various vocal utterances normally sung by the male to the female during precopulatory courtship. It has been the subject of intense investigation since Immelmann's (1965c, 1967) pioneering studies into its development. Domesticated Australian Zebra Finches have become the principal focus of investigations into development, control, perception and function of song, and, recently, variation and structure of song in wild birds has also been investigated. Unfortunately, no studies of singing have been made on wild Lesser Sundas Zebra Finches, although some laboratory work has been conducted in recent years.

Zebra Finches sing a soft song that sounds squeaky and cheerful to the human ear and, perhaps, more mechanical than musical. Parts have a strange ventriloquial quality that makes singers hard for humans to localise. Song of the Lesser Sundas subspecies is softer and noisier than that of the Australian subspecies. Singing can be heard in wild Australian Zebra Finches throughout the year, irrespective of whether they are breeding or not. However, wild Lesser Sundas Zebra Finches sing less often; for example, Clayton *et al.* (1991) rarely heard them sing during their field study that spanned part of the breeding period. In Australia, some parts of the song can be heard more than 20 m away, depending on the weather, but one needs to be much closer to hear the complete song. Distance Calls, both loud and soft, are embedded among the notes that make up the song, but these cannot be distinguished in the songs of the Lesser Sundas subspecies. Careful listening will also allow one to distinguish songs of different individuals since each is unique, an observation first made by Morris (1954).

The sound and organisation of Zebra Finch song is similar to that of the Double-barred Finch (Hall 1962) and the Masked Finch (Zann 1976b). Both species have harmonically rich elements and there is no interruption between the phrases. In contrast the Long-tailed and Black-throated Finches have a number of tonal elements in their songs and pauses between phrases.

Structure

In both subspecies of Zebra Finches notes occur in sequences that are repeated throughout each song performance and constitute phrases, or verses. Sonograms of two wild Australian Zebra Finches are shown in Figure 10.6 and three captive-bred Zebra Finches from Timor shown in Figure 10.7.

Unfortunately, investigators of Zebra Finch song have not standardised their terminology. The smallest components, or units, are called elements (syllables, notes). On sonograms elements appear as morphologically discrete black tracings that are separated from their neighbours by intervals without tracings, namely gaps of silence that range from 5 to 10 ms in duration. In some instances elements are not separated by gaps of silence but by sudden changes in amplitude (Williams and McKibben 1992). Elements are not merely descriptive units but functional ones as well, since Cynx (1990) has shown that singing is never interrupted in the middle of elements, but only in the silent intervals. Consequently elements as depicted by tracings on sonograms are the vocal units of production in Zebra Finch song. However, when elements are copied the element itself and the silent interval preceding it are copied as one unit

(a)

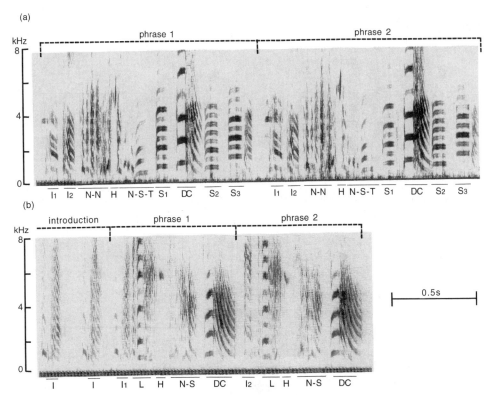

Fig. 10.6 Sonograms of songs from two wild Zebra Finches from different parts of the Australian distribution: (a) Alice Springs, and (b) northern Victoria. Both songs are of the undirected type in which a female is not involved, and show two successive song phrases. The start of singing is shown in male (b) where a couple of introductory notes precede the first phrase. The letter code beneath each note or element indicates its type, and the number refers to the version within that type: I, Introductory; L, Ladder; H, High; N-S, Noise-structure; DC, Distance Call; N-N, Noise-noise; N-S-T, Noise-structure Tone; S, Stack.

(Williams and Staples 1992), hence, strictly speaking, the element and its preceding silent interval are the unit of song in the Zebra Finch.

Each male sings a number of different elements in a set sequence and together these constitute the song-phrase (Hall 1962; Immelmann 1965a, 1969) ('song-unit', Price 1979; 'motif', Sossinka and Böhner 1980; 'song' Cynx *et al.* 1990). Within one performance, males may repeat the song-phrase a number of times to form a song ('bout' Price 1979; 'strophe' Sossinka and Böhner 1980), the first phrase of which is usually preceded by several identical elements, which form the song introduction. Some of these introductory elements may be incorporated into the phrase itself (e.g. Figure 10.6b). Intervals of approximately five seconds separate the

(a)

(b)

(c)

kHz

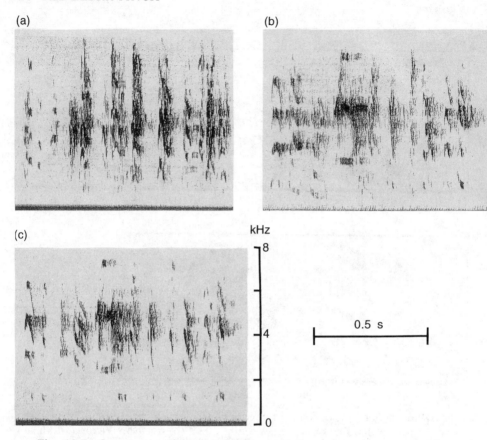

8

4

0.5 s

0

Fig. 10.7 Sonograms of songs of three semi-domesticated Lesser Sundas Zebra Finches whose ancestors came from Timor. Only one song-phrase is shown for each male (a, b, and c).

end of one song and the start of the next during which the male resumes a normal perching posture. The song-phrase is the natural unit of investigation of Zebra Finch singing and is used by most researchers, rather than song, since the number of song-phrases sung per song depends on motivation and varies according to whether the song is directed or undirected (see below).

Song-phrase macrostructure (Table 10.2)

Considerable variation exists in the gross features of the song-phrase among individuals, populations and regions (Zann 1993a,b); variation also exists between domesticated and wild forms (Slater and Clayton 1991; Zann 1993a), and between the two subspecies (Clayton *et al.* 1991). In wild Australian Zebra Finches, the number of elements sung

Table 10.2 Comparison of song-phrases from Australian and Lesser Sundas Zebra Finches showing medians and interquartile ranges; after Clayton *et al.* (1991) and Zann (1993a).

Parameter	Wild Australian	Domesticated Australian	Semi-domesticated Lesser Sundas
Number of elements	7 (6–8)	6 (5–8)	12 (11–15)
Elements/s	7.9 (7.1–8.7)	10.0 (8.8–11.6)	14.1 (12.1–15.3)
Duration (s)	0.86 (0.71–0.97)	0.66 (0.54–0.80)	0.88 (0.81–1.1)
Repeats (% males)	73	89	66
Frequency (kHz)[a]	1.0 ± 0.42	?	1.6 ± 0.49
n	402	46	12

[a] Fundamental frequency (mean ± *s.d.*)

per song-phrase ranges from 3 to 14 with a mean of 6.7, which is not significantly different from that of domesticated birds. However, the latter sing their song-phrase more rapidly (10.0 elements per s *vs.* 7.9), consequently the mean duration of the song-phrase of domesticated birds is correspondingly shorter than that of wild ones (0.66 s *vs.* 0.86 s) (Zann 1993a). Domesticated Zebra Finches incorporate fewer Distance Calls into the song than wild birds (Slater and Clayton 1991). Semi-domesticated Timor Zebra Finches have song phrases with more elements than wild Australian birds (12.0 *vs.* 6.7); they sing them at a faster rate (14 elements per s *vs.* 7.9), and at a higher pitch (fundamental 1.6 *vs.* 1.0k Hz) (Clayton *et al.* 1991).

Types of song-phrase elements (Figure 10.8)

Although the average Australian male sang around seven elements in its song-phrase, fourteen distinct categories or types were identified in an analysis of sonograms from 402 wild Zebra Finches from locations in southeastern and central Australia (Zann 1993a). Most elements sung by males were different types although 73% of males contained 'repeats', that is elements of the same type, but usually different versions (e.g. $I_{1,2}$, and S_{1-3}; Figure 10.6a).

Broadly speaking, the structure of the song-elements can be divided into call-like and non-call-like ones. Three elements were similar in structure to the three most common calls given by Zebra Finches, and together constituted 67% of all elements sung, namely

(1) the Introductory Element, whose structure strongly resembled that of the Tet Call, constituted 24% of all elements, and was found in the song-phrases of 90% of males;

(2) the Stack Element, whose structure strongly resembled that of the Stack Call, constituted 17% of all song-elements, and was found in 80% of males;

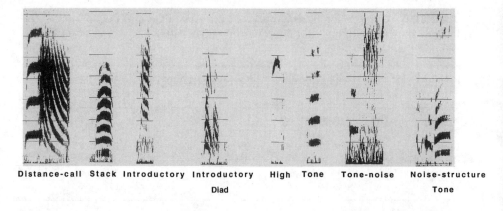

Distance-call Stack Introductory Introductory High Tone Tone-noise Noise-structure

Diad   Tone

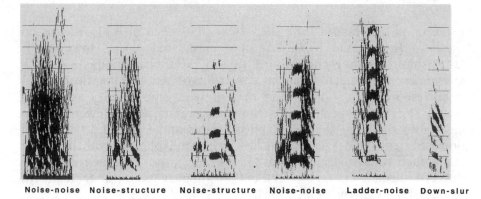

Noise-noise Noise-structure Noise-structure Noise-noise Ladder-noise Down-slur

Distance-call   Distance-call

Fig. 10.8 Fourteen types of song-elements sung by wild Australian Zebra Finches. Horizontal lines show 1 kHz intervals. (After Zann 1993b.)

(3) the Distance-call Element, whose structure matched that of the Distance Call, constituted 16% of all song-elements, and was sung by 87% of males; in 50% of males the Distance-call Element was the same version used as the Distance Call, but much softer.

Immelmann (1962a) was the first to identify the presence of call-like elements in the song phrase, and Price (1979) later distinguished three types, short, medium and long that correspond respectively, with the Tet, Stack and Distance-call Elements.

The remaining 33% of song-elements were unique to the song, having no direct structural resemblance to any calls (except for the Introductory Diad), and were composed of eight different types, of which the High Element formed 10%, and occurred in 67% of males. This distinctive element was by far the briefest (median duration 8 ms) and highest-pitched (median fundamental frequency 6.4 kHz) sound given by Zebra

Finches, and was the only sound that did not have numerous harmonics (see Figures 10.6 and 10.8). Its fundamental frequency was five times greater that of the next-highest-pitched element in the song and Williams *et al.* (1989) believe its structure has arisen from suppression of lower harmonics. By contrast, the remaining eight types of song-elements contained various combinations of unstructured noise and harmonic spectra, and their occurrence was variable among males. Production of these compound elements was less stereotyped than other elements and varied within and among individuals; consequently, unambiguous classification is more difficult. Criteria for identifying these less-diagnostic elements, are given in Zann (1993a).

The number of element types sung by Lesser Sundas Zebra Finches is unknown. No comparisons have been established between the call repertoire, which itself is still largely undescribed, and the song elements, although some males appear to produce elements that faintly resemble Distance Calls.

Sequence of elements

Each male sings its song-elements in a fixed order and the pattern is fairly consistent among individuals. With Australian males it is possible to recognise three sections to the stereotyped song-phrase: a start, middle and an end. The first section consists mostly of Introductory Elements; a single High Element is the centrepiece of the middle section, and is preceded and followed by compound elements; the end section consists mostly of the Stack Element and the Distant-call Element, in that order (e.g. Figure 10.6b), but not always (e.g. Figure 10.6a). The Distance-call Element is much louder than the rest (Zann 1993a), and the most resistant to attenuation, so that it is often the only element heard clearly from a distance, or in strong wind. With its position at the end of the song-phrase the Distance-call Element serves as a 'coda' or 'punctuation' that separates one phrase from the next.

The organisation of the song-phrase into three sections of elements appears to have some functional basis. When production of song in domesticated Zebra Finches was experimentally disrupted, Williams and McKibben (1992) found that elements disappeared, and were later recovered, not singly, but in 'chunks'. Furthermore, during song learning there was a tendency to learn elements from the song tutor, or model, in chunks. These chunks correspond to the sections identified in the song-phrase of wild Zebra Finches

Evolution of the song-phrase

The sequence of elements in the song-phrase of the Australian subspecies shows similarities with the series of calls given around take-off, so conceivably, the ancestral song may have originated from such a sequence (see above). Unfortunately good sonograms of the take-off calls are difficult to obtain because the exertions involved in lift-off distort call structure (Zann 1984). In general, the same sequence occurs in the song-phrase, but inserted between the Introductory Elements and the Stacks is the middle section of non-call-like song-elements, namely those that are original and exclusive to the song, and include the High Element and its neighbouring compound elements (Figure 10.9).

The Introductory, Stack and Distance-call Elements may have been 'borrowed' from their counterparts in the call repertoire, but the origins of the non-call-like elements are not obvious. Nevertheless, all ten elements have probably been derived from the Distance Call through systematic repetition, modification and omission of the tonal and noise components of the call (Zann 1993a). Thus, in this theoretical scheme the Distance Call has a dual role in the evolution of the elements of the song-phrase—it not only provides a sound that is used unmodified in the end section of the song, but it also serves as a source from which the non-call-like elements in the middle section have been derived. One can only speculate on how such an arrangement might have arisen. One possibility is that the original ancestral song consisted of three unlearned calls given at take-off, after which the female-like Distance Call was subjected to sexual differentiation during which it came under control of higher vocal centres. Subsequently, during its ontogeny the Distance Call became open to auditory input from an external model, perhaps from the father, and the derived elements were a by-product of this process. The last step would require all the song elements to be copied from the external model in one process, thus emancipating the entire song-phrase from the lower vocal control centre.

The changes that have produced the song phrase of the Australian Zebra Finch may be of fairly recent evolutionary origin. The incorpora-

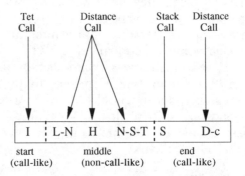

Fig. 10.9 Schema showing sections of the song-phrase of the Australian Zebra Finch and the possible evolutionary source of the elements from the call repertoire. I, Introductory Element; L-N, Ladder-noise Element; H, High Element, N-S T, Noise-structure-Tone Element; S, Stack Element; D-c, Distance-call Element; TC, Tet Call, DC, Distance Call; SC, Stack Call.

tion of call-like elements into the song does not occur with the Lesser Sundas subspecies. Conceivably, the call-like elements have been lost since geographic isolation, but it is more likely that they only evolved in the Australian subspecies after isolation. Absence of sexually dimorphic Distance Calls, the source of non-call-like elements, suggests they never existed in the Lesser Sundas subspecies. Finally, none of the three species of *Poephila* have sexually dimorphic distance calls, nor do they incorporate them into their songs (Zann 1976b).

Song variation

While there are fairly clear rules of organisation of the song-phrase in wild Australian Zebra Finches, each individual sings a unique, stereotyped phrase. Recording of songs from the same individual up to one year apart show phrases identical in element number, structure and sequence, and singing duration (Zann 1990). Similar stability exists in domesticated Zebra Finches (Nordeen and Nordeen 1992). Thus, there is the potential for the song-phrase to serve as a permanent individual signature.

Geographic variation in the song-phrases of 402 males from 33 colonies in two distant parts of Australia was investigated using canonical discriminant classification of 31 variables (Zann 1993b). Song-phrases from southeastern Australia were clearly distinguishable from those of central Australia. While there were no differences in the duration of the phrase, there were significantly more elements in those from central Australia (7 *vs.* 6); moreover, they were sung more rapidly (8.5 elements per s *vs.* 7.4) and had a higher fundamental frequency. The composition and sequence of non-call-like elements in the middle section of the phrase also differed significantly between the two geographic zones. When analysis was restricted to southeastern Australia, variation in the middle section again distinguished song-phrases from different colonies, and their position in the classification was more or less proportional to the geographic distance separating them and the estimated likelihood of dispersal. When analysis was focused down to a few colonies, significant variation in the middle section of the song-phrase was found from one year to the next and depended on the origin of the large influx of immigrants (Chapter 8).

In summary, variation in the song-phrase of the Australian Zebra Finch occurs principally in the non-call-like elements that constitute the middle section. Whereas the High Element is almost a permanent fixture, the occurrence of the nine compound elements is quite variable, and these form the most labile part of the phrase. By contrast, the call-like elements, at the start and end sections are stereotyped. Similarly, the macrostructural features, (phrase duration, element number, speed of singing) are not highly variable across colonies, and so must be more rigidly specified. This was confirmed by (Clayton 1990b) in hybridisation and cross-fostering experiments between the two subspecies. Macrostructure of hybrid song-phrases was intermediate between the two subspecies, moreover, there was no significant difference in macro-structure between normally reared and cross-fostered males. Both results suggest that at the macrostructural level the song-phrase is unresponsive to a wide range of social experiences (see below).

Directed and Undirected Songs

Adult male Zebra Finches, like those of most estrildines (Hall 1962), sing in two contrasting situations:

(1) during the stage 2 courtship dance (Chapter 9) in which the song is directed at the female less than 20 cm away;

(2) when alone, or in a variety of social contexts, where the song is not oriented towards any individual; here it is not accompanied by any dance or movement and not followed by any overt sexual behaviour (Morris 1954, Immelmann 1962a).

The first type is called Directed (or Courtship) Song and the second, Undirected (or Solitary) Song.

In Zebra Finches and other estrildines, the relationship between Directed and Undirected Song has long been an enigma. Whereas a female elicits the Directed Song in a sexually aroused male, her presence will inhibit the performance of the Undirected Song, which is sung when the male does not seem aroused. The two types of singing also differ in details of performance.

The basic structure of the song-phrase in both song types is identical, but there are significant quantitative differences in overall singing performance in both wild-caught and domesticated Zebra Finches that indicate that Directed Song is a more intensive performance than Undirected Song. Directed Song has (a) more introductory elements, (b) more song-phrases per song, and (c) the elements are sung faster than in Undirected Song (Sossinka and Böhner 1980). There is no difference in loudness between the two types (R. Zann, unpublished observations). In domesticated birds Bischof *et al.* (1981) found a continuum in the above measures of song performance that increased from Undirected Song to Directed Song; the intensity of song corresponded with the increasing releasing value of the test stimuli provided.

Visual components also differ between the two types of song. Birds singing Undirected Song adopt a range of postures from weak to intense (Figure 10.10), whereas those singing Directed Song normally adopt only the full display. In its weakest expression, a male simply extends its neck vertically from a resting position during the song-phrase. With increasing intensity, the posture becomes more upright and plumage erected; the head feathers first, then the ear coverts, flanks, and finally breast. Therefore, some extreme Undirected Song postures resemble those adopted

Fig. 10.10 (a) High and (b) low intensity Undirected Song postures.

during Directed Song except for the 'angular head' (Figure 9.5c). No differences are detected in the head movements during both types of singing. As each phrase is sung the bill opens and closes once or twice and the head pivots to and fro to one side, swinging about 45° from the forward position; it pivots to the other side for the next phrase.

Intensity differences in the acoustic and visual components of singing have led a number of authors to conclude that Undirected Song and Directed Song have a common control mechanism in which Undirected Song is simply the sexual display with the lowest threshold, whereas Directed Song is the one with the highest threshold (Morris 1957; Kunkel 1959; Immelmann 1962a; Güttinger 1970). In contrast, Caryl (1981) concluded that Directed and Undirected Songs did not share a common control mechanism because he found no correlation between rates of Directed Song and Undirected Song, and the sexual stimuli important for Directed Song were not important for Undirected Song.

Both types of song are strongly influenced by gonadal hormones, consequently there is a reduction in singing if males are castrated, and a restoration of singing if there is replacement with gonadal androgens (Pröve 1974; Arnold 1975a; Harding *et al.* 1983). Frequency of songs was positively correlated with levels of circulating testosterone, but threshold concentrations were significantly smaller for Undirected Songs (Pröve 1978; Pröve and Immelmann 1982). Nevertheless, analysis of how androgens affect the production of Directed and Undirected Songs indicates that they have separate controls and are not simply at different points on the same unitary song system. Walters *et al.* (1991) discovered that Directed Singing is oestrogen dependent and causes songs to be sung faster and to be directed at females, whereas Undirected Song is more androgen dependent and is either independent of oestrogen entirely, or extremely sensitive to small quantities. How these hormones affect the control circuitry in the brain is yet to be determined.

Functions of song

As with most estrildines, song appears to have no territorial function in Zebra Finches (Morris 1954; Kunkel 1959; Hall 1962; Immelmann 1962a, 1965a). They possess no territory and males take no aggressive action towards singing males, nor do singing birds behave aggressively before or after singing (Dunn 1994).

Function of Directed Song

It was generally assumed from its context in the courtship display that Directed Song was a pre-copulatory signal that elicited female solicitation (Morris 1954), however, it was not until experiments by Clayton and Pröve (1989) that this assumption was confirmed. Semi-domesticated females of both subspecies of Zebra Finches and Bengalese Finches were implanted with oestradiol to heighten their sexual response and then placed alone in sound-shielded boxes where they were subjected to songs played back through a loudspeaker. Female responsiveness towards the songs was measured by the number of copulatory solicitations (Tail-quivers). Females responded strongest to songs of conspecific males, preferring them to those of heterospecifics; moreover, the Zebra Finch females preferred the song of their own subspecies, thus confirming the hypothesis that Directed Song on its own serves as a means of recognising species and subspecies. Cross-fostering experiments between subspecies showed that this preference is learned from exposure to the songs of their fathers before 35 days of age (Clayton 1990b). Females also discriminated among songs within subspecies. They preferred long song-phrases over shorter ones, that is, phrases with more elements rather than phrases with fewer elements. Presumably, a male with a long complex song-phrase would not only elicit a copulation solicitation more rapidly from his mate, but would more be successful with extra-pair females as well, and probably be preferred at pair formation over males with simpler songs. In choice-tests females are attracted to males that sing frequently (ten Cate and Mug 1984; Collins *et al.* 1994). Provisional results from other captive studies suggest that females have a higher reproductive success if they can pair with males whose songs they prefer (Williams *et al.* 1993). Both studies support the hypothesis that sexual selection plays an important role in the evolution of estrildine song.

Directed Song is also used for individual and kin recognition, a suspicion long held by Immelmann (1968, 1969) on the basis of the large individual variation in song structure in domesticated and wild-caught birds. Playback experiments with domesticated Australian Zebra Finches showed that females could recognise songs of their mates and preferred them to similar, yet familiar songs of other males (Miller 1979a). Further experiments showed that daughters also preferred the songs of their fathers over that of other males even after many months of separation (Miller 1979b). The ability to memorise the father's song for the purpose of recognition and discrimination first occurs in both males and females between day 25 and day 35, and the ability extends to a least six months of age (Clayton 1988a). Finally, it is likely that adult females quickly learn to recognise the song of new partners.

Function of Undirected Song

In contrast to Directed Song, the functions of Undirected Song are less obvious. Immediate overt responses are not observed among potential recipients of Undirected Song in captive and wild Zebra Finches. Moreover, the absence of detailed field studies and experiments has, until recently, limited knowledge of its function to mere speculation. Immelmann (1968) for example, concluded that it was basically functionless and resembled subsong and was purely an indication of a 'very tranquil mood'. Recently, however, Dunn (1994) focused investigations specifically on the contexts of Undirected Song in the wild and conducted experiments on both wild and captive birds. He concluded that Undirected Song not only has costs in terms of energy expended and risks taken but that it has multiple functions. Undirected Song occurred at two main locations in the Danaher colony—around the nest during the breeding season and at flock feeding sites. Undirected Song occurred in the flock throughout the day with slightly more males singing in the morning than the afternoon, but within males rates were similar. Males sang Undirected Songs in the flock throughout the year but a greater proportion sang during the non-breeding season than during the breeding season, but the individual rate of singing was no higher. Although testosterone levels fell in the non-breeding season this did not affect the rate of Undirected Song in the flock. When female partners were experimentally removed the rate of Undirected Song increased significantly and fell again when the partner was returned, or when the male re-paired. The finding suggests that the female partner inhibits, but does not stop, male Undirected Song performance in the flock, and that males use Undirected Song to attract females for pairing purposes or to advertise their availability for extra-pair copulations; they may also advertise their quality so that they might rapidly re-pair should they lose their current partner.

During the breeding season, Dunn (1994) found that Undirected Song occurred most frequently during the nest-building and early egg-laying stage. Typically, the male sang when the female was in the nest and he had just left her. If the male sang immediately the female entered the nest it significantly increased the time she spent inside. This enhanced the male's ability to guard the female from extra-pair matings during her fertile period, and also had the advantage of preventing brood parasites from dumping eggs, or other pairs from taking over the nest. Thus, a collateral function of Undirected Song at the nest site is mate- and nest-guarding. Undirected Song tends to decline during incubation but increases after young hatch. Ten Cate (1982) found that Undirected Song increases steeply in domesticated Zebra Finch fathers from day 8 to day 24, after which singing occurs at sustained levels.

Development of song

Like most song birds Zebra Finches learn their songs early in life. Immelmann (1965c, 1967, 1969) made this discovery in pioneering experiments in which he manipulated the early social life of young males by exposing them to different auditory experiences extending from the egg stage up until sexual maturity. Some males he cross-fostered to Bengalese Finches in sound-shielded boxes where they were isolated from all sounds made by Zebra Finches. He prevented others from hearing any songs at all until maturity, by either hand-raising them or by having females raise them unaided. Finally, he had others cross-fostered to Bengalese Finches, where they could hear and see other Zebra Finches. His results can be summarised as follows:

(1) Zebra Finches must learn the details of their song-phrase since only a rough version exists without learning (see below);

(2) they can learn the complete song of the Bengalese Finch foster-father which completely masks the unlearned Zebra Finch framework;

(3) elements of the song can be learned and memorised as early as day 25, about the time they start performing subsong, but after the full song is performed, around day 80, new elements cannot be learned— a sensitive phase for song-learning exists in the first three months of life;

(4) Zebra Finches learn their songs from the male that feeds them, so it is usually the father or foster-father who serves as their song tutor;

(5) if there is no father or foster-father, young males show a bias towards learning the songs of Zebra Finches in preference to those of other species.

Immelmann (1969) concluded that wild Zebra Finches would be likely to learn songs of their fathers, and an early end to the sensitive phase was necessary in order to prevent accidental learning from heterospecific estrildines that they might encounter.

In due course this fascinating investigation inspired a long series of intriguing follow-up experiments by other researchers, in particular, by P. J. B. Slater and co-workers in the last decade, who used song learning in domesticated Zebra Finches as a model for teasing apart the subtle interactions involved in the development of behaviour. A detailed review of progress made in understanding Zebra Finch song development from Immelmann's initial work up until 1988 can be found in Slater *et al.* (1988). An early assumption of Slater's approach to song learning was that wild Zebra Finches would be unlikely to have their father available as a song tutor after 35 days of age because bonds between them would be broken at about this time. It was reasoned that wild young become independent around 35 days (Immelmann 1962a) and the re-nesting father would tend to drive them away. Consequently, his group initially focused on how young males chose a song tutor other than the father and what factors delimit the sensitive phase. My studies at the Danaher colony show that although nutritional independence is almost complete by 35 days, roosting bonds can continue with the parents until around 50 days and association in feeding flocks is likely beyond that age, especially when one considers the abilities of kin to recognise one another (Chapter 9). Also, no observations have been made of the father driving away his young. Consequently, contact with the father and his song is highly likely beyond 35 days providing opportunity for the father to serve as tutor in wild Zebra Finches.

Physical development of singing

The first subsong begins in wild birds between 28 and 35 days of age when still in the company of their parents and siblings, but it is so soft that when one observes from as little as a few metres distant, the singing posture is the only clue to its performance. The upright stance, pivoting of the head and open and closing of the bill follow the adult display except that the plumage is uniformly fluffed. Up close, quiet bursts of toneless sound are heard which are punctuated with irregular squeal-like warbles. Within a week or so subsong gets louder and is often preceded and interrupted by a series of Distance Calls, during which the plumage sleeks again. The calls get softer before the transition to subsong and some are incorporated into the rambling unstructured performance. After day 35, males are often seen singing alone on the tops of bushes. In captivity, there is great individual variation in the onset and frequency of subsong.

Arnold (1975b) detailed the changes in song structure from 40 days of age to adulthood. At first elements are poorly formed and highly variable in structure and sequence, and no song-phrases are discernible—the song has a rambling or babbling quality. By day 50, element variability is reduced and there is a vague resemblance to adult song in element form and organisation, and by day 60 almost all elements of the final song are recognisable, but their production is not as stereotyped as that of adults. A sequence of elements is recognisable by day 60 although slight changes may subsequently occur. By 90 days of age the song contains elements that are sung with the stereotyped form and sequence, typical of adults more than four months of age, although a small proportion of elements can change slightly after this age (Morrison and Nottebohm 1993; Slater *et al.* 1993). The song-phrase shortens slightly after day 90 due to a decrease in the interval between the elements; in addition, the number of song-phrases per song increases. In summary, Zebra Finch song development is like that of most species of song birds where three phases are discernible: (a) the rambling subsong phase between 25 and 50 days that bears little resemblance to the adult song; (b) a plastic song phase between 50 and 80 days in which some features of adult song are present and during which the elements of the adult song gradually 'crystallise' out to form (c) the final song phase. It is not clear whether Zebra Finches overproduce song elements at the onset of plastic song and gradually lose all but those reproduced in the adult song via a process of selective attrition (Marler and Peters 1982), or the final version crystallises out some other way.

Motor development of song in Zebra Finches is temporally compressed in comparison with other species of song birds that have been studied. Consequently, the two stages necessary for development of song, in the Zebra Finch (namely, memorisation of the acoustic structure of the tutor's song and the correct development of the sensory–motor integration that permits reproduction of that song structure) overlap in time. In other species of song birds, such as the Swamp Sparrow *Melospiza georgiana*, the two processes are temporally discrete (Marler and Peters 1982).

Songs without tutoring

In his preliminary experiments Immelmann (1967, 1969) showed that males deprived of the opportunity to hear other adult finches from the nestling stage to maturity sang aberrant, self-improvised elements. Nevertheless, these untutored songs still retained the species-typical temporal patterning that formed the macrostructural organisation of song-phrases and song-bouts; in other words, the contents were abnormal, but the packaging still fairly normal. This was subsequently confirmed by Price (1979), Clayton (1990b) and Morrison and Nottebohm (1993). Price concluded that the conservative macrostructural organisation of song in the Zebra Finch is due to a 'neuromotor constraint'. Strangely, not all Zebra Finches showed this constraint; the songs of birds visually and acoustically isolated by Eales (1985) after day 35 displayed no phrasing at all, and had many repeated elements. Conceivably, the early exposure to the father's song in Eales' study somehow disrupted the unlearned framework, or there may have been important differences in breeding stock from which experimental animals were sampled.

Element structure is abnormal and variable in untutored males, and mostly consists of non-call-like elements (Price 1979). Untutored elements are abnormally high in frequency and long in duration, and include many with upward inflections, and others that are click-like (Morrison and Nottebohm 1993; Williams *et al.* 1993). If untutored males are held in groups they tend to learn from each other (Slater *et al.*

1993). Call-like elements are not found in songs of isolates although their Tet and Stack Calls develop normally. It appears that males cannot form song-elements out of their own calls, but must learn them from other individuals. Wild Zebra Finches incorporate the father's Distance-call Element or his Distance Call into their song-phrase (Zann 1990), and possibly his Tet and Stack Calls as well. Similarly, untutored males raised in isolation with their mothers, or foster-mothers, will learn the female Distance Call and incorporate it into their aberrant songs (Immelmann 1967, 1969; Price 1979, Eales 1987a); a phenomenon also found in Bengalese Finches (Clayton 1987b). Price (1979) concluded that incorporation of developmentally conservative calls into the song-phrase was a means by which element formation could be rigidly specified, and so preserve the species song structure from accumulating learning errors that might occur with the developmentally labile non-call-like elements. The developmentally conservative Tet and Stack Calls could serve this purpose, but not the Distance Call, whose noise component must be learned, and whose absence from many Distance-call Elements of domesticated Zebra Finches is a manifestation of its developmental lability and high susceptibility to copying errors.

Sensitive phase for song learning

Both domesticated and free-living Zebra Finches mainly learn their songs in the second month of life, specifically between days 35 and 65. This was first shown by Eales (1985) who removed domesticated young males from their father's cage at day 35, 50 or 65, and either held them in visual and acoustic isolation from all adults until day 120, or held them in a cage next to a pair of adults Zebra Finches, the male of which served as a new song tutor. Those given the new tutor at day 35 learned all their song from him, while those given the new tutor at 65 days kept the father's song and learned nothing; those switched at 50 days learned a hybrid song. In the isolated group the proportion of song-elements learned from the father was in direct proportion to the time spent in his cage, and by day 65 all his song had been learned. However, those isolated at 35 days sang aberrant, amorphous songs, but when given access at six months of age to normal male Zebra Finches in an aviary, they learned a stereotyped, species-typical song-phrase. Strangely, when this last experiment was replicated by Slater *et al.* (1993) the majority showed signs of learning from the father and only modified theirs songs slightly when exposed to adult male tutors after 120 days. When Eales (1985) had males raised alone by their mothers they sang abnormal, unstable songs, but subsequent exposure to normal male tutors at 65 days of age showed that they were still capable of learning his song. When they did, it replaced the earlier version based on female calls (Eales 1987a); this also happens with Bengalese Finches (Clayton 1987b).

These results indicate that domesticated Zebra Finches mostly learn their song between 35 to 65 days of age, but if no acceptable tutor is available during that interval the ability to learn can be extended until one comes to hand, if not, songs learned before day 35 can be used. The important point is that it is experience, in combination with age, that determines the end of the sensitive phase. On the basis of their findings on untutored adult males, where song learning could be extended well into adulthood, Morrison and Nottebohm (1993) concluded that the closing of the sensitive phase probably results from social interactions with the tutor which lead to the acquisition of a 'stable motor memory of song'. Experiences such as singing a stereotyped song, hearing and copying elements from others, and development of particular types of song were considered to be less important factors in the closing of the sensitive phase. Therefore, Zebra Finches are clearly not 'open-ended' learners like canaries that can learn a new song each breeding season (Nottebohm 1993), but they are not strictly 'age-specific' learners either.

Most wild Zebra Finches will also learn their song during the 35–65 day sensitive phase. This was shown at the Danaher colony by comparing songs of fathers and their sons. The sons, who were caught during the second month of life and caged in an aviary within the colony were, in effect, socially isolated from the father (Zann 1990). There was a significant positive correlation between the time of exposure to the father outside the aviary and the proportion of elements shared. Those sons not caged until 65 days of age showed a strong match with the father's song, whereas those caged soon after day 35 showed no resemblance at all.

Learning before day 35

By swapping nestlings between nests of different males, Arnold (1975b) showed that no learning occurred before fledging. After fledging, the father's song is completely memorised by the son before day 35, but whether he sings it or not depends on what happens during the second sensitive phase. If he receives inadequate tutoring during the second sensitive phase, or it is disrupted to some extent so that there is a mismatch with the visual and auditory stimulation he received before day 35 he will produce the song learned before day 35 from the primary tutor. For example, if deprived of auditory, visual and physical contact with other Zebra Finches he will sing the song learned before day 35 just as completely as controls kept continuously with the father for 100 days (Böhner 1990). Similarly, if he cannot see the tutor, or cannot interact vocally with him (Eales 1989), or is exposed to a tutor of a different species to his rearing father (Clayton 1987c), or to a different morph to that of his rearing father (Mann et al. 1991, Slater and Mann 1991), or exposed to two tutors of different species, either successively (Clayton 1987d), or simultaneously (Clayton 1988b), he will sing the father's song heard before 35 days of age. Finally, if the acoustic structure of songs and Distance Calls of the new tutor diverge too strongly from those of the father they will be rejected (Slater and Jones 1995).

Exactly how Zebra Finches learn their father's song between fledging and day 35 is unknown and requires investigation.

Learning from the father

Immelmann (1967, 1969) found that domesticated male Zebra Finches learned the song of their rearing fathers even when cross-fostered to Bengalese Finches and able to hear, see and interact with conspecifics during the sensitive phase. Other semi-naturalistic aviary studies found that sons copy their father's song in preference to other Zebra Finches (Arnold 1975a; Schwab 1986; Williams *et al.* 1993), but not in every case (50%—Mann and Slater 1995), and in one study sons showed no preference at all for the father's song (Williams 1990).

In an attempt to determine if wild birds sing their father's song, I compared song-phrases of 40 sons with those of their 20 rearing fathers during a three-year study at the Danaher colony (Zann 1990). In order to detect evidence of copying, sonograms of song-phrases each of the two individuals in question were examined element by element for matching, and a per centage score of matched elements compiled. Conceivably, two phrases could have elements matching by chance, that is, without any copying or transmission from one individual to the other. To determine what level of matching occurred by chance in this population I compared the song-elements of 57 dyads chosen at random from 55 unrelated mature adults, and found that a median of 20% of elements matched, with a range of 0–57%. Next, I determined how high a matching score had to be for it to be statistically unlikely to have been drawn from the population of chance matches, and found that the probability of matching scores of $\geq 54\%$ arising by chance was less than 2.5%. Therefore, scores greater than this were unlikely to occur by chance and may have arisen by copying. I estimated that some 61% of sons matched the songs of their respective fathers and presumably copied from them. Some sons matched their father's song in every element, whereas others, even in the same brood, showed poor matches, some not exceeding chance levels (e.g. Figure 10.11, son (b) *vs.* son (c)). In one family, it was possible to compare song phrases across three generations, a grandfather, father and four grandsons. Matching scores reached criterion in all but two grandsons, one of whom was confined to the aviary during his sensitive phase and effectively excluded from learning the father's song (see Figure 3 in Zann (1990)).

Slater and Mann (1990) have argued that it is conceivable that a son did not actually learn from the father himself, but from a tutor that happened to sing like the father. Unless this tutor was a relative of the father's it is unlikely that such an individual would be present in the colony since the probability of a chance matching with the father is only

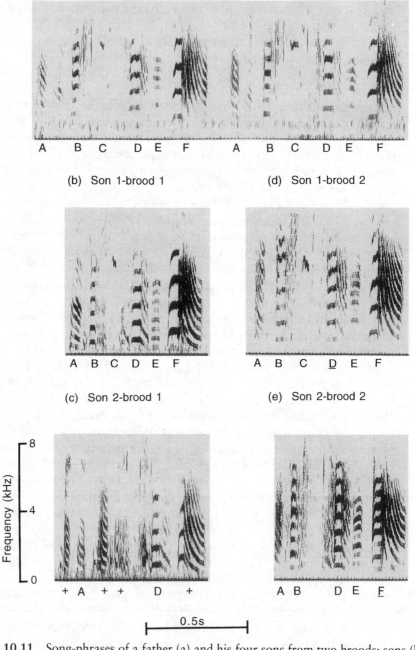

Fig. 10.11 Song-phrases of a father (a) and his four sons from two broods; sons (b) and (c) were from one brood and (d) and (e) from another. Two song phrases of the father are shown and one of each son. Elements of sons that match those of the father have the same letter; underlined letters are good, but not perfect matches, and a + sign indicates that the element has no matches. The percentage of a son's elements that matched those of the father are: (b) 100%, (c) 33%, (d) 83%, and (e) 92%. (After Zann 1990.)

2.5%. High song diversity at Danaher is probably due to the large influx of immigrants each year (Chapter 8). Moreover, the probability that sons in successive broods and in successive generations would encounter this same individual that happened by chance to sing the family song and learn from him are remote. Finally, I showed that Distance-call Elements and Distance Calls match between fathers and sons and some matches cross four generations—further evidence for sons learning from the father tutor. Therefore, the most parsimonious explanation is that sons whose songs matched those of their fathers actually copied from them. This is consistent with much of the laboratory work. What is difficult to explain is why about 40% of sons learned from other free-flying males, rather than their own father despite his availability as a song tutor. This can even happen in the one brood where one sib will accurately copy the father and another will only show chance levels of matching, and presumably copied someone else. This presents a new problem that can only be answered in the laboratory, namely, how are song tutors chosen?

How is the father chosen as a tutor?

In the laboratory a young male will learn the father's song in preference to that of another male, providing that the father is present during the second sensitive phase. This will occur whether the youngster is reared with the father for the first 35 days of life (Böhner 1990) or reared by a single mother and only encountering the father for the first time after day 35 (Eales 1987b). Experiments indicate that a range of cues are used to recognise the father.

Immelmann (1967, 1969) noticed that Zebra Finches tend to learn songs from those males that feed them, and this individual is usually the father in domesticated birds (and exclusively so in wild birds). However, in some domesticated stock, males other than the father will feed young and serve as song models (Williams 1990). Thus, the father–son nutritional bond may be one way a son recognises the father as a suitable song tutor; the occurrence of Undirected Song before and after feeding of young may be important for the learning process. Since provisioning care ceases before the main sensitive phase begins, this might not be the proximate mechanism that leads a son to chose the father as a model after day 35.

Both Böhner (1983) and Eales (1987b) noticed that fathers in cages behave aggressively to their young, and since aggression is believed to have an important influence on sexual imprinting in the Zebra Finch (ten Cate 1984) they thought that males may choose to copy the song of the most aggressive male they encounter. Indeed, when two non-fathers were offered as tutors Clayton (1987a) found that the one most aggressive towards the young male was the one he chose to learn from. However, recent experiments indicate a preference to learn from males aggressive to other individuals, and not necessarily the one aggressive to the young male seeking a song tutor (P.J.B. Slater, pers. comm.). In a small breeding cage this aggressive individual would normally be the father. As mentioned previously, aggression by the father towards his young has not been observed in the wild (Chapter 9) and could be an artefact of confinement, but young males frequently witness their father chasing and supplanting other Zebra Finches that come too close. A study by Slater and Richards (1990) found that in pairs where nest boxes were removed and re-nesting prevented, males learned more of the father's elements than if re-nesting occurred. One possible explanation might be that young have closer contact with a non-re-nesting father than a busy re-nesting one, and so have opportunity to learn more thoroughly. This finding with caged birds again stands in contrast to that found with wild birds where re-nesting had no significant effect on the amount of song learned from the father (Zann 1990). However, Slater and Richards (1990) also found that levels of aggression by the father to the young were no different if they re-nested or not, a finding which concurs with the situation in the wild.

Further meticulous experiments by Mann and Slater (1994) later showed that young males have at least two other methods of identifying their father from other males: they prefer to learn from the male with whom they were housed before the sensitive phase, and they prefer to learn from the male that is paired to the female that raised them. Thus, both parents influence tutor choice; the father's influence is direct, and the mother's indirect.

How are other song tutors chosen?

Laboratory experiments have revealed that the quality of a tutor's song, and the ability to interact visually and vocally with him are of critical importance (Eales 1987a). When offered an equal and simultaneous choice of two Zebra Finch tutors after day 35, a young male normally choses one, rather than producing a hybrid song from several tutors (Slater and Mann 1991). However, a hybrid song will occur if two tutors are offered in succession, in which case the male tends to learn sequences of elements from each (ten Cate and Slater 1991; Slater and Jones 1995), but with a preference for that heard later (Slater et al. 1991). During simultaneous choices the amount of singing by potential tutors does not affect choice, unless it is exceptionally low (Böhner 1983; Clayton 1987a). Song familiarity though, does affect choice of a tutor. If a young male is offered a choice of two tutors in which one sings a phrase similar to that of the father and the other does not, Clayton (1987a) found that the former is preferred. Young males also prefer to learn a song from a male that is mated rather than an unmated one (Mann and Slater 1994). If a tutor is somehow inferior (e.g. a female) males can still learn from it and produce a song-phrase, albeit an aberrant one, but if a normal male is subsequently encountered they will learn his song even if it means copying over the previous song (Eales 1987b). How they decide that their first song (and tutor) is inadequate is unknown.

Although Zebra Finches can learn the songs of other species such as Bengalese Finches (Immelmann 1969) or Red Avadavats *Amandava amandava* (Price 1979), several experiments, including Immelmann's (1965c, 1967, 1969) initial ones, indicate that there is a bias towards learning the song of their own species. When cross-fostered to Bengalese Finches to day 35 and then allowed full contact with a Zebra Finch tutor throughout the sensitive phase, but only limited contact (visual and vocal) with the Bengalese Finch foster-father, Eales (1987a) found that Zebra Finch males could learn the conspecific song and overcome the tendency to learn the rearing father's song if he was a heterospecific. In contrast, if the rearing father was a conspecific, full contact with a heterospecific did not override the conspecific song, indicating an own-species bias. Further evidence of bias was provided by Clayton (1988b), who limited exposure to conspecific and heterospecific tutors after day 35, to visual and vocal contact only. She found that normal-reared Zebra Finches only learned from the Zebra Finch tutor, and males cross-fostered to Bengalese Finches incorporated Zebra Finch elements. In a further experiment, Clayton had Zebra Finches cross-fostered to mixed parents, one a Zebra Finch (male or female) and the other a Bengalese Finch (male or female) and on exposure to the two species of tutors after 35 days of age found that hybrid songs were sung in which elements from both tutors occurred, but more from the tutor that was a Zebra Finch; again evidence for bias. The exact cause of bias is unknown but may relate to ten Cate's (1982) finding that significant differences exist in the quality and timing of parental care between parents of the two species, a factor assumed to be responsible for an own-species imprinting bias (Chapter 11). If the male of the mixed foster-pair was a Zebra Finch then a foster-reared male might be expected to prefer a Zebra Finch tutor but why it should do so when its foster-father was a Bengalese Finch is a puzzle.

Young males also prefer to learn from the song tutor that has the same visual appearance as their own species even when this tutor sings the incorrect song. In a clever experiment Clayton (1988c) exposed Zebra Finches with a range of rearing backgrounds simultaneously to two song tutors, a Bengalese Finch that sang a Zebra Finch song, and a Zebra Finch that sang a Bengalese Finch song. All males learned from the conspecific tutor, indicating that visual appearance was paramount. In a similar experiment Mann *et al.* (1991) offered tutors belonging to two different colour morphs and found a preference towards that of the father's morph. In a further experiment Clayton (1988c) exposed males to two Zebra Finch tutors, one that sang a Zebra Finch song and one that sang a song composed of Bengalese Finch elements, but the young males did not discriminate between them. She concluded that both visual and vocal interactions are important for selection of song tutors, but Zebra Finches have no preferences for copying Zebra Finch elements as such, but probably have a preference for tutors that have Zebra Finch calls and Zebra Finch song macrostructure (Clayton 1989a).

The need for visual and vocal interactions between young male Zebra Finches and their song tutor was impressively demonstrated by Eales (1989) who exposed males to (a) tutors they could see and with whom they could interact vocally, (b) tutors they could not see but could interact with vocally; and (c) tutors they could hear, but not see or interact with vocally. Males in group (a) all learned from the tutor, while none did so in group (c). Nevertheless, in group (b) half learned from the tutor they could not see which suggests that the visual components of the song tutor are not always essential for song learning. This aspect has been followed up recently by Adret (1993) who placed young male Zebra Finches in a Skinner box and trained them to peck at a key that resulted in the broadcast of a Zebra Finch song from a tape via a loudspeaker placed inside the cage. The test male could hear, but not see a control bird, a male sibling companion that could also hear the taped song, but had no switches to control its production. Test males not only found the song reinforcing and worked the key in order to hear the taped song, but they also memorised it and sang it at maturity. In contrast, the control males barely learned any elements of the song. Adret concluded that operant conditioning with song as a reward strongly influences song learning, that is, the young male is not some passive 'sponge' soaking up the song of the selected tutor, rather it exercises some control over the learning process. Exactly how this might work in the natural context is unknown. In a similar Skinner box experiment ten Cate (1991b) also conditioned test males to song playback, but was unable to demonstrate song learning, perhaps because the operant response (perching) was less appropriate than that in Adret's study (1993)

Ten Cate (1986a) noticed that foster-reared male Zebra Finches often sat very close to the singing Bengalese Finch foster-father and appeared to 'peer' at, or 'listen' intently to him. Listening such as this is extremely rare in normally raised Zebra Finches but is common among young mannikins (Morris 1958). Ten Cate believes that listening is a means of attending carefully to the song performance in order to copy it, and may be essential when the song to be learned is soft, complex and not sung very often, which is often the case with mannikins.

Overview

Both field and laboratory findings suggest that sons are inclined to learn their fathers' songs. There appears to be an overlap in the mechanisms that promote learning of the father's song, and additional mechanisms to ensure that they at least learn the songs of Zebra Finches, rather than those of heterospecifics. All details of the father's song are learned in the first sensitive phase during early post-fledging and this is the primary backup system, an insurance against failures to learn the conspecific song during the main sensitive phase between 35 and 65 days of age. During the second month of life, field data suggest that young Zebra Finches still have tight social bonds with the parents, and the early end to the sensitive phase fosters the learning of song from the father. Furthermore, if an incorrect song is learned, the sensitive phase remains open until a satisfactory tutor is encountered and a normal song learned. During the 35–65 day interval, the young male is inclined to learn from the individual that should be his father; moreover, the fact that he prefers to learn from a tutor that sings like his father appears to be another means of guiding his song towards the paternal version. Finally, imprinting on the siblings and/or parent is a means of ensuring, at least, that a Zebra Finch will be the tutor.

Böhner (1990) proposed a two-step learning process. The first stage occurs between fledging and day 35 during which all the details of the father's song are memorised. After day 35, the onset of the second stage of learning, no new song will be acquired if the father continues to be present during the main sensitive phase, but if he is absent a new tutor is chosen. This new tutor is not chosen at random, but is guided by the visual and vocal experiences gained before day 35.

Given the redundancies built into the song learning process that appear to promote learning of the father's song in preference to that of other males it is puzzling that only about two thirds of wild males appear to learn the father's song at all, and some of their copies are not very accurate. Even in cage studies, where only a few tutors are present, inaccurate copying is a common feature of song learning, although one benefit is that it enables individuals to develop unique versions that can be used for individual recognition (Böhner 1983). Variability in the learning process in the wild, especially among siblings, is unexplained, but recent findings by Jones and Slater (1993) raise the possibility that females, other than the mother, also influence what song a male develops. In this way males arrive at a song that females find attractive yet is not necessarily the same as their fathers'. This situation may be comparable to 'action-based' learning found in some other species of songbirds (Marler and Nelson 1993).

In his initial study, Immelmann (1967, 1969) found that the Bengalese Finch and the African Silverbill *Euodice cantans* also learned their songs during an early sensitive phase and could learn from heterospecifics in a manner similar to that of the Zebra Finch. Subsequent investigations by Clayton (1987b) showed that song development in the Zebra Finch and Bengalese Finch is almost identical in all details, except that the latter does not include call-like elements into its phrase. Unfortunately, few studies of song development have been made on other species of estrildines to determine whether all estrildines possess a short sensitive phase early in life or whether it only occurs in those that have particular ecological adaptations. In considering the biological advantage of a short and early sensitive phase in Zebra Finches, Immelmann speculated that it was a means of ensuring that only the conspecific song would be learned, thus precluding accidental learning from any heterospecific a young male might encounter in mixed breeding colonies or in mixed feeding flocks later in life. Too little is known about the biology of free-living Silverbills and the wild ancestors of Bengalese Finches to determine if this explanation also applies to them. Moreover, with Zebra Finches, the risk of accidental learning from heterospecifics appears low under present ecological conditions, since mixed colonies and mixed feeding flocks are rare (Chapter 4). Conceivably, during periods of intense aridification, Zebra Finches may have been forced to move out of much of the interior of Australia to seek refuge in areas where close living with other species of estrildines occurred and so heightened risk of learning incorrect songs. The provision of numerous, widely dispersed, artificial water points since European settlement may have limited the extent to which Zebra Finches are now forced to disperse, and consequently their exposure to other estrildines has become limited in recent times. Unfortunately, too little is known of the comparative breeding biology and song development of wild estrildines to proceed beyond the speculation stage.

Control of song

Canaries and domesticated Australian Zebra Finches are the two species of songbirds used as experimental models for investigations of brain function, in particular, the neural basis of learning and mechanisms leading to sexual differentiation in the brain. Investigations of the vocal control system in these species offer 'a unique opportunity to study the interplay of hormonal, environmental, and developmental forces that mould a neural circuit responsible for complex vertebrate behavior' (Arnold 1992). Although conceptually discrete, mechanisms for the development, production, and perception of song are difficult to separate in practice. Neural control of song in the Zebra Finch is an area of vigorous research activity and only the barest outline of the main findings will be presented here. Fortunately, F. Nottebohm (1993), a leading researcher in the neurobiology of song learning for the past 25 years, has recently written a masterly review of the important developments. He compares the neural mechanisms of song learning in the Zebra Finch, an 'age-limited'/'experience-limited' learner, with that of the canary, a classical 'open-ended' learner, highlighting similarities and differences and summarising the main questions for future research.

When Price (1979) deafened Zebra Finches late in the sensitive phase they failed to develop normal song. This is believed to have arisen because they needed to be able to hear what version of song they were actually producing in order to compare it with the memory of the tutor's song they previously heard and were trying to copy. The unlearned elements of song plus the components learned from the tutor during the sensitive phase are believed to combine to form a type of neural representation, or template, which the male attempts to copy.

Song production and neural control

Vocalisations in birds originate from the syrinx, a muscular organ located at the junction of the two bronchi. Each half of the syrinx has a membrane that vibrates under muscular tension when air is expired from the lungs so producing sound (Greenewalt 1968). The pattern of muscular tension on the membranes is caused by nerve impulses which travel from the brain stem along paired tracheosyringeal nerves to each side of the syrinx. The production of these nerve impulses is controlled by the song control circuitry in the forebrain. In addition, the circuitry controls the respiratory patterns that produce the timing of the song. The structure and function of this circuit have been under investigation for nearly three decades and an outline of its components (nuclei, or clusters of neurones) and their connections via nerve projections, as it is currently understood, is shown in Figure 10.12.

Basically, there are two circuits—the HVC–RA–nXXts circuit controls the production of song, and the 'recursive loop' is important for learning of song. Control of the syrinx in the adult emanates from the HVC which is the pattern generator of the fine detail of the song (Nottebohm et al. 1990). The HVC and the RA are not connected at hatching, but as the nestling grows the HVC sends nerve projections to the RA so that the performance of the first subsong coincides with the initial connection between the two key control nuclei. Subsequent development of song goes hand in hand with an increase in the size of the HVC and the strengthening of its link with, and control over, the RA (Nottebohm et al. 1990). If any one of the control nuclei or their projections are experimentally destroyed or damaged, the production circuit ceases all function and song is lost. New neurones are not only continually added to the HVC during maturation, but are also added in the adult Zebra Finch as well.

The recursive loop also originates at the HVC and ends at the RA but detours via three control nuclei, area X, DLM, and LMAN. This indirect circuit is believed to be involved in song learning and perception. Experimental destruction of components of the recursive loop does not stop production of song in adults, but if destroyed before or during the sensitive phase it prevents normal song development (Bottjer et al. 1984). Where song learning has been experimentally delayed in untutored males, destruction of the LMAN prevents learning in adulthood (Morrison and Nottebohm 1993). Electrical recordings in the song control circuits found that auditory stimuli, especially vocalisations, stimulate neurones in both production and recursive circuits, with Field L (Figure 10.12) being a major source of auditory input to the HVC (Katz and Gurney 1981; Williams 1989); these stimuli are transmitted to the brain stem

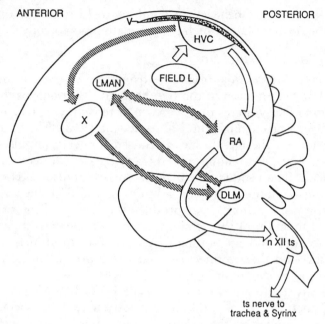

Fig. 10. 12 Schematic diagram of the neural system that controls the acquisition and production of song and learned vocalisations in the Zebra Finch. A sagittal view of the right side of the brain is shown. Open arrows show the main motor pathway from the forebrain down to the brain stem and the nerves that go to the syrinx. The dark arrows show an important alternative pathway, the recursive loop, which is involved in song learning. Abbreviations of the major control centres, or nuclei, connected by nerve projections (arrows) as follows: HVC, hyperstriatum ventrale, pars caudale, (or 'higher vocal centre'); RA, robust nucleus of the archistriatum; nXII ts, tracheosyringeal part of the hypoglossal nucleus; DLM, medial portion of the dorso-lateral thalamic nucleus; LMAN, lateral part of the magnocellular nucleus of the anterio neostriatum; V, lateral ventricle. Field L provides auditory input to the HVC but is not part of the motor pathway. (From Nottebohm *et al.* 1990.)

and syrinx (Williams and Nottebohm 1985). The recursive circuit responds more to the bird's own song than to those of other individuals. Since the neural circuit for song is reduced in females it is likely that their discrimination of song occurs in a different part of the brain (see below).

One hypothesis that attempts to explain aspects of song perception, learning and production in male Zebra Finches proposes that during the sensitive phase, songs or other sounds, such as Distance Calls, have a direct effect on the development of the actual control circuits themselves, via the HVC, and connections to the RA are formed during song learning (Nordeen and Nordeen 1988). While growth in the neurones of the HVC and RA is related to vocal learning there is a concomitant loss of neurones in the LMAN (Arnold 1992). Deafening is believed to have two effects on song performance: it denies access to any song model, and it denies auditory input to guide the ontogeny of the production circuit (Nottebohm *et al.* 1990). Song learning in the Zebra Finch is conceived as an interaction between auditory and motor components in which a song is analysed by the auditory system, which then compares it with the stored auditory model of the target song, and any discrepancy between the two triggers a change in the motor pattern that produces the song—a negative feedback loop (Williams 1989).

Auditory feedback is necessary for adult song

When Price (1979) deafened adult Zebra Finches by removing the cochlea he found that only minor changes occurred in the song-phrase following surgery. He concluded that adult song had crystallised, namely, the sensory–motor development phase was complete and a permanent central motor program formed, so that hearing was no longer necessary for song production. This was confirmed by Bottjer and Arnold (1984), who found, in addition, that proprioceptive feedback from the syrinx was also unnecessary for song production. Recently, however, Nordeen and Nordeen (1992) found that deafening did have significant effects on adult song, but the effects were not fully developed until about four months after surgery. Element structure and temporal organisation were grossly distorted by deafening, implying that song crystallisation does not eliminate the need for auditory input for maintaining the neural circuitry controlling learned songs. This is consistent with the finding that the song control system in adults is sensitive to auditory input (Williams 1989) and is also consistent with the discovery that the HVC continues to incorporate new neurones in adulthood and these may need the guidance of auditory feedback for the formation of new synapses (Nordeen and Nordeen (1988). However, as Nottebohm (1993) has recently pointed out, just because neurones in the song control circuit of adults are sensitive to auditory input does not mean that the function involves learning, it could be communication.

The effects of deafening on the structural integrity of the male Distance Call are unknown, but that too may retain its neural plasticity after it has been produced in the adult form.

Other evidence suggests that the adult song retains a degree of structural plasticity. When Williams and McKibben (1992) injured, to varying degrees, the right tracheosyringeal nerve to the syrinx, there were short-term deficits in element production but these disappeared after the nerve regenerated. Subsequently, however, there were permanent changes in the temporal patterning of the song-phrase irrespective of whether the nerves regrew or not. Non-call-like elements tended to be lost in 'chunks' or strings, and new elements, mainly Stacks and Distance-calls, were added. The song-phrase became shorter as gaps closed up where strings of elements were lost. The changes in temporal patterning are believed to result from a reorganisation of the production circuit that changes the central program. Williams and McKibben (1992) hypothesised that when the nerve injury disrupted function of the syrinx, proprioceptive and/or auditory feedback to the control circuit detected the deficit and initiated a slow, but permanent, reorganisation of the temporal patterning of the song-phrase. Another part of the control circuit added new elements, but these were limited to call-like ones. This experiment supports the view that there are two roles played by the control circuit, one for the production of elements via the tracheosyringeal nerves and another that controls the respiratory musculature that does not involve innervation by the tracheosyringeal nerves.

Right-side dominance of song control

Mirror image duplicates of the song control circuitry exist in each side of the brain and each innervates the ipsilateral half of the syrinx via the left and right tracheosyringeal branches of the hypoglossal nerve. In seven species of song birds investigated to date, it has been found that destruction of part of the left nerve causes a profound loss of song elements because the left half of the syrinx atrophies and is rendered useless; in contrast, when the right side is damaged there is little or no effect on song. Therefore, the control circuit for production of song located on the left side of the brain is considerably more important than that on the right and the phenomenon is known as 'left hypoglossal dominance' of song control (Nottebohm 1980).

However, when similar experiments were conducted on Zebra Finches, Williams et al. (1992) found that song production was disrupted more after a section of the right nerve was cut and removed than after one from the left. For example, all high elements disappeared after damage to the right nerve, but only half disappeared when the left nerve was rendered inoperable; yet in neither case was the temporal organisation of the song affected. Similarly, when the right HVC was destroyed, more of the song was lost than when the left HVC was destroyed, and disruption was more severe than that produced by destruction of the right tracheosyringeal nerve. This is the first reported instance of right hemisphere dominance in birds and contrasts, interestingly, with findings on the Java Sparrow, the only other estrildine investigated to date, where song control was found to be left-side dominant (Seller 1979).

Sexual differentiation of the brain

Adult female Zebra Finches do not sing even if treated with the male hormone, testosterone (Arnold 1992). One reason is that their brains have not completed the song control circuits. The song control nuclei in females are insignificant in volume relative to those of males. Furthermore, the key connection between the HVC and the RA is greatly reduced or absent in the female brain (Williams 1985). The brain of

Zebra Finches is the most sexually dimorphic of any song bird studied to date (Arnold 1980; Simpson and Vicario 1991a). Sex differentiation of the brain in Zebra Finches begins during the first few weeks of post-hatch development, before subsong and the sensitive phase begin. Sex differences in the song-control centres arise either from differential growth and addition of neurones in males, or from differential atrophy and cell death in females (Arnold 1992). Either way the result is the same: volumes of the neural centres increase in males and regress in females. These changes lead to detectable sex differences between days 15 and 30 post-hatch.

Sexual dimorphism in the brain of Zebra Finches is believed to arise from hormonal action in the days immediately following hatching. However, the exact stages are uncertain because, paradoxically, there are no sex differences in levels of plasma and brain steroids; nevertheless, the oestrogens that masculinise the song control circuit in males are synthe-sised in the brain itself from androgens produced from the adrenals (Schlinger and Arnold 1992; Arnold 1993), and this may possibly begin while the male is still in the egg (Arnold 1992). A number of investigators have given estradiol to female Zebra Finches immediately after hatching and found that the brain anatomy changes to that of males (see review by Arnold (1992)). These females are capable of learning the fathers' songs and if implanted with androgens at adulthood give normal males songs (Pohl-Apel and Sossinka 1982, 1984). Recently, Simpson and Vicario (1991b) elicited songs and male Distance Calls in females by early oestro-gen treatment alone, without the androgen treatment in adulthood. These females learned both songs and calls from the father and were indistin-guishable from those of normal males. This is the first demonstration that hormone treatment can affect calls as well as song in female Zebra Finches. Furthermore, lesions to the tracheosyringeal nerves caused loss of element morphology in the song and loss of the male features of the Dis-tance Call, just as they do in males (Simpson and Vicario 1990). There-fore, early oestrogen treatment masculinises the song control circuit of female Zebra Finches to a fully functional state and enables them to learn and produce complex male-specific vocalisations. Adkins-Regan and Ascenzi (1987) found that oestradiol implants in gonadectomised adult females had the same effect on oestradiol-treated nestlings. These 'mas-culinised' females sang both Directed and Undirected Song and gave other male behaviours, such as the courtship dance and nest solicitation. By contrast, males were demasculinised by the procedure.

Gonadal hormones are necessary for the normal development of singing in Zebra Finches. Arnold (1975b) and Adkins-Regan and Ascenzi (1990) found that castrated nestlings still learned their father's song, but sang less frequently and more slowly than controls; however, there is some uncertainty whether castration removes all traces of androgen (Bot-tjer and Hewer 1992). Although Pröve (1983) found peaks in testosterone production during the sensitive phase, these could not be detected by Adkins-Regan and Ascenzi (1990) so that correlations between hormones and song learning are still unclear at this stage (Nottebohm 1993).

In adult male Zebra Finches, testosterone affects the production of song. It is selectively concentrated in certain cells of the control circuit, and it causes the syrinx to increase muscle mass (Luine *et al.* 1980). However, the effects of steroids on song production are more equivocal, as contrasting effects of castration on singing have also been found: Arnold (1975a) detected losses in frequency of singing and tempo whereas Bottjer and Hewer (1992) could find no detectable differences. One possible explanation for the discrepancy is that Arnold measured Directed Song and Bottjer and Hewer measured Undirected Song. However, Walters *et al.* (1991) demonstrated that Directed Song is oestrogen dependent, whereas Undirected Song is more androgen dependent.

Song perception

It has been proposed that the song control circuits are not only important for the learning, development and production of song, but also have a significant role in the perception of song. By means of operant techniques it has been shown that male Zebra Finches are better at discriminating between familiar songs than females, especially when one is the male's own song (Cynx and Nottebohm 1992a). Discrimination in this instance is centred around the bird's own song so that aspects of perception may be related to whether the bird can or cannot produce it, but this requires verification. Unexpectedly, ability to perceive this way varies seasonally, being better in summer than in winter, and is thought to result from fluctuating levels of plasma testosterone (Cynx and Nottebohm 1992b). Possibly, testosterone results in a general improvement in memory, rather than just improvements in memorising song. The left hemisphere of the brain is better at this discrimination task than the right, although the right hemisphere is better at another task where minute versions of unfamiliar song need to be discriminated (Cynx *et al.* 1992). The evidence suggests that males and females perceive Zebra Finch songs differently, and that males perceive familiar songs differently from unfamiliar ones. Blood plasma concentrations of testosterone in wild Zebra Finches are significantly higher in the breeding season (Dunn 1994), so males might be better at discriminating their songs from others at this time of the year, but the biological benefits of doing so are not apparent. Perhaps this ability is related to kin recognition, which is high among brothers (Burley *et al.* 1990) and may lead to the formation and maintenance of groups in breeding colonies. Similarly, if females prefer first cousins as potential mates, then the ability to discriminate among similar songs would also be an advantage to them. Contrasts in male and female perception of song would be worth pursuing, especially those aspects that might indicate male quality, and are most attractive to females, since discrimination among these has a clear biological function for females, whereas none has been suggested for males.

Auditory perception

Zebra Finches do not differ significantly in basic auditory sensitivity from other species of song birds studied to date (Dooling *et al.* 1992). Using positive reinforcement techniques, Hashino and Okanoya (1989) established that the range of best hearing for domesticated Zebra Finches is 1–6 kHz (Figure 10.13). This is the same for most other species of birds that have been tested so far, and is much inferior to that of humans (Okanoya and Dooling 1987). Thus, Zebra Finches are fairly insensitive to sounds above and below this range, but within it they are most sensitive to sounds around the 4 kHz level which is slightly higher than that for other songbirds. When Hashino and Okanoya (1989) examined the frequencies of the Distance Calls they found that most of the sound energy was between 2 and 5 kHz with a distinct peak at 4 kHz. Coincidence of frequencies of best hearing and those produced in the vocal utterances is a common finding among avian species studied to date (Dooling 1982).

The ability to resolve different frequencies against a background of noise is a more realistic or natural test of perceptual ability, and is measured by the critical band or signal-to-noise ratio. This increases monotonically at a rate of 3 dB per octave in Zebra Finches, and is the same as that found for all songbirds and other vertebrate species mea-

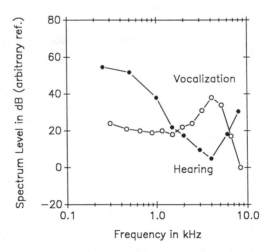

Fig. 10. 13   Average auditory sensitivity of five domesticated Zebra Finches to pure tones, and the power spectra of 14 Distance Calls. Sensitivity thresholds were established by operant reinforcement procedures. (From Hashino and Okanoya 1989.)

sured to date (Okanoya and Dooling 1987). Of course, Zebra Finches are more sensitive to their own sounds than those of other species when both are mixed in a sequence together (Okanoya and Dooling 1988, 1990, 1991a,b; Dooling *et al.* 1992). Furthermore, in one of the few psychophysical demonstrations of sex differences in auditory perception in higher vertebrates, male Zebra Finches were found to discriminate complex heterospecific calls better than females. This may be related to the fact that males learn to produce Distance Calls and songs and females do not, and is consistent with the finding of Cynx *et al.* (1992) described above on sex differences in song discrimination. How specialised perception for conspecific calls is acquired is unknown, but the method of detection is probably a frequency-based filtering system rather than a 'matched-filtering' one involving a signal match with an internal 'template' (Okanoya and Dooling 1991b).

The minimum detectable threshold for temporal acuity of Zebra Finches is 2.5 ms for a broad spectrum noise. Within the band of best hearing, resolution is lowest (3.0 ms) on frequencies centred on 5 kHz and highest (7 ms) on those centred on 1 kHz (Okanoya and Dooling 1990). As with most species there is a trade-off in resolving power between frequency and time.

Timbre is a source of much between-individual variation in Distance Calls. In addition, Williams *et al.* (1989) established that the timbre of specific elements in the song is under active control of the vocal organ and accounts for much variation in song structure within and between individuals. Zebra Finches can perceive differences in timbre and be trained to discriminate between song elements that vary by as little as 6 dB in just one harmonic out of nine (Cynx *et al.* 1990). They are much better at the task when the element is embedded in its normal place in the song phrase. Inter-hemispheric perceptual differences of Zebra Finch song have also been detected (Nottebohm *et al.* 1990).

In summary, temporal and frequency thresholds of Zebra Finches are about the same as that of other species of songbirds so far studied. However, their auditory perception is tuned to variations in timbre, especially those elements given in the song, which is appropriate for a species with a rich production of harmonics in its vocalisations.

Summary

In addition to the song, Australian Zebra Finches have 12 distinct vocalisations, of which the Distance Call is the most important. It is the only long-distance vocal signal emitted by the species. It differs between the subspecies and sexes, and shows strong geographical and individual variation. Domesticated male Zebra Finches give abnormal Distance Calls due to learning errors. The Distance Call in males is learned from the father early in life, but that of females is not learned. Only four vocalisations of the Lesser Sundas Zebra Finches have been described, but further work is needed.

A song contains repeatable phrases, each of which consists of a mean of six elements or notes in the Australian subspecies and 12 in the Lesser Sundas subspecies. Songs of wild birds differ geographically, and differ quantitatively from those of domesticated Zebra Finches. Elements are sung in a fixed sequence in the phrase and fall into two types: those that resemble calls and those that do not. Song is believed to have evolved from a sequence of calls given at take-off. Non-call-like elements are thought to have originated from the Distance Call. Two types of song are performed: Directed Song, a pre-coital signal directed at the female, and Undirected Song which is not directed at any individual, but may have a number of possible functions, including sexual advertisement and mate-guarding.

Subsong begins soon after fledging and the plastic song crystallises into the full adult song around 90 days of age. Laboratory studies of domesticated Zebra Finches show that they learn most details of their songs in the first two months of life, but they do not need to learn the macrostructural components. They learn from the father in the first month of life, mostly after fledging, but will copy song from other males during the main sensitive phase (35–65 days of age) depending on the amount of social interaction they have with their father and with other males. A range of developmental strategies exist which appear to bias learning from the father. Nevertheless, only about two thirds of wild males have a similar song to that of their fathers. In some wild families, song traditions can be detected over three generations.

Song production is controlled by neural circuits located in the forebrain. During the first month of life, oestrogens cause the brains of males to develop song circuits; none develop in females, but if they are treated with oestrogen, they too develop circuits, and can learn to sing the father's song and give his Distance Calls. The circuit on the right side of the brain is more important than the circuit on the left side, and is the opposite to that found in other species of birds investigated so far. For song to retain its structural integrity, the bird must be able to hear itself sing, even when adult. This indicates that song remains plastic after it has crystallised into the adult form. Song control circuits are not only involved in the learning and production of song but may have a significant role in perception of song. Males and females perceive song differently and use different parts of the brain. Male perception may depend on levels of circulating androgens and vary seasonally.

Auditory perception in Zebra Finches is similar to that of other small birds. Best hearing is between 1–4 kHz, and they are most sensitive to frequencies around 4 kHz; sensitivity coincides with the frequencies of the sounds emitted. The minimum detectable threshold for temporal acuity is 2.5 ms.

11 Sexual imprinting and mate choice

'These results indicate that early experience seems to have a crucial influence on the later choice of the sexual and social partner. The acquisition process shows some characteristics typical for sexual imprinting.'

K. Immelmann (1969).

Imprinting

In the mid-1960s Immelmann began to investigate puzzling accounts from aviculturalists who reported that when they fostered out eggs of rare estrildines to pairs of Bengalese Finches for rearing, the young, on reaching adulthood, often formed breeding pairs with Bengalese Finches. Consequently, in a long series of pioneering experiments, Immelmann (1969, 1972a) systematically cross-fostered eggs of Zebra Finches belonging to both domesticated and wild stocks to pairs of Bengalese Finches that raised the young to independence. Young were then held individually until maturity whereupon sexual preferences were tested in 'free choice experiments'. Males, which were given a choice of a female Zebra Finch or female Bengalese Finch at each end of a three-compartment cage (the 'double-choice test'), overwhelmingly directed their courtship towards the latter. Holding conditions after independence had no effect on preference; they all preferred females of the foster-parent species, even if they could hear and see their own species. The number of siblings in a brood had no effect on the direction of imprinting, so Immelmann concluded that young imprint on the individuals that fed them, namely their parents, not on their siblings.

The sexual preference was highly stable. It persisted in many retests even after imprinted males had been forced to pair and breed with Zebra Finch females for periods of up to four years and during which many broods were successfully reared. Yet, when removed from their partner and given a choice of strange females in the double-choice cage they preferred the Bengalese Finch. Immelmann (1972a) concluded that imprinting was 'absolutely irreversible' since it appeared to last for life no matter what experiences the male subsequently had.

Clearly, these sexual preferences for the foster parent species began before independence, and by switching nestlings at different ages from their natural parents to the foster-parents, Immelmann determined that the onset of sexual imprinting occurred between 15 and 20 days post-hatch. Similarly, by switching independent young at different ages from the foster-parents back to Zebra Finches he found that it was possible to reverse the sexual imprinting on the foster-parent species before 40 days post-hatch, but not afterwards. He concluded that sexual imprinting ends before 40 days post-hatch. Immelmann also believed that there was an own-species bias since reversal of sexual imprinting could only be made from Zebra Finches to Bengalese Finches if it occurred before day 25 whereas it could occur up to day 40 from Bengalese Finches to Zebra Finches. Furthermore, if nestlings were foster-reared by a mixed-pair, one a Bengalese Finch and one a Zebra Finch, 17/21 males became sexually imprinted on Zebra Finches. However, preferences here were not as clear cut and showed signs of 'double-imprinting' (see below).

In contrast to males, female sexual preferences were more difficult to determine, not only because of their more passive and subtle courtship behaviour, but their reaction to the stimulus males in the preference tests were qualitatively different from those displayed by their male counterparts (Chapter 9). Nevertheless, in the double-choice test the male of the foster-parent species received the first courtship greeting, although subsequently the Zebra Finch male received some greetings if he courted vigorously. Females also spent most of the time perched near the Bengalese Finch. From this small sample ($n = 5$) Immelmann tentatively concluded that females also sexually imprint, but have a stronger own-species bias than males.

Immelmann's inspirational series of experiments encouraged other researchers, principally J. Kruijt and C. ten Cate at Groningen University, and N.S. Clayton at Bielefeld University to investigate the problem of sexual imprinting in more detail, especially the causal mechanisms that lead to the formation of the preference and how this subsequently affects mate choice. Cross-fostering was again the principal method of exposing young to the rearing species and while the double-choice test was not always used it was found to be a reliable method for testing sexual preference (ten Cate *et al.* 1989).

Is there an own-species bias?

In follow-up experiments using larger samples, Sonnemann and Sjölander (1977) showed that female Zebra Finches fostered by Zebra Finch pairs had a very strong sexual bias towards Zebra Finch males, whereas those fostered by Bengalese Finch pairs also showed sexual preferences, but these were fairly evenly divided between the two species. These differences were interpreted as evidence for an own-species bias and an hypothesis for an unlearned, or innate preference was proposed where the effect was stronger in females than males (Immelmann 1972a,b). Sonnemann and Sjölander (1977) also thought that their results might reflect the different processes the sexes use to select a partner.

Ten Cate considered the possibility that sources of bias other than a species effect might also exist in the imprinting and testing processes and began a series of subtle experiments in which he tried to tease apart the various factors involved. First, he found that Zebra Finch foster-parents interact with their nestlings more intensely before 30 days of age than do Bengalese Finch foster-parents, and in mixed foster-pairs the Zebra Finch parent fed the young more than the Bengalese Finch parent and was also more aggressive towards them. However, the Bengalese parents allo-preened and clumped with the young more after day 30 (ten Cate 1982). Thus, parental care by Zebra Finches and Bengalese Finches is different in kind, timing and quantity, and these differences may bias the strength of imprinting towards the Zebra Finch parent. Second, in the double-choice test, sexually imprinted females could be biased towards Zebra Finches because Zebra Finch males have a more vigorous style of courtship than Bengalese Finch males. When the courtship vigour of the Zebra Finch male was experimentally lowered to that of the Bengalese Finch male, the imprinting bias towards the Zebra Finch was strongly reduced (ten Cate and Mug 1984). This finding is consistent with Immel-mann's (1959) mate recognition investigations (Chapter 9). Furthermore, with males raised by mixed pairs ten Cate and Mug (1984) found that male sexual preference for the female Zebra Finch was greater if the fos-ter-mother was a Zebra Finch than if she was a Bengalese Finch, which suggests that familiarity with female Zebra Finches is a another source of bias in the double-choice test (see below). Third, males foster-reared by mixed pairs of Zebra Finches and Bengalese Finches had their sexual bias towards Zebra Finches shifted experimentally towards Bengalese Finches when the caring behaviour of the Zebra Finch parent was experimentally reduced (ten Cate 1984). Fourth, during the interval of isolation from the foster-parents to the first double-choice test, ten Cate and Mug (1984) found a steady drift in preference from Bengalese Finches towards Zebra Finches, consequently, this might also be another important source of bias towards Zebra Finches.

In summary, experiential biases during the sensitive phase and biases during testing, make it almost impossible to verify the existence of an own-species sexual and social bias in Zebra Finches. Moreover, Clay-ton's (1990d) extensive studies on cross-fostering between subspecies of Zebra Finches failed to reveal any sub-species biases in sexual imprinting.

What characteristics of the sexual object are learned?

Visual characteristics of the rearing parents are learned, but behaviour, especially social interaction, is also important. If raised by another colour morph, males prefer females the same colour as that of the rear-ing parents (Walter 1973), and if released into aviaries both males and females quickly pair up and breed with a partner the colour morph of the rearing parents (Immelmann et al. 1978). The importance of visual cues is consistent with the finding that imprinting in nestlings does not begin until visual perception and discrimination develop at around day

16 (Bischof and Lassek 1985). During the sensitive phase of imprinting, offspring identify the sex of their respective parents by means of behaviour and subsequently develop a preference for the morphological characteristics of the opposite sex parent (Immelmann 1985). Thus, a son imprints on visual features of his mother and subsequently directs sexual preferences towards individuals with those features; conversely, he directs aggressive responses towards those individuals that have the same features as his father (Vos *et al.* 1993; ten Cate *et al.* 1993). Similarly, the daughter develops a preference for the visual features of the father and directs her sexual preferences to individuals with those features (Weisman *et al.* 1994).

Behaviour of the rearing parents also influences the development of sexual preferences. This was demonstrated by ten Cate *et al.* (1984) who removed normally reared males from their parents at 31 days of age and caged them with Bengalese Finches under a range of conditions whereby the amount of social contact was regulated. The extent to which Zebra Finches tended to direct a (small) proportion of their courtship towards Bengalese Finches in the double-choice tests depended on the degree of social behaviour directed at them by Bengalese Finches during earlier exposure. Both aggressive and non-aggressive behaviours directed at young Zebra Finches (not vice versa) influenced their sexual preference, but non-aggressive ones were more important. Finally, the first ten days of contact with Bengalese Finches were more effective in stimulating a change to a sexual preference towards Bengalese Finches than contact after that time. This confirms Immelmann's finding that the sensitive phase does not end before day 40, but challenges his finding that first exposure to heterospecifics after day 30 does not influence sexual imprinting.

Among Zebra Finches fostered by mixed pairs, or exposed to Bengalese Finches before the end of the sensitive phase, were some males that courted both species in the double-choice tests. Ten Cate (1986b) called these males 'dithers' and found that their dithering persisted in subsequent tests which indicated that they had become 'double-imprinted' on the two species. Follow-up experiments, where males were tested for preference with females of Bengalese Finches, Zebra Finches and hybrids of the two species, showed that dithers preferred the hybrid. This suggested to ten Cate (1987) that dithers had learned both features of the rearing species and that one image, or internal representation, was synthesised from attributes of both species, rather than two separate images, one for each species, existing side by side; however, recent experiments have not confirmed this and the question remains open (Vos *et al.* 1993).

While visual characteristics of the bill and plumage, and the type and intensity of behaviour have been found to be important, the role of vocalisations has yet to be determined. Recognition learning of the song

(Clayton 1988a) and Distance Calls (McIntosh 1983) occurs between 25 and 35 days of age, so it is likely that they also provide significant cues in the preference tests. Furthermore, the relative importance of these characteristics in the exposure period may be different between males and females. This appears to be the situation as far as the double-choice test is concerned, since males appear to more strongly impressed by the visual features of the stimulus birds (Immelmann 1959; Vos 1994), whereas females are more strongly impressed by the courtship behaviour of the males, although in the first instance they attend to visual cues (Immelmann 1959; Weisman et al. 1994). Ten Cate (1985a) argues that actual sex differences in imprintability have not been proven, since the sexes differ in the way cues are used in the final testing procedures and this is an outcome of different strategies used at pair formation. Indeed Clayton (1990b) showed that when the behavioural responses of the birds being offered for choice are controlled, there are no sex differences in the strength of the initial sexual preference in either subspecies of Zebra Finch.

Do siblings affect sexual imprinting?

Effects of siblings are not clear, and depend on the detail of experimental procedures. Immelmann (1969, 1972a) could find no effect, whereas Kruijt et al. (1983) found that males foster-reared to Bengalese Finches preferred females of that species in the double-choice tests more strongly if they had fewer siblings. However, neither sibling number nor their sex affected sexual preference in females fostered by Zebra Finches or by Bengalese Finches (ten Cate and Mug 1984), and the effect was of marginal significance in males (ten Cate 1984). On balance, the effect of siblings on sexual preferences is limited in comparison to that of the rearing parents. The only instance where siblings have a pronounced effect on sexual imprinting is when they are raised by hand; if raised with their siblings they will not imprint on humans, but on each other (Immelmann 1969). Therefore, it seems that siblings only imprint on each other when the parental object is completely aberrant.

Sexual preference—a two step process?

Although Immelmann (1969, 1972a) was initially impressed by the irreversibility of sexual preference in adulthood, he was later impressed by the extent of individual variation in the strength of the preference, especially where a few immatures fostered to Bengalese Finches could have their sexual preference switched back to Zebra Finches if exposed to conspecifics after 40 days of age (Immelmann 1975a, 1979; Immelmann and Suomi 1981). This was confirmed by ten Cate (1984) and Clayton (1987b). Other observations led to suspicions that particular experiences in early adulthood could change the direction of the preference and when this problem was investigated experimentally, both Immelmann et al.

(1991) and Kruijt and Meeuwissen (1991, 1993) produced almost identical results.

Both groups of investigators found that foster-reared Zebra Finches would alter their sexual preferences as a consequence of breeding experience in adulthood; however, this effect was blunted if given a short, prior exposure to a female of their foster-mother's species on reaching sexual maturity. The results clearly indicate that, first, the sensitive phase does not end before day 40 as originally proposed, but can extend into adulthood, and, second, the preference for the foster-parent species acquired early in development is not completely stable since it can be altered in adults. However, if adult males are given just a very brief exposure to the foster-parent species, such as that provided in preference tests, it stabilises the initial preference. Both Immelmann *et al.* (1991) and Kruijt and Meeuwissen (1991, 1993) proposed that under these experimental conditions, sexual imprinting in Zebra Finches had two stages: (a) a sensory stage that operates before 40 days of age during which information on the characteristics of the sexual object are learned and first preferences developed, and (b) a verification stage in which this information is subsequently verified during the first courtship to adult females. During verification, the initial preference is either consolidated if the female is the same species as the rearing parents, or it is modified if it is a different species.

The species of the first female courted by the males is the critical factor in the verification stage of sexual imprinting. This fact was established by Bischof and Clayton (1991) who controlled which species of adult female was first encountered by cross-fostered male Zebra Finches at 100 days of age. If a Bengalese Finch was courted first, the Zebra Finches all showed a strong preference in the double-choice test for the Bengalese Finch, since the characteristics of the initially preferred species were confirmed. However, if confined first with a female Zebra Finch, the males varied strongly in their preference owing to the conflict between the characteristics of the female first courted and those learned in the early sensitive phase. Consequently, this had a significant effect on the final sexual preference. Further experiments showed that the strength of the first preference established during the rearing phase depended on how much males begged to the foster-parents and how much they were fed by them. Moreover, the effect of the second exposure to adult females on the final sexual preference depended on the amount of courtship singing a male directed to either species. Sample sizes were quite small in this follow-up study and it requires verification. Notably, when the foster-parents were Zebra Finches, first exposure to female Bengalese Finches had no effect on the final preference; all males tested in this way ($n = 10$) preferred Zebra Finches, which suggests an own-species bias (cf. Kruijt and Meeuwissen 1993), since first courtship to Bengalese females did not modify the final preference. Nevertheless, a few individuals did court the

Bengalese Finch female, but their begging, feeding and song rates were not monitored, so that the issue of an own-species bias remains open.

These experiments have considerably advanced our understanding of the processes involved in the development of sexual preferences, but the picture is still not complete. The experiments of Bischof and Clayton (1991) need to be repeated with female Zebra Finches to ensure that the two-stage mechanism also applies to them. Also, a comparative study of male and female Bengalese Finches, in the first instance, would indicate the generality of the process, at least as far as estrildines are concerned. Additional studies are also needed that relate behavioural interactions with parents, siblings and with those first courted to the strength of the final preference. Finally, it is difficult to envisage the operation of the two-step development of sexual preference to the situation in the wild where processes must merge. There must be considerable overlap in the timing of the sensory stage and the verification stage since young begin to court before day 40 in the wild (Chapter 9). Moreover, a 'first courtship' event would difficult to pinpoint because the full sequence develops gradually and does not culminate in the song and dance until around day 60 or later, with all degrees of completion in between. In addition, individuals first courted are often immature so do not match the characteristics of the rearing species learned in the sensitive phase; exactly what is being verified and when is uncertain.

Function and evolutionary implications of sexual imprinting

Immelmann (1972b, 1975a,b) concluded that sexual imprinting was adaptive because it enabled birds to recognise members of their own species and thus ensure that sexual and pair formation behaviour were restricted to conspecifics of the opposite sex. The advantage of imprinting as a species recognition system is that it ensures that whatever characteristics are provided by the rearing parent they will be the ones preferred by their offspring at maturation. The early end to the sensitive phase and the precision and stability of the preference ensure that the species of the rearing parents are imprinted upon and preferred and the risk of 'misimprinting' on alien species is unlikely. This is especially important in species such as the Zebra Finch that have early maturation and pair formation.

Immelmann surmised that imprinting would reduce the risk of pairing with alien species; he considered this a distinct possibility because Zebra Finches form mixed-species flocks with other species of estrildines (Immelmann 1962b) and breed in mixed colonies. However, while incidences of these mixed flocks and breeding colonies are currently rare (Chapter 4), it does not mean they were not more prevalent in the past (Chapter 10).

Immelmann (1975a,b) argued that sexual imprinting was particularly important in estrildines and other groups of rapidly evolving species in which adaptive radiation produced several closely related species similar in appearance and occupying the same habitat. The importance of sexual imprinting for positive mating assortment between the two geographically isolated subspecies of Zebra Finches was beautifully demonstrated by Clayton (1990d). Most species of Australian estrildines appear to have relied on divergence of colour patterns of bill and plumage to establish species isolation, rather than differences in behaviour and vocalisations (Morris 1958). For example, colours and patterns of bill and plumage are very different between Zebra Finches and Double-barred Finches yet their vocalisations and behaviour are similar. Sexual imprinting would maintain sexual isolation between them.

In a rapidly evolving species where behaviour and appearance can change due to mutation, sexual imprinting enables sexual preferences to be altered from one generation to the next (Immelmann 1972b, 1975b). Ten Cate and Bateson (1988) have proposed that sexual imprinting could be an important component of sexual selection by leading to the evolution of conspicuous characteristics if the preferred partner is slightly novel relative to the appearance of the parent that was learned during imprinting. Indeed, the same attributes of Zebra Finch song are used for sexual selection (Clayton and Pröve 1989) and sexual imprinting (Clayton 1990d) and provide a means of investigating the relationship between the two.

Strong parallels have been drawn between song learning and sexual imprinting in Zebra Finches (reviewed by Clayton (1989b, 1994) and ten Cate *et al.* (1993)). First, both processes require a special kind of memory that is sensitive to information available over a brief period and once acquired remains stable. Second, social interactions are important in determining what is learned, from whom, and when. Initially, the sensory phase of sexual imprinting was believed to precede (and possibly guide) that of song learning, but new evidence indicates considerable overlap so that the two must be considered distinct processes, not different manifestations of the same developmental system. Moreover, the rearing mother and father have opposite roles: sons imprint on their mother's appearance and develop a sexual aversion for that of the father, yet at the same time acquire a preference for his song (ten Cate *et al.* 1993).

Positive assortative mating between subspecies

When unpaired adults of both Australian and Lesser Sundas Zebra Finches from semi-domesticated stocks were released into a single large aviary, Böhner *et al.* (1984) and Clayton (1990a) found that there was a significant tendency for birds to pair with members of their own subspecies, that is, to assort positively. Reproductive isolation occurs despite the fact that if forced to breed with one another the two subspecies produce fertile hybrids, whose offspring are also fertile (Clayton 1990d). How the subspecies discriminate between one another and why the capability should exist in the first place, since the nearest populations of the two subspecies are currently separated by at least 600 km of ocean, are questions Clayton investigated in a series of complex experiments (Clayton 1990a,b,c) at Bielefeld University and whose results are well summarised in Clayton (1990d).

Both subspecies are morphologically and behaviourally distinct (Chapters 3 and 10). Playback experiments showed that females preferred songs of their own subspecies (Clayton and Pröve 1989) and mate-choice experiments demonstrated that females preferred males with the correct breast band pattern (Clayton 1990a). Thus, at pair formation, females could use song, size and plumage to discriminate among males of the two subspecies. Clayton (1990b) set out to determine the relative importance of these attributes by means of cross-fostering experiments.

When females that had been fostered to consubspecific or heterosubspecific pairs to day 35 were subsequently implanted with oestradiol at adulthood and tested for song preference in playback experiments they preferred the song of the foster-father's subspecies and did not discriminate between males that were foster-reared or not (Clayton 1990b). Therefore, females must have learned the macrostructural features of their foster-father's song before day 35. This shows the importance of paternal influence on the learning of the female song preference and is generalised to songs of other males of the same subspecies at adulthood. Additional experiments showed that females also learned the songs of male siblings or aviary mates and this too influenced their song preferences in adulthood.

In multiple mate-choice tests, normally raised females of both subspecies preferred normally raised males of their own subspecies over those that had been cross-fostered. Females must have discriminated between them using either behavioural cues, or a combination of visual and vocal cues (Clayton 1990d). How this preference developed was subsequently examined in cross-fostered and normally reared birds where the sexual preference was tested in the multiple choice apparatus and subsequently confirmed in pair-formation trials conducted in aviaries (Clayton 1990a). Preference of the normally reared group was stronger than the foster-reared group so that during pair formation preferences of the former prevailed over those of the latter, especially in females, since they have a more decisive role in pair formation than males (Chapter 9). The important finding is that the behavioural differences in subspecies sexual preference have arisen from sexual imprinting, and this is probably the factor that maintains sexual isolation between Australian and Lesser Sundas Zebra Finches in captivity. Plumage and macrostructural song cues are both learned for subspecies recognition, but neither are altered by rearing experience.

Clayton (1990d) considered that the cues used for assortative mating of subspecies might also be available for mate choice within species. For example, macrostructural differences in element number are used to distinguish subspecies but the same cues are also used to discriminate among singers of the same subspecies in playback experiments (Clayton and Pröve 1989). Thus, the same cue is used for both sexual imprinting and sexual selection. Bill colour and breast band size may also fall into this category.

Mate choice

At pair formation, Zebra Finches usually have a choice of unmated individuals with whom they can pair and it is critical that they make the correct choice, especially since the pair bond lasts for life, and it is rare for individuals in the wild to outlive more than three or four partners (Chapter 6). This means choosing and pairing with the highest 'quality' individual available in order to maximise fitness. Presumably, phenotypic indicators of genetic quality and reproductive and rearing competence should be the cues used to attract partners. Optimising choice of a partner should be a concern for males as well as females, since males make an almost equal contribution to the reproductive effort of the pair, and, indeed, this is the case (Chapter 9). Nonetheless, Zebra Finches are sexually dimorphic indicating that some form of sexual selection is operating. For this reason they have been chosen as subjects by evolutionary biologists, notably by N.T. Burley and her students in the United States, who are interested in mate choice and sexual selection in monogamous species.

Choice of mates and copulating partners

Sex ratios in flocks and breeding colonies fluctuate from month to month in response to changing patterns of mortality and immigration (Chapter 8) so that the extent of choice varies significantly from day to day and may even be affected by seasonal factors. Occasionally, through lack of available options, some individuals may be forced to pair with partners of lesser quality. Under these circumstances it is adaptive to seek matings outside the pair bond with individuals that are superior to the current partner and this presents birds with another choice, namely, which member of the opposite sex should be chosen for these extra-pair copulations. Opportunities for extra-pair copulations will be high for males and fairly continuous throughout the long breeding season, except for the fertile period of their own partners when they are constrained by supply of sperm and mate-guarding duties (Chapter 9). Benefits of extra-pair copulation are obvious for males in that they can increase their reproductive success without any additional parental investment. Females, in contrast, have fewer opportunities for extra-pair matings given the close guarding by the partner and the fact only about one to three fertile periods occur each breeding season. A number of benefits have been hypothesised for female extra-pair copulations, including the most likely ones of 'good genes', and 'genetic diversity' (reviewed by Birkhead and Møller (1992)), but the pattern of occurrence of extra-pair copulations from the Danaher colony do not support either of these (Birkhead *et al.* 1990). The relative importance of the two types of mate choice should also be kept in perspective: DNA finger printing at the Danaher colony found that only 12% of females had extra-pair offspring and these constituted only 2.4% of all offspring sampled (Chapter 9). Thus, the choice of a mate is a critical decision for both male and female Zebra Finches.

Laboratory experiments on domesticated Australian Zebra Finches have investigated both types of heterosexual choices: partners for the pair bond and partners for extra-pair copulations. Choice of pair bond partners has been investigated in multiple-choice arenas that allow two-way interaction between the choosing individual and those offered for choice. Preferences are normally measured by timing how long the choosing bird spends in front of the cages of the various individuals on offer and subsequent experiments have shown that this measure of attractiveness is a true indication of the pairing preference (Burley *et al.* 1982; Clayton 1990a,b); similarly, scoring of Directed Song is another means of determining pairing preference (ten Cate 1985b). There are three important questions: is there agreement on which individuals are the most preferred, and if so, what cues make them attractive, and are these cues reliable indicators of quality? So far researchers have concentrated on female preferences for male Zebra Finches.

Female Zebra Finches require males of a minimum standard of quality but their preference for a mate is relative to recent experience (Collins 1995). That is, if a female had recently experienced a high quality male she will not accept a lesser quality one shortly afterwards.

Females prefer males with deep red bills

Females prefer those males with the reddest, brightest bill colours, and males that had the bill painted to a 'super red' shade were preferred over males that were painted with deepest naturally occurring shades (Burley and Coopersmith 1987; but see Sullivan (1994) and Weisman *et al.* (1994) where orange-billed males were preferred). Hence, for females, the redder the bill the better, whereas males prefer females with the standard orange-red colour and avoid females whose bills are too red or too yellow. In a similar series of preference tests, in which bill colours of males were not artificially altered, Houtman (1992) likewise found a positive relationship between bill redness and attractiveness. Rate of singing was also positively related to attractiveness, and could compensate for an unattractive bill colour (Collins *et al.* 1994).

Is bill colour an indicator of male quality? Colour of the bill in wild and domesticated Zebra Finches varies among individuals but is consistent within them, although it changes during the course of reproduction (Burley *et al.* 1992). Bill colours of domesticated and wild Zebra Finches have a similar range and distribution of shades; they are redder and darker (higher colour scores) in males and more orange and paler (lower scores) in females, but there is some overlap between them. Bill colour scores of wild birds decline over the breeding season in populations from central Australia and northern Victoria (Burley *et al.* 1992; Zann 1994b). Among laboratory birds, significant declines in bill colour and weight occur over a five-week breeding cycle in both sexes, and colours quickly recover if breeding is prevented. Other factors also affect bill colour. For example, at the end of a drought in central Australia all birds had pale bills, but these increased in colour when birds were held in brief captivity on an enriched diet. Given the above effects, Burley *et al.* (1992) concluded that bill colour is an indicator of quality but is affected by physical condition which is lowered by reproductive effort and harsh living conditions. However, contrary to expectation, bill colour in wild females was positively correlated with ectoparasite loads, but experiments are needed to determine how parasite load affects the physiology, bill colour and mate choice before this finding can be fully interpreted (Burley *et al.* 1991). Despite the dynamic quality of bill colour, the optimally preferred colour for members of the opposite sex is attained just before, and during egg-laying, that is, during the female's fertile period. This is the time when both partners need to be at their most attractive; the male needs to keep his mate from seeking extra-pair copulations, but if she does, she may need to attract the best quality male.

In an aviary study, Burley and Price (1991) found that females invited, or tolerated, extra-pair copulations from only a subset of those males available, namely those with redder bills and larger breast bands than possessed by the female's own mate. Bill colour is strongly heritable in domesticated Zebra Finches, but evolution is only proceeding slowly towards optimal sex-specific colours because it is constrained by low dominant genetic variance and opposing selection in males and females (Price and Burley 1993).

Females prefer males with high song rates

In multiple choice tests, Houtman (1992) found that females preferred some males over others and there was agreement on which males were most attractive. These were males with high rates of directed singing. Later these same females invited extra-pair copulations from males that had higher rates of singing than their own partner, while those with lower rates were ignored or deterred. High song rate was not a consequence of female preference (Collins 1994). In follow-up cross-fostering studies Houtman (1992) found significant heritability in singing rates between father and son and between brothers. Consequently females that copulated with frequently singing/courting males could expect to have sons that also sang frequently and were attractive to females. Whether it is the song component of the courtship display and/or the dance component that makes males attractive is unclear. Playback experiments by Clayton and Pröve (1989) showed that females prefer long complex song-phrases. However, in these experiments song rate was controlled.

Singing rate, like bill colour, is high in males in good condition. Houtman (1992) found singing rate was positively correlated with the amount of fat stored around the clavicle, which is one indication of physical condition. Another correlation was found between singing rate and the mean weight of the singer's offspring at independence. Presumably, only competent fathers in good condition can produce heavy offspring to independence.

Sheridan (1985) attempted to determine which component of courtship females found the most attractive, the song or the dance. Despite efforts to court and sing, castrated and muted males were not as attractive as intact males which courted more frequently and gave the full display.

In a semi-natural aviary study, Ratcliffe and Boag (1987) could find no correlation between song rates and male mating success; instead, aggression (supplantings) was strongly correlated with success. Their experiments were designed to heighten aggression and competitiveness among males and this may have obscured any effect of singing.

Effects of hormones on male attractiveness

Rate of singing and bill colour were positively correlated in Houtman's (1992) study. This is not surprising because both depend on levels of testosterone in the blood. In laboratory and wild Zebra Finches rates of Directed and Undirected Song were positively correlated with testosterone levels (Chapter 10), theoretically enabling females to assess the underlying physiological state of males they encounter. Predictably, when males are castrated they lower their rates of singing and courtship; if subsequently implanted with testosterone, the effects are reversed (Arnold 1975a; Harding *et al.* 1983). Similarly, after castration the bill colour changes from red to the orange colour typical of females, but testosterone implants quickly restore the red colour typical of males (Cynx and Nottebohm 1992b). Sheridan (1985) enhanced the attractiveness of males by implanting androstenedione (an androgen), but could find no increases in singing and dancing performance that could be serve as cues for the females; unfortunately, he did not examine changes in bill colour, which may be more sensitive to hormones than sexual behaviour and could have been the cue used by females in this experiment. In both sexes there is a need to examine in detail the nature of interactions among hormones, physical condition, bill colour and courtship behaviour.

Male choice of females

In aviary studies, males preferred extra-pair copulations with females that had the standard orange-coloured bills (Burley and Price 1991) and oddly, in both males and females, there was a strong tendency for individuals that had high reproductive success within the pair to be engaged in extra-pair courtships and copulations. That is, individuals with a high reproductive success were preferred as partners in extra-pair copulations. Double-choice tests show that male choice of good-quality females can be just as discriminating as female choice of males (Wynn and Price 1993).

Are close relatives preferred as mates?

When quadruplets of domesticated Zebra Finches were offered a choice of pairing with very-close relatives or non-relatives Slater and Clements (1981) found, unexpectedly, that most pairings were with relatives. Specifically, there was a strong tendency for preferential pairing between mothers and sons, a weaker one between fathers and daughters, and a slight tendency for pairing between siblings. Two groups of investigators attempted, unsuccessfully, to confirm this intriguing result. They used relatives other than parents and offspring because the close temporal association between parents and offspring during rearing could bias pairing tests. Schubert *et al.* (1989) offered females the opportunity to pair with siblings, first cousins, or unrelated males and found that there was no significant preference or aversion for any type of pairing. Similarly, Fetherston and Tyler Burley (1990) found no significant bias towards pairing with siblings among males or females in their study. In a separate study, where perching preference was tested in a multiple-choice arena, Burley *et al.* (1990) also found that there was no discrimination between siblings and non-kin; however, there was a significant preference in females, but not males, for first cousins over unrelated males (Chapter 9). It is likely that this is a pairing preference.

Schubert *et al.* (1989) and Fetherston and Tyler Burley (1990) also found that the reproductive success of sibling pairs was inferior to other pairs in some measures, but not others, and the differences were neither large nor consistent. This suggests that sib–sib breeding has slight negative effects on reproductive success, but the extent to which this is an outcome of initial inbred aviary stock is unknown.

All four pairing studies concur in the finding that domesticated Zebra Finches do not have a mechanism for avoiding incest and inbreeding, but Slater and Clement's (1981) conclusion that birds prefer to pair with very close relatives is not supported. This is consistent with the study of wild Zebra Finches at the Danaher colony: no parent–offspring pairings were detected over four breeding seasons, although one sib–sib pairing was discovered (Chapter 6). While high mortality produces a continuous supply of unmated wild Zebra Finches the extreme mobility of members of even the most permanent colonies makes it fairly unlikely that first order relatives will pair up very often. Even if relatives form pairs, it seems that penalties are slight as far as reproductive success is concerned, but there is a need for an extensive series of laboratory experiments on wild-caught and first-generation aviary-bred birds to verify this.

Band colours affect mate attractiveness

Placement of coloured plastic leg bands in unique combinations on Zebra Finches and other species of birds is a standard procedure researchers and aviculturalists use for identifying individuals from a distance, and it was assumed that bands would not alter behaviour. In 1978, when establishing a new colony of domesticated Zebra Finches, Nancy Burley noticed that individuals that had bands of certain colours tended to pair and breed first, while those with certain other colours paired up last and had poor breeding records (Burley *et al.* 1982). On the basis of this astute observation she established a long series of preference experiments using a multiple choice arena that allowed two-way interaction, and tested the hypothesis that leg bands of certain colours are attractive to Zebra Finches and other colours are unattractive. She found that females preferred males with red bands the most, and those with light-green bands the least, and did not distinguish between unbanded birds and those with orange bands. Males preferred females wearing black bands the most and those with light-blue bands the least and did not distinguish between unbanded birds and those with orange. Bands of other colours were neutral in attractiveness. Burley's (1981a) initial report of these findings attracted criticism on the methodology and interpretation of the results from Immelmann *et al.* (1982) and Thissen and Martin (1982). These criticisms were initially answered, in part, by Burley (1982), and later in more comprehensive fashion when the full investigation was published (Burley *et al.* 1982). Effects of band colours on attractiveness for birds of the same sex went in the opposite direction to the heterosexual preferences: females avoided those with black bands and males avoided those with red bands (Burley 1985b). In another series of experiments males and females were given heterosexual tests in which the stimulus birds were either unbanded or had either red or light-blue bands. Males and females preferred the those with red bands the most and those with blue bands the least (Burley 1986a).

Preference for red band males was also shown by recently captured wild females (Burley 1988b) and although males were not tested it was assumed their colour band preferences were no different from their domesticated counterparts. Therefore, under captive conditions colour bands bias the sexual attractiveness of the wearers. Since this fact became known researchers have been careful when investigating mate choice and other behaviour in Zebra Finches to avoid bias and have omitted red and light-green bands on males and black and light-blue bands on females. The finding also stimulated field ornithologists to examine, in retrospect, whether colour banding biased behaviour of their wild subjects, and new experiments were established to investigate possible effects.

In considering why these band colours affect attractiveness in Zebra Finches, Burley *et al.* (1982) noticed that the attractive colours were present on the bill of the male (red) and the plumage of both sexes (black), whereas the least preferred colours (light-blue and light-green) were not represented, and concluded that the attractive colours probably enhance pre-existing preferences for these features in the opposite sex. The ultimate causes of these preferences and aversions are unknown, but band colour experiments with Double-barred Finches suggest that colour band biases reflect naturally occurring differences in species colouration. Male and female Double-barred Finches preferred members of the opposite sex wearing light-blue leg bands and avoided those with red bands; there were no sex-specific preferences (Burley 1986a). The bills of the sexually monomorphic Double-barred Finches are a bluish-grey (Immelmann 1965a) and Burley (1986a) postulated that their preference for blue bands and their aversion for red bands and the corresponding preference of Zebra Finches for red bands and their aversion for blue bands is not a coincidence but is related to selection for preference for their own species. Each species is biased towards colours specific to themselves and biased against colours specific to their congeners and these biases may have reinforced differences in plumage and bill colour that arose during the speciation.

Burley (1986a) hypothesised that male aversion for males with red leg bands was possibly a means of avoiding more competitive or sexually attractive individuals. However, in an aviary study designed to verify effects of band colour on male competitive ability, Ratcliffe and Boag (1987) could find no significant difference between males with red bands and those with light-green bands in their ability to compete with one another for nest sites and females. While conceding that Zebra Finches may be sensitive to band colours, they concluded that any effects are obscured, or overridden, by intra-sexual competition and preferences for other attributes, especially behavioural ones.

Females prefer symmetrically banded males

Recently Swaddle and Cuthill (1994a) showed that female Zebra Finches can detect and respond preferentially to males wearing symmetrically placed colour bands over males wearing asymmetrically placed ones. In a multiple-choice apparatus, females were given a choice of males, each wearing two orange and two pale green bands, but each with the bands placed in one of the six possible combinations. Ten females were tested and found to significantly prefer those males that had the two combinations that were symmetric. Since each of the six combinations had precisely the same number of bands of the same colours it is the bisymmetric pattern that is preferred.

Female preference for symmetrically banded males is believed to be a manifestation of a general female preference for bilateral symmetry in the sexual characteristics of males (Møller 1990). It is hypothesised that males of superior phenotypic and genetic quality have ornamental traits that are more symmetrical in pattern than those of inferior males because of their superior ability to buffer themselves against environmental or genetic stress during development. With Zebra Finches there are two sets of paired secondary sexual characteristics found in males, namely the chestnut-coloured ear coverts and the flank marks, and it is reasonable to hypothesise that females would prefer males with larger and more symmetrically shaped and positioned markings. Recently, Swaddle and Cuthill (1994b) have demonstrated that female Zebra Finches prefer males with symmetrical patterns to the fine black bars on the chest and foreneck over those with asymmetrical patterns.

Females mated to green-banded males have more extra-pair copulations

Females are active in their choice of extra-pair copulatory partners. Recent aviary studies showed that females mated to unattractive (light) green-banded males participate in more extra-pair courtship and extra-pair copulations than do females mated to attractive red-banded males (Burley *et al.* 1994). This is not due to differential sexual attractiveness, since males did not discriminate, but is related to the willingness of females mated to green-banded males to participate in extra-pair relations and the corresponding unwillingness of those females mated to red-banded males. The latter are believed to reject advances from extra-pair males in order to increase the paternal confidence of their attractive mates and thereby obtain assistance with rearing the offspring. Unattractive green-banded males are doubly disadvantaged: they have a low rate of success in extra-pair copulations yet their partners have a high rate of participation in extra-pair copulations.

Effect of band colours on physical condition and mortality

After establishing that Zebra Finches are sensitive to certain colours of plastic leg bands worn by the opposite sex, Burley (1985a) began to investigate the implications for the breeding biology and initiated a series of long-term (15–22 month) breeding experiments in aviaries. In the banded-male experiment, 24 males were allocated either all red bands, all light-green bands or all orange bands at random so that when 24 unbanded females were released into the aviary they had the opportunity to pair with males with attractive bands, unattractive bands, or bands neutral in attractiveness, respectively. In the complementary banded-female experiment, 24 unbanded males could pair and breed with 24 females banded with either all black bands, all light-blue bands or all orange bands, thus presenting males with females that were attractive, unattractive or of neutral attractiveness respectively. Over the experimental period there was a significant effect of band colour on mortality rates: red-banded males lived longer than light-green-banded males, black-banded females lived longer than light-blue-banded females, and orange-banded birds of both sexes had intermediate life spans. Causes of mortality could not be clearly established, but Burley hypothesised that they were related to reproductive stress arising from reproductive effort. Before birds died they showed signs of physical decline (weight loss, loss of bill colour, poor feather condition); unfortunately, physiological symptoms of stress, such as the level of blood plasma corticosteroids, were not examined.

I attempted to verify the effects of leg band colours on survivorship and physical condition of wild Zebra Finches trapped at the Cloverlea colony in northern Victoria (Zann 1994b). One day each month for 32 successive months birds were captured by walk-in trap. In addition to their uniquely numbered metal band, unbanded males were allocated, at random, either three red plastic bands (attractive), or three light-green plastic bands (unattractive); females were given either black bands (attractive) or light-blue bands (unattractive). At subsequent recaptures birds were scored for physical condition (wing moult, weight and bill colour). Scores taken before and after intervals of breeding showed that bills faded in colour in both sexes, but there was no significant effect of band colour on fading, nor were there any significant band colour effects on weight and rate of moult in either sex. Furthermore, recapture rate over a 12 month interval (23%, $n = 44$) did not differ significantly between the sexes nor between the categories of band colours within the sexes. I concluded that under these field conditions any differential reproductive costs that arose as a consequence of band colours had no discernible effect on physical condition, and consequently, no effect on mortality.

Irrespective of any colour band effect, I predicted that costs of reproduction would be higher in the wild population than in Burley's (1985a) captive birds, since energetic demands of foraging, nest building, provisioning of young, predator avoidance and thermoregulation, and so on, would be greater in free-flying birds. However, Burley's birds bred continuously for 15–22 months whereas my birds bred for a median duration of three months (range 2–8 months). The breeding season at Cloverlea extends for eight months and the estimated number of breeding attempts, based on the nearby Danaher colony, would range from one to six with a mean of two (Chapter 7), although some birds could have bred elsewhere before and after their tenure at Cloverlea. Therefore, while costs of reproduction would be higher per breeding attempt in the wild, the continuous uninterrupted episodes of breeding in captivity would not only prevent recovery, but successive breeding episodes would become increasingly devastating so that symptoms of stress would suddenly set in and death would follow quickly (Burley 1988a).

It is also possible that wild birds are more resistant than domesticated birds to any debilitating effects of reproduction since aviculturalists are advised not to let domesticated Zebra Finches breed excessively so as not to 'weaken the birds' (Immelmann 1965a). A comparison of reproductive costs of continuously breeding wild-caught and domesticated Zebra Finches and the associated physiological causes would be worth investigating.

Effect of band colours on reproductive success

In the banded-male experiment and the banded-female experiment, attractive males (red banded) and attractive females (black banded) respectively, reared about twice as many young to independence than their unattractive counterparts (light-green banded males and light-blue banded females) (Burley 1986b). Given the effort involved in rearing these additional young, one might expect birds with attractive bands to suffer higher costs and die sooner, but they lived longer. Burley (1986b) proposed two non-exclusive hypotheses to explain the phenomenon: (a) the 'differential access hypothesis' in which birds with attractive bands have the opportunity to chose superior-quality members of the opposite sex, which have a high reproductive output and are capable of tolerating high reproductive costs without much stress; and (b) the 'differential allocation hypothesis' in which an unattractive individual pairs with a bird with an attractive band and compensates by taking a disproportionate share of the parental investment and produces more offspring per breeding attempt. This unattractive individual thus incurs heavy costs and associated reproductive stress, which reduces life span. Indeed, time-budget estimates showed that parental investment of attractive birds was less than that of unattractive birds (males 40% *vs.* 51%, females 52% *vs.* 64%; Burley 1988a). Moreover, four red-banded males restricted their parental investment sufficiently to allow them to pair and breed with a second female. The fact that the second female was attracted to an already mated male is evidence for the differential access hypothesis. Further evidence for this hypothesis came from the finding that reproductively successful attractive females in the banded-female experiment had males that did not have increased mortality which indicates that these males were of better quality.

In an attempt to discover if reproductive effort of wild Zebra Finches was also affected by band colour, I carried out a retrospective analysis of the 'lifetime' reproductive effort of 144 pairs of Zebra Finches studied during their tenure at the Danaher colony over a four-year investigation (Zann 1994b). Here each individual had a unique combination of three colour bands and one metal band, and I compared the reproductive effort of pairs where males had one red (attractive) band ($n = 35$) in their combination with those having one light-green (unattractive) band ($n = 35$) and those having neither red or light-green (neutral attractiveness) bands ($n = 34$). Males with both red and light-green bands ($n = 9$) in their combination were excluded from the analysis. Reproductive effort of females with a black (attractive) band ($n = 42$) were compared with those without black bands (neutral attractiveness) in their combination ($n = 63$). No birds at Danaher wore light-blue bands (unattractive on females) as they were not approved by the Banding Scheme for allocation.

Females paired to red-banded males laid significantly more eggs, had larger clutches and more clutches than females paired with light-green banded males and males that had no red or light-green bands (Zann 1994b). There were no band colour effects among females except that those pairs where the female had a black band fledged significantly fewer young. Obvious biases in the allocation of band colours were not found, nor were any found in the detection of breeding attempts of pairs belonging to different band colour categories. The results, if true, are remarkable because only one red band produced the effect in males and 66% of them were estimated to have been paired before banding so that females must have adjusted their reproductive effort after the male received the red band. This finding is consistent with that of Burley (1986b) for domesticated males and supports the differential allocation hypothesis. Further evidence in support of this hypothesis has recently been found in Barn Swallows *Hirundo rustica* where females respond to changes in male ornamentation (De Lope and Møller 1993).

At the very least, the band colour study at Danaher indicates that female sensitivity to red bands does exist in free-living Zebra Finches and despite the fact that they do not affect reproductive success, female physical condition or mortality, it would be prudent to omit red bands from male band combinations in future. It would also be prudent to allocate all combinations in a truly random fashion.

Effects of band colours on sex ratio of offspring

In her original paper, Burley (1981a) reported findings consistent with the hypothesis that pairs of Zebra Finches biased offspring production in favour of the sex of the more attractive parent. This intriguing idea was supported by the more extensive data set generated by the banded-male and banded-female experiments (Burley 1986c). The sex ratio of nestlings is believed to be modified by brood reduction soon after hatching. Nestlings are thought to be sexed by Begging Calls and those belonging to the sex of the less attractive parent are starved or removed from the nest. Burley cites unpublished research by R. Balda and J. Balda in which sex differences in the frequency of utterance of 22 different types of nestling Begging Calls differed between the sexes. Nestlings of first and second generation offspring of wild-caught Zebra Finches also have sex differences in Begging Calls; specifically females have a greater range of call variants than males, but no calls are unique to one sex (Roper 1993).

Brood reduction does occur in wild Zebra Finches (Chapter 9), but there is no evidence for or against facultative manipulation of numbers of nestlings in broods. However, if resources are limiting and it is necessary to reduce the size of the brood, selection should favour a mechanism to bias the sex ratio towards that sex that would achieve the highest fitness. Under most conditions the sex of offspring likely to inherit the attractive characteristics of the most attractive parent should be favoured, although in locations where there is a local imbalance of adults of one sex (see Burley *et al.* 1989) fitness would be higher by favouring the nestlings of the sex in most demand.

Until a cheap molecular technique is available to sex recently hatched Zebra Finch nestlings it is difficult to envisage field investigations of sex ratio bias comparable to those performed by Burley. High rates of nest predation and fledgling mortality would make gathering of such data arduous at best.

Summary

Zebra Finches are monogamous and form lifelong bonds in which both sexes participate in all aspects of parental care. Consequently, choice of the best quality partner is of major importance for the fitness of males and females. Members of the opposite sex are also chosen for participation in extra-pair copulations. Females prefer males with deep-red bills for both pair formation and for extra-pair copulations. Redness of the bill is inherited and indicates physical condition, but is subject to fading as a consequence of reproductive effort and harsh environmental conditions. Females are also attracted to those males that show courtship vigour and sing frequently; these males are preferred for extra-pair copulations. A high song rate is heritable and is also correlated with good physical condition. Both red bills and high song rate are also correlated with high levels of testosterone so that both attributes may indicate good physical condition, strong courtship vigour and high intra-male competitive ability.

Despite visual and auditory cues for kin recognition, Zebra Finches have no biases for, or against, pairing with relatives, and a brother–sister mating has been found in free-living birds. High mobility and dispersal makes sib–sib pairing infrequent. Inbreeding depression in domesticated Zebra Finches has not been clearly established.

Wild-caught and domesticated Zebra Finch females are attracted to males wearing red bands and avoid males wearing light-green bands; males are attracted to females wearing black bands and avoid those with light-blue bands. Colour preferences may be based on biases towards their own species-typical colours and against species-typical colours found in closely related species. Females prefer males that have bands placed in bilaterally symmetrical arrangements. It is hypothesised that males of superior quality have their secondary sexual characteristics arranged in a more symmetric fashion than do inferior individuals and that band preferences are an expression of this. In domesticated Zebra Finches, coloured leg bands affect mate selection, extra-pair copulations, parental investment, reproductive success, sex ratios of offspring and longevity. Experiments on wild Zebra Finches did not detect colour band effects on mortality or physical condition, perhaps because breeding seasons were not as long as those possible in captivity. However, in wild pairs where males have red bands, there was a significant increase in the reproductive success (eggs, clutches). In order to prevent behavioural artefacts in field and laboratory studies, red bands are not recommended for male Zebra Finches.

12 Life history and adaptations

'The large fluctuations of the Zebra Finch population at Jindi Jindi resembles that of the grasshoppers and suggests that both breed prolifically when conditions are good; most die but a few survive over the dry years. Autumn breeding is presumably an advantage in a species that can breed at three months of age because, when numbers are very low, the survival of even a few will boost the potential of the breeding population in the spring.'

S. J. J. F. Davies (1986).

The primary aim of the final chapter is to integrate the various anatomical, physiological and behavioural attributes of Zebra Finches in order to understand the principal thrusts of their life histories and adaptations.

Both subspecies of Zebra Finches must be considered successful in their respective environments because both are relatively numerous, wide-spread in distribution, and persistent in a rapidly changing environment. Understanding their life histories is the key to understanding why they are successful. Unfortunately, few details of even the most basic aspects of the field biology of the Lesser Sundas subspecies are known. Consequently, little can be said about its microevolutionary adaptations to a highly seasonal environment. Whereas Immelmann (1962a, 1965a) described the basic breeding ecology of almost all of the Australian estrildines, little has been published on their demography, with the exception, of course, of the Zebra Finch. Demography is the key to understanding life histories (Stearns 1992), but unfortunately demographic data are only available from just a few locations within the continent-wide range of the Zebra Finch so that the extent of local adaptations in demographic and other traits in different environments are only vaguely suggested at this stage.

The demographic traits listed in Table 12.1, measured principally from the Danaher colony in northern Victoria, show that the Zebra Finch life history is distinguished by an exceptionally early age of maturation, high levels of pre-adult mortality, and multi-broodedness.

Precocial breeding

Age of maturation is a pivotal trait in life history evolution because fitness is more sensitive to this trait than any other (Stearns 1992). The benefits of early maturation are shorter generations and higher survival

Table 12.1 Demographic traits significant in the life history of Australian Zebra Finches.

Trait		Value or Range
Breeding nests suffering predation		66%
Juvenile mortality (fledglings lost before 35 days of age)		67%
Life expectancy at hatching		51 days
Fledglings that reach breeding age (percentage banded that reach day 80)		22%
Adult recapture rate[a]	Northern Victoria	14–19 months
	Alice Springs	2–7 months
Annual adult survival		4–28%
Age at first breeding in both sexes		2–3 months
Maximum life span	Northern Victoria	5 years
	Alice Springs	1.3 years
	Mileura[b]	3–5 years
'Life-time' reproductive success at the Danaher colony[c]		
	males	1.7 ± 2.5 (0–13)
	females	1.7 ± 2.5 (0–11)
Clutch size ($\bar{x} \pm s.d.$)		5.0 ± 0.98
Number of breeding attempts per season[d]	males	1.9 ± 1.7 (1–6)
	females	1.7 ± 0.9 (1–4)
Interval between successful clutch initiations		52 ± 16 days
Secondary sex ratio (percentage males at independence)[e]		44–60%
Tertiary sex ratio (percentage males at adulthood)[e]		52%

[a] Time (months) last 10% of birds recaptured after banding.
[b] Davies (1986).
[c] Mean ± *s.d.* (range) number of young raised to independence at the Danaher colony.
[d] Mean ± *s.d.* (range)
[e] Northern Victoria, Alice Springs, Top End, Mileura.

to maturity because the juvenile period is shorter. Selection on age of maturation is strong in populations that are growing rapidly and selection on survival and fecundity is relaxed.

Zebra Finches at the Danaher colony, where the breeding season is extended to seven to nine months, can produce at least two generations of descendants before the end of the season. Zebra Finches in central Australia, where continuous breeding can occasionally extend for up to 15 months (Chapter 7), could theoretically produce four to five generations in one breeding episode, but data have not yet been gathered to confirm this. Breeding at two to three months of age appears to be a widespread trait in Zebra Finches, and is not restricted to any geographic location—it has been observed in all three widely scattered populations where breeding has been studied: York (Western Australia), Armidale (New South Wales) and Wunghnu (Victoria). Precocial breeding also occurs in domesticated Zebra Finches as well as in offspring of wild-caught birds. Precocial breeding can make an important contribution to population growth, for example, at the Danaher colony, precocial breeders constituted 44% of the pairs breeding in the second half of the breeding season and their reproductive success was not significantly different from that of older pairs (Chapter 6). However, intergeneration and intrageneration effects of precocial breeding on survival and reproduction need to be investigated; findings on other species of birds consistently show that older or more experienced adults produce more young than novice breeders (Rowley 1983; Newton 1989).

The fact that precocial breeding only occurred at the Danaher colony among those young hatched in the first half of the season implies trade-offs between current reproduction and survival (and future reproduction). Young hatched in the second half of the season appeared to mature as rapidly as those hatched in the first half of the season, except for males that hatched very late in the season (Chapter 3). Individual variation in age of first breeding among those hatched in the first part of the season was pronounced (see Figure 6.5), but the reasons for this are unknown. Food supply and other determinants of parental and juvenile mortality are likely to affect decisions about whether to make precocial breeding attempts or not. The experimental supplementation of abundant supplies of dry seed did not stop the postponement of breeding by those hatched in the second half of the season. This suggests that survival and condition of potential first-time breeders may not be the deciding factor in whether to delay breeding or not. However, availability of half-ripe seed in the second half of the season could be the crucial difference, since it may enable potential breeders to predict levels of juvenile mortality likely to be incurred by a late breeding attempt. We can only speculate on what are the proximate mechanisms involved here.

Although comparative data for other species of Australian estrildines are few, it appears that precocial breeding may be exclusive to Zebra Finches. Immelmann (1962a, 1965a) mentions instances of precocial breeding among wild Gouldian and Painted Finches, but this has not been confirmed for wild or captive birds (S. Tidemann, pers. comm.). Failure to make precocial breeding attempts may be a consequence of sexual immaturity or lack of suitable environmental conditions. In captivity, where conditions for precocial breeding should be optimal, aviculturalists report that other species of Australian estrildines, including both subspecies of the Double-barred Finch, first breed at six to nine months of age (Immelmann 1965a; Queensland Finch Society 1987; R. Kingston, pers. comm.).

During histological and physiological investigations of testes maturation in a range of tropical species of estrildines, Sossinka (1975, 1980b) found that only *Amadina erythrocephala*, an arid-adapted African species, had a very rapid rate of maturation comparable to that of the Zebra Finch; other species took from four to nine months to reach full maturation. Sossinka (1975, 1980a) found that early sexual maturation occurred in both domesticated Zebra Finches and offspring of wild-caught birds. There was little individual variation and maturation rate was not affected by a range of captive conditions, including day length. Sossinka concluded that early maturation was 'internally programmed' and unresponsive to environmental conditions, at least the ones he tested. Therefore, failure to make a precocial breeding attempt by birds hatched in the second part of the breeding season at the Danaher colony is unlikely to be due to immature gonads. Sossinka believed that other species of estrildines were not physiologically capable of accelerated sexual maturation because of the energetic demands of moult, although there is no evidence for this assertion. Given the very rapid gonadal maturation of *A. erythrocephala* reported by Sossinka (1980b) one would predict that both it and its congener, *A. fasciata,* would be precocial breeders, but there is no evidence to date that this occurs in the wild.

But for the Zebra Finch, it is unlikely that many opportunities would be available for young of most species of estrildines to breed in their season of hatching. Precocial breeding in the Zebra Finch depends on two physiological conditions and one environmental one. First, rapid gonadal maturation is a consequence of a greatly abbreviated juvenile refractory period in the maturation of the testes in males (Sossinka 1975), and an accelerated growth of a few ovarian follicles in females (Sossinka 1980a). Secondly, rapid growth enables subsequent surplus energy to be allocated to reproduction and the slow continuous method of moulting (Chapter 3) may spread the energetic cost so that it does not interfere with breeding. Finally, early sexual maturation will be fruitless unless environmental conditions are still suitable for breeding, namely a ready supply of ripe and half-ripe grass seeds for feeding young. The fact that Zebra Finch nestlings do not require insect food means that breeding is not tied to those times of the year when insects are available. While breeding occurs in most months of the year in many parts of the distribution, the extent to which any one breeding episode extends for more than four or five months may be quite rare. Cultivation in some regions has permitted breeding episodes of eight to nine months on a regular annual basis, but, without irrigation, grass seed production and the breeding season would be truncated in all but the exceptionally good seasons. Nevertheless, precocial breeding in Zebra Finches has been reported for non-irrigated areas, namely York and Armidale.

Patterns of mortality among adults and juveniles are important factors in selection for early maturation and may possibly account for the fact that other species of estrildines have not attained sexual maturity as early as the Zebra Finch.

Mortality patterns

Nest predation (66%) is quite high among Zebra Finches breeding at the Danaher colony in northern Victoria (Table 12.1) and is comparable to that found among breeding birds at Alice Springs (R. Zann and N. Burley, unpublished observations). In general, intensity of nest predation of passerines is high in Australia and variation in reproductive effort of the 'old Australian endemic' species of passerines occurs through number of breeding attempts rather than clutch size (Rowley and Russell 1991). High re-nesting rates in Zebra Finches may be a response to predation levels, and the fact that their eggs are smaller than expected for estrildines of their mass (Chapter 6) may be an adaptation for this.

Nest predators in different parts of the range appear to be well adapted to exploiting Zebra Finches. They not only take a heavy and constant toll of eggs and nestlings from permanently established colonies that have regular breeding seasons, but in regions where breeding is unpredictable in time and space predators quickly track and find nesting colonies and rapidly exploit them.

The next most distinguishing feature of the Zebra Finch life cycle, after the early age of maturation, is the extremely high rate of juvenile mortality. The proportion (22%) of fledglings that appear to reach breeding age (day 80) is well below the 42–86% range for a number of different passerine species summarised by Newton (1989). Risk to juveniles is highest between fledging and day 35, the age of nutritional independence (67% losses), but after independence they grow out of risk so that losses between independence and day 80, the age of sexual maturation, fall to 27%, which is not distinguishable from that of adults (see Figure 8.2). Losses before day 35 are mostly mortalities since dispersal by young before that age is unlikely; however, estimates of losses from independence to sexual maturity confound disappearances due to dispersal with those due to mortalities. Nevertheless, mortalities before maturation must be very high and 'instantaneous juvenile mortality' is believed to be the major selective force in promoting and maintaining early maturation (Stearns 1992).

In modelling the relationship between environmental conditions and early maturation, Stearns and others assumed that growth rate of juveniles was a potential cue for monitoring prevailing demographic conditions: when growth slowed down there were increases in juvenile mortality rates and this was compensated by delayed sexual maturation. However, early sexual maturation in Zebra Finches appears be a fixed trait across its geographic range and, although demographic data exist only for the Danaher colony, growth rates were not noticeably affected by environmental conditions, although there was a delay in the onset of moult in the second half of the breeding season (Chapter 3). Further study is needed to determine if the growth rate of potential breeders is a cue to undertake precocial breeding or not.

Among estrildines, early maturation is the derived trait and later maturation the ancient trait fixed in the lineage, hence early maturation of Zebra Finches must be an adaptation to unique ecological conditions. In estrildines other than the Zebra Finch, delayed maturation must reduce the rate of juvenile mortality sufficient to compensate for the increased risk of having to survive for a longer period before reaching breeding age. The environment of the Zebra Finch is distinguished by long breeding seasons or episodes and high levels of juvenile mortalities and both factors would have selected for early sexual maturation. Juvenile mortalities probably arose mainly through extrinsic factors—those not sensitive to changes in reproductive decisions. Starvation and predation are the main sources of extrinsic mortalities in Zebra Finches, and the effects of reproductive costs on ageing constitute the main intrinsic source of mortalities. I argued in Chapter 7 that reproductive costs in Zebra Finches are not normally important, except when many successive broods are raised without interruption, such as in captivity, or during exceptionally long breeding episodes that may occasionally arise in central Australia; otherwise, cessation or decline in breeding conditions stops further breeding attempts and minimises long term costs. After early maturation evolved, other selection pressures maintained it by reducing the level of juvenile mortality, principally through high reproductive competence of young novice breeders.

What prevents Zebra Finches from attaining sexual maturation earlier than they do already? Energetic constraints imposed by moult are probably the main factors. Moult into adult body plumage begins around day 35 and is not complete before about day 70. Failure to develop complete foraging skills may be another factor in preventing the advancement of breeding age, because techniques needed to take rearing food, such as half-ripe seeds from standing heads of grass, may take some time to develop. Fortunately, the more complex hunting skills needed to take flying insects are unnecessary.

Longevity and breeding opportunities

Life span, based on recapture data suggests that Zebra Finches are very short-lived birds (see Table 8.1), but high mobility confounds accurate estimates of adult mortality using recapture data. Nevertheless, mortality of Zebra Finches must be very high given their high reproductive capacity, especially their early age of sexual maturation, which is characteristic of short-lived birds (Newton 1989).

Although elderly Zebra Finches, even those five or six years old, may still breed regularly in captivity (Immelmann 1965a) there is no information on how rate of reproduction changes with age in wild birds. High mobility and high mortality make such data difficult to collect. The ability to breed at a relatively old age should be strongly selected by environmental conditions that prevail throughout the interior of Australia. Annual breeding is frequently impossible, hence it is essential for those lucky individuals that manage to hang on until good conditions come again to be capable of breeding early and efficiently. These individuals will have the highest reproductive value. Davies (1986) emphasised the importance of longevity in desert-adapted species, such as the Zebra Finch, in relation to the pattern of rainfall. Populations can only be maintained in desert areas if survivors can make it through until the next fall of rain, whereupon food supply recovers and breeding can occur. This is the key to their survival in a non-annual environment. Thus, longevity in Zebra Finches should be adjusted to exceed the interval between successive breeding episodes. Most estrildines appear to have a lifespan of around five years, consequently, selection in Zebra Finches and their ancestors has simply maintained the ancestral trait rather than selected for it in the first place.

Life-time reproductive success and sex ratio

Values for 'life-time' reproductive success in Table 12.1 are likely to be underestimates, given the high mobility among the Zebra Finches at the Danaher colony. This problem confounds estimates of life span, which is the principal demographic determinate of breeding life span in birds (Newton 1989). Two other problems obscure accurate estimates of life-time breeding success of wild Zebra Finches: (a) extra-pair copulations, and (b) intra-specific brood parasitism. Both factors inflate estimates for some individuals and underestimate them for others.

With Zebra Finches at the Danaher colony, survival between fledging and nutritional independence probably accounts for the greatest effect on life-time reproductive success. However, environmental fluctuations in time and space are likely to be the major determinants in life-time reproductive success of most populations inhabiting the dry interior of Australia. Where fluctuations are most extreme, the entire breeding lives of only a few individuals will fall within a period of good food supplies, whereas the vast majority will probably not be able to breed at all, or have very limited opportunities.

Mean and variance in life-time breeding success for male and female Zebra Finches at the Danaher colony are identical (Table 12.1). However, given the strong sexual dichromatism, one might expect a greater variance in males than females and extensive DNA fingerprinting data may reveal this. No sex differences in life span were detected but more data are needed from a range of colonies. The adult sex ratio was slightly male-biased at all study colonies, but failed to reach statistical significance. Thus, despite pronounced sexual dichromatism, sexual selection had not produced strong sex differences in life-time reproductive success, lifespan or numbers of adults. These findings are consistent with those of other monogamous species where there are few opportunities for polygynous matings.

Adaptations and pre-adaptations to unpredictable arid environments

Many traits possessed by Zebra Finches that adapt them to their arid and semiarid environment are also possessed by other species of estrildines that do not live in such environments (Table 12.2). These 'pre-adaptive' traits became fixed in the lineage of the subfamily early in its evolution. On the one hand, such traits may still have an adaptive function yet no genetic variation exits for their expression in extant populations upon which microevolution can work. On the other hand, these traits may confer no advantage to species living in mesic environments yet they persist due to phylogentic inertia.

To understand the adaptations of Zebra Finches to their environment it is necessary to distinguish traits derived from local adaptations from those ancestral ones of the lineage. In theory this can be achieved by examining the pattern of the traits upon the lineage or clade of the

Table 12.2 Physiological, morphological and behavioural traits of Zebra Finches considered adaptive to arid environments. 0 = ancestral trait of estrildines (plesiomorph); 1 = derived trait of Zebra Finches (apomorph).

Traits	Origin	Source
Granivory	0	Heidweiller and Zweers 1990 MacMillen 1990
Tolerance of saline water	1	Oksche *et al.* 1963; Immelmann 1970
Tolerance of dehydration	1	Cade *et al.* 1965; MacMillen 1990
Bill-down drinking	1	Heidweiller and Zweers 1990
High rate of evaporative cooling	1	Calder and King 1963
Low metabolic rate	0	Marschall and Prinzinger 1991
High body temperature	0	Calder 1964
High zone of thermal neutrality	0	Marschall and Prinzinger 1991
High thermal conductance	0	Marschall and Prinzinger 1991
Early sexual maturation	1	Immelmann 1962a; 1963b
Constant gonadotrophic activity	1	Immelmann 1963b; Sossinka 1975
Slow continuous moult	1	Zann 1985a
Simultaneous moulting and breeding	0	Immelmann 1963b; Zann 1985a
Strong sexual dimorphism	1	Immelmann 1959; Burley 1981a
Nest roosting	1	Immelmann 1962a
High mobility and dispersal	1	Zann and Runciman 1994
Opportunistic breeding	1	Immelmann 1963b
Permanent pair bond	0	Immelmann 1963b
Multiple broods	?	Zann 1994a

tribes of Australian estrildines. Fortunately, Christidis (1987a,b) has constructed suitable cladograms of the Poephilini and its sister-group, the Lonchurini, based on chromosomal and molecular characters, which are, presumably, non-adaptive. The physiological and behavioural traits thought to be important for survival of Zebra Finches in arid, unpredictable environments can be mapped on the cladogram and their pattern determined by means of character optimisation methods (Brooks and McLennan 1991). The success of such an exercise depends on the breadth of comparative data for other members of the Zebra Finch tribe and the sister tribe.

Granivory and water relations

Granivory may be associated with an adaptive shift to arid ecosystems in small birds, such as the estrildines, although given the dryness of seed this idea appears counter intuitive. However, the greatest efficiencies in production of metabolic water from seeds arise in birds of estrildine size (< 24 g), in particular, those of Zebra Finch size, which can achieve a positive water balance at ambient temperatures below 30°C (Chapter 5). In addition to the size advantage, a subfamily trait, Zebra Finches have an extraordinary capacity to conserve water by reducing losses through evaporation (Cade *et al.* 1965; Lee and Schmidt-Nielsen 1971), excretion (Skadhauge 1981) and egestion (Calder 1964), consequently they can achieve a positive water balance at even higher temperatures, and this allows captive birds to survive indefinitely without drinking. It is likely that wild Zebra Finches can withstand long periods without drinking providing they can obtain good supplies of seed. This undoubtedly helps Zebra Finches survive drought by allowing them to travel long distances to water as well as freeing them totally from the need for drinking under certain conditions. The ability to endure extremes of thirst and dehydration is a reserve capacity utilised in times of emergency rather than employed for every day use. These adaptations to aridity have arisen through a refinement of subfamily characteristics, namely a coincidence of a granivory and small size.

Estrildines and heat regulation

Some physiological properties of Zebra Finches that enable them to survive in desert environments may also be fairly general attributes of estrildines, although in others they may not be developed to the same degree as in the Zebra Finch. For example, low basal metabolic rates, high rates of thermal conductance, and high thermoneutral zones have also been demonstrated in tropical species of estrildines drawn from all three tribes, including some that inhabit mesic environments (Marschall and Prinzinger 1991).

Drinking

The tip-down method of drinking in the Zebra Finch and seven other Australian species is believed to be an advantage in arid environments (Chapter 5). Heidweiller and Zweers (1990) postulated that the complex tip-down drinking evolved from the tip-up method found in all other species of estrildines and that its evolution was associated with an ecological shift in habitat from mesic to semiarid and arid environments with an associated switch to a diet of grass seeds. The anatomical and behavioural adaptations for dehusking grass seeds served as pre-adaptations for tip-down drinking, but why this should have evolved in the Australian estrildines and not African ones is unknown. Immelmann and Immelmann (1967) believe it may be associated with the greater insectivorous diet in the latter, where some arid-adapted species (e.g. *Uraeginthus granatina, Pytilia melba*) can go for months without drinking, providing insects are plentiful.

Phylogenetic analysis of the method of drinking is somewhat ambiguous, but it suggests that tip-down drinking evolved twice in the Poephilini (Figure 12.1). It is an ancient character that originated early in the evolution of the tribe some time after the ancestor of the Painted Finch separated off from the common ancestor of the other members and therefore it is the ancestral condition as far as Zebra Finches are concerned.

Nest roosting

Desert environments are renowned for low nocturnal temperatures and Immelmann (1962a) suggested that roosting in nests is one means of conserving heat among desert inhabitants. In the dry season in tropical northern Australia, temperatures also fall sharply at night. Indeed, Davies (1986) believes that low temperatures in Australian deserts and arid regions provide considerable environmental stress for animals. Estrildines, with their high zone of thermal neutrality, are particularly susceptible to cold, and Zebra Finches can make significant energy savings by roosting in nests (Chapter 6). The incidence of nest and perch roosting in the subfamily was extracted from the literature (Goodwin 1982; Immelmann *et al*. 1963, Immelmann 1965a), and although nest roosting is also found in some estrildines outside the Poephilini, especially in some African genera (*Spermestes* and *Amadina*), character optimisation in both the Lonchurini and Estrildini showed that perch roosting is plesiomorphic for the Poephilini. However, phylogenetic analysis shows that nest roosting is not a recently evolved trait in the Poephilini, but arose in an early ancestor shortly after the speciation of *Emblema pictum* (Figure 12.1).

There is some congruence in the patterns of distribution of the roosting and drinking characters when mapped onto the phylogeny of the

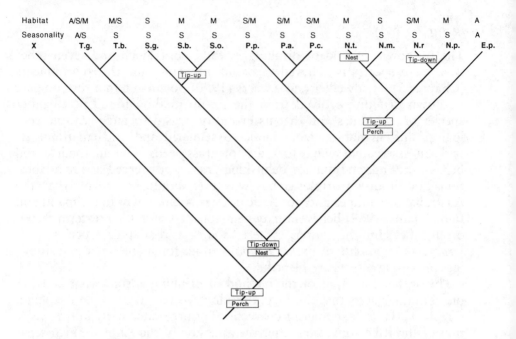

Fig. 12.1 Method of drinking and roosting mapped onto a phylogenetic tree of the Poephilini constructed by Christidis (1987b) using chromosomal and biochemical characters. Drinking and roosting were polarised using the sister group (X), the Lonchurini (see Fig. 1.1), as the outgroup. Tip-up drinking and perch roosting are ancestral traits (plesiomorphies) in the Poephilini, and tip-down drinking and nest roosting are derived traits (apomorphies). T.g. = *Taeniopygia guttata*; T.b. = *T. bichenovii*; S.g. = *Stagonopleura guttata*; S.b. = *S. bella*; S.o. = *S. oculata*; P.p. = *Poephila personata*; P.a. = *P. acuticauda*; P.c. = *P. cincta*; N.t. = *Neochmia temporalis*; N.m. = *N. modesta*; N.r. = *N. ruficauda*; N.p. = *N. phaeton*; E.p. = *Emblema pictum*. Habitat: A = arid, M = mesic, S = semiarid. Seasonality: A = aseasonal, S = seasonal.

Poephilini. Tip-down drinking and nest roosting evolved in an early ancestor of 12 of the 13 species. This suggests that the common ancestor of the tribe occupied an environment that became increasingly arid and that isolation of one population led to the speciation of *E. pictum* and the ancestor of the remaining species, which initially remained in an arid environment. Subsequent vicariance events, such as those proposed by Cracraft (1986) probably led to isolation of populations in environments that underwent climatic change.

Breeding potential

In addition to precocial breeding, Zebra Finches possess a number of traits that give them a high breeding potential which is utilised when suitable conditions arise. Some are unique to the species, whereas others are found to varying degrees in other estrildines. The ability to breed at anytime of the year is a consequence of a tonic gonadotropic activity of the hypothalamo-hypophysial system which maintains the gonads in a semi-permanently active state (Chapter 7). To date this has not been demonstrated for any other estrildine. In contrast, other characteristics of Zebra Finches that facilitate opportunistic breeding may be pre-adaptations, in the sense that they may have evolved in an ancestral species. For example, the ability to moult and breed at the same time is also an advantage for an opportunistic breeder, but may not have arisen as a microevolutionary adaptation in the Zebra Finch, since it occurs in at least six other species of Poephilini that live in semiarid habitats (Schoepfer 1989; Tidemann and Woinarski 1994). Similarly, the tight, life-long bond is an advantage for an opportunistic breeder since it expedites ovulation and accelerates breeding. However, a life-long bond appears to be a characteristic trait of the subfamily, although only a few, the *Poephila*, the *Amadina* and the *Uraeginthus* species, have a pair bond that is as tight year-round as that of the Zebra Finch. Interestingly, all of these species are arid-adapted and appear to have developed the trait independently. Finally, slow wing moult may not be a microevolutionary adaptation of Zebra Finches that facilitates mobility and dispersal since it occurs in the Double-barred Finch (Schoepfer 1989), and also in Masked and Long-tailed Finches (Woinarski and Tidemann 1992) that are not highly mobile. However, the ability to moult continuously (Chapter 3) has not been found in any other estrildine and may be another trait that adapts Zebra Finches to their unique predominantly aseasonal life cycle.

Sexual dimorphism

Of all estrildines, few species are as strongly sexually dimorphic as the Zebra Finch. Size dimorphism is extremely rare in estrildines, and only occurs in the Pin-tailed Parrot-Finch *Erythrura prasina*, where the tail of the female is only about half as long as the male's (Goodwin 1982). Although the sexes of Zebra Finches differ in size, the differences are only slight statistical ones (Chapter 3), and the species should be considered sexually monomorphic in the strict sense. Where secondary sexual characteristics exist in estrildines, there are differences in colouration in plumage, bill and soft parts; that is, they are sexually dichromatic.

Table 12.3 shows that Zebra Finches are the most sexually dichromatic of estrildines, with four to five male-specific plumage markers having evolved in the Australian subspecies (ear, flank, chest, throat, and possibly abdomen), and three in the Lesser Sundas (ear, flank, chest). The location of these sex-specific markers makes them highly conspicuous when viewed from the front or side, yet they are invisible from

Table 12.3 Levels of sexual dichromatism in estrildines based on illustrations and species descriptions of mature adults taken from the literature (mainly Immelmann *et al.* 1965a; Goodwin 1982)

Sexual chromatism	Description	Examples
0. Monochromatism	External phenotype of adult male and female are qualitatively and quantitatively indistinguishable	*Stagonopleura bella, Neochima temporalis, Erythrura psittacea,* most species of *Lonchura* and *Spermestes*
1. Slight dichromatism	Slight quantitative differences in size of colour markings or intensity of colouration exist between the sexes, but such characters are only relative and only evident when mated partners are compared	*Taeniopygia bichenovii, Stagonopleura guttata, S. oculata, Poephila personata, P. acuticauda, P. cincta, Erythrura trichroa, E. tricolor, Pytilia phoenicoptera, Estrilda astrildid, E. troglodytes, E. melopoda, Uraeginthus angolensis*
2. Moderate dichromatism	Sexes similar but male colouration is brighter or more extensive than in the female; adults can be sexed on sight whether in pairs or not	*Chloeba gouldiae Heteromunia pectoralis, Pytilia afra, Uraeginthus cyanocephala, Lagonosticta rhodopareia, Mandingoa nitidula*
3. Strong dichromatism	Adult male has one large coloured plumage marker not found in the female	*Neochmia modesta, N. ruficauda, N. phaeton, Emblema pictum, Amadina fasciata, A. erythrocephala, Pytilia melba, Lagonosticta senegala, Estrilda melanotis*
4. Extreme dichromatism	Males have more than one sex-specific colour marker	*Uraeginthus bengalus* (2 markers), *Erythrura prasina* (2–3), *Amandava amandava* (2), *Taeniopygia guttata* (4–5)

above. Thus, the sex-specific plumage markers provide sexual identity to conspecifics approaching from close to intermediate distances. Moreover, during courtship males display their markers to females to maximum advantage, especially during the song and hop-pivot dance in stage 2 courtship. As in most Australian estrildines the greyish plumage on the dorsal surface makes them cryptic to aerial predators, and this may be the reason that eclipse plumage has not evolved in Zebra Finches (Immelmann 1959).

In addition to these visual cues to sexual identity, the Australian Zebra Finch is one of the few species of estrildines that has sex-specific Distance Calls. These frequently uttered calls enable conspecifics to identify the sex of unsighted individuals from 50 to 100 m away and provide the means by which individuals can be brought into visual contact for confirmation of sexual identity. Acoustic identification of sex from Distance Calls is problematic in the Lesser Sundas subspecies, where the calls of the sexes are identical but for slight statistical differences. Also, this subspecies is less sexually dichromatic than the Australian subspecies: it lacks the throat barring and the chest band is small or absent. Hence, it is likely that identification of sex from a distance is slightly more difficult. One can only speculate on the origin of these differences. Possibly, the Lesser Sundas subspecies lost some of the more conspicuous sex-specific characteristics when it invaded more densely vegetated habitats with different predators, or perhaps these characteristics evolved only recently in the Australian subspecies after geographic isolation of the Lesser Sundas form.

Australian Zebra Finches obviously have considerably redundancy in their system of signalling sexual identity. In contrast, the closest living relative of Zebra Finches in Australia, the Double-barred Finch, does not have sexually distinct distance calls and strangers probably require several minutes, or possibly hours of behavioural interaction or 'interrogation' in order to determine each other's sex. Certainly, males (but not females) of the three species of *Poephila* grassfinches have difficulty identifying the sex of strangers in captivity and usually court everyone they encounter until they learn their sex (Zann 1976a). Immelmann's (1959) investigation of social and sexual behaviour of the white colour morph beautifully illustrates the difficulties Zebra Finches have when there are no sexually dichromatic features (Chapter 9). He concluded that male-specific markings of Australian Zebra Finches act as a 'frozen display' which, on the one hand prevents males from courting and allopreening one another, and on the other hand stimulates the initial approach of the female at the onset of courtship and prevents her from fleeing. Therefore, sexual signalling in the Zebra Finch brings opposite sexes together from a distance, prevents misidentification of sex and subsequent time wasting and aggression, so that rapid pair formation is effectively promoted and reproductive activity is stimulated. Such urgency in these short-lived birds would be necessary in certain arid environments where breeding conditions can arise suddenly without much warning and must be exploited as soon as possible. High mortalities also select for rapid re-pairing in order to maximise breeding opportunities. In highly mobile Zebra Finch populations where there is much turnover in membership of colonies, there may not be sufficient time for unpaired individuals to become personally acquainted with one another's sex in pre-breeding flocks in the gradual manner of the Double-barred Finch. However, there is a need for more comparative studies on captive grassfinches in order to compare rates of pair formation between sexually dichromatic and monochromatic species.

Conspicuous sex differences in appearance are not expected in permanently monogamous species, such as estrildines, since competition among males for females is expected to be low. Indeed, the demographic data for Zebra Finches and their patterns of parental investment are consistent with this view. However, evidence for intersexual selection comes from both captivity and field studies where females have been shown to discriminate among males for the purposes of pair formation and extra-pair copulations (Chapters 9 and 11).

What behavioural and/or ecological factors selected for and maintain sexual dichromatism in Zebra Finches and a handful of other estrildines, and what has prevented other species of estrildines evolving strong sexual dichromatism? Phylogenetic analysis suggests that strong sexual dimorphism evolved independently three times among the Poephilini: (a) in the Zebra Finch, (b) in the ancestor of the *Neochmia* clade (monochromatism evolved back in *N. temporalis*), and (c) in the Painted Finch (Figure 12.2). With the exception of *E. prasina*, estrildines with strong or extreme sexual dichromatism live in arid environments, but only the Zebra Finch and the Painted Finch have largely unpredictable breeding seasons. Possibly, the ancestor of the *Neochmia* was also an inhabitant of arid environments.

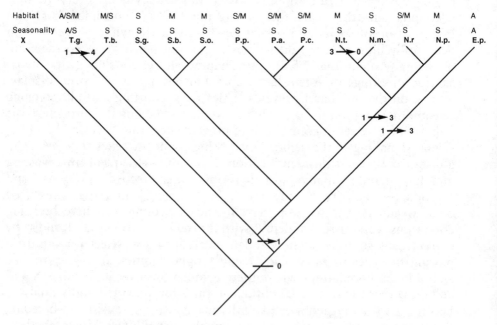

Fig. 12.2 Degree of sexual dichromatism mapped onto a phylogenetic tree of the Poephilini constructed by Christidis (1987b) using chromosomal and biochemical characters. Sexual dichromatism was polarised using the sister group (X), the Lonchurini (see Fig. 1.1), as the outgroup. 0—sexual monochromatism, 1—slight dichromatism, 2—moderate dichromatism, 3—strong dichromatism, 4—extreme dichromatism (see Table 12.3). Abbreviations of species names and habitats as in Fig. 12.1.

Burley (1981b) predicted that sexual indistinguishability is an advantage in monogamous, social species, especially those that flock in the breeding season, or breed colonially. A literature and museum survey of 150 species supported her prediction, but there were three notable exceptions, the Flock Pigeon, the Budgerigar and the Zebra Finch. The need for a rapid breeding response was thought responsible for selection for strong dichromatism in these arid-adapted species. Burley argued that it was an advantage for individuals that bred in colonies to conceal their sex in order to reduce the level of sexual interference competition and so enhance fitness. Sexual interference occurs in Zebra Finch colonies but seems limited to only a few days in the fertile period: copulating males may be knocked off by others and rates of extra-pair courtship and forced copulations suggest that male interference of females can be quite high (Chapter 9). It is surprising that Zebra Finches have not adopted nest copulation where it would be free from interference from other colony members since it has evolved a number of times in at least a dozen species of estrildines from all three tribes (Immelmann and Immelmann 1967). Interestingly, most species of nest copulators are highly social and sexually dichromatic ones. Sexual indistinguishability does not necessarily prevent sexual interference and promote social harmony. For example, white Zebra Finches are much more socially disruptive than wild-type Zebra Finches, although this arises from incompatibilities between external appearance and behavioural responses. Similarly, field observations of *Poephila cincta*, a species where the sexes are barely distinguishable, shows that defence of the mated partner from sexual competitors is high (Zann 1977) and suggests that where permanent colonies are established, sex is difficult to conceal even when the sexes appear indistinguishable to humans.

Conservation

Human activities in Australia and Indonesia have increased the distribution and abundance of the species, although protection from trapping in the latter country may be needed. The extent of available habitat has increased through clearing of forests and woodlands and changes in land use and cultivation have increased the availability of food plants (Chapter 2).

Since European settlement of arid Australia, thousands of artesian bores have been drilled for the pastoral industry and these have inadvertently provided drinking water for dispersing Zebra Finches and allowed them to establish semi-permanent populations in areas not previously occupied (Davies 1977a). Currently however, there is a program to conserve the artesian supplies by systematically capping many degraded and wasteful bores not used by stock; this will exclude Zebra Finches and other native wildlife that require drinking water from some localised areas.

In southwestern and southeastern Australia where clearing of woodlands and mallee for cultivation has extended Zebra Finch habitat, lack of suitable indigenous nesting bushes is limiting in some regions especially where thorny exotic species, such as boxthorn are subjected to eradication.

Grazing by exotic herbivores in the vast mulga shrublands of arid Australia has degraded much of the chenopod understory. Water previously used by chenopods has become available for growth of annual grasses which now predominate the understory (Davies 1977a). Accordingly, there has been a great increase in the distribution and abundance of seed food for Zebra Finches and other granivorous species (Reid and Fleming 1992). Nonetheless, overgrazing by cattle and rabbits of these grasses especially those growing on floodplains and riverine environments has led to serious degradation of vast areas of rangelands. In some cases highly valuable drought refugia, centred on rare 'islands' of nutrients and moisture, have been irreversibly destroyed by overgrazing, and in other cases this has allowed woody weeds to take over and prevent regrowth of grasses. Thus the carrying capacity for grazers and granivores alike has been permanently reduced in some places. Heavy, sustained use of rangelands by cattle, rabbits and kangaroos is an important conservation problem in Australia and hopes of restoration rest mainly on the elimination of rabbits and lower stocking rates of cattle, especially during drought.

Thankfully, however, the conservation of Zebra Finches is not a current concern, nor likely to be one in the near future for either subspecies.

Summary

Many of the traits that enable Zebra Finches to thrive in an unpredictable arid environment have arisen in distant ancestors and are lineage-specific adaptations that have not originated as a consequence of microevolutionary selection on Zebra Finches or their immediate ancestors. Many physiological (high body temperature, low metabolic rate, high thermal conductance), morphological (small body size) and life history (graminivory, life long pair bond, life span) traits characteristic of the Estrildini are adaptive to arid environments. This suggests that the proto-estrildine probably evolved in an arid or semiarid environment. In addition to these pre-adaptations other arid-adapted traits probably evolved in the ancestor of the Poephilini (nest roosting, tip-down drinking, slow moult, breeding while moulting) and were maintained in descendent species during the climatic changes in Pliocene–Pleistocene epochs. Finally, there are those unique traits that arose in Zebra Finches as microevolutionary adaptations to an arid, unpredictable environment. Early sexual maturation, precocial breeding, constant gonadal activity, opportunistic breeding, ability to withstand extremes of dehydration, high mobility and high dispersal are the distinguishing adaptations of Australian Zebra Finches. These traits enable survivors of perennial droughts to find and exploit any favourable conditions, and by means of a prolific capacity for breeding, populations can be quickly restored before the next devastating drought. Field work is needed on the Lesser Sundas subspecies of Zebra Finch in order to determine how these adaptations to aridity have been modified in the monsoonal seasonal environment that prevails throughout its range.

Appendix 1

List of the principal domesticated colour morphs of Zebra Finches in Australia, Europe and the United States. Combinations of the main mutations can be found in Kloren (1982) and the poster by van Keulen (Postbus 86, 7440 AB Nijverdal-Holland, van der Bergsweg 25).

Mutation	Inheritance mode	Year of discovery
1. fawn (cinnamon)	sex-linked recessive	1927
2. light-backed	sex-linked recessive	1955
3. cream-backed		
4. chestnut-flanked (marked white)	sex-linked recessive	1948
5. agate		
6. orange-breasted		1978
7. white	autosomal recessive	1921
8. pied	autosomal recessive	
9. crested	autosomal dominant	1976
10. grizzle	autosomal recessive	
11. Florida fancy		1971
12. yellow-billed	autosomal recessive	
13. isabel		
14. phaeo		
15. recessive silver	autosomal recessive	1960
16. dominant silver	autosomal dominant	
17. dominant cream	autosomal dominant	
18. recessive cream	autosomal recessive	
19. penguin	autosomal recessive	
29. frizzled		
21. saddle-back	sex-linked recessive	
22. cheekless		
23. grey-cheeked		1967
24. fawn-cheeked		
25. black-cheeked		1984
26. black-backed		
27. black-breasted	autosomal recessive	1968
28. black-fronted		
29. black-bodied		
30. black-faced	sex-linked recessive	1959

Appendix 2

List of seed species eaten by Zebra Finches in Australia. Names in bold have been identified as items of major importance in the diet in quantitative studies.

Species	Location and source[a]	Growth habit[c]	Flowering[d]
POACEAE			
Amphibromus neesi	SE[4]	P	Sp–A
Aristida contorta	WA[2], C[7,] EA[3]	A/P	Sp, A
Aristida armata	EA[3]	P	A
Aristida inaequiglumis	C[7], K[5]	P	?
Aristida muricata	EA[3]	P	A
Astrebla lappacea	EA[3]	P	
Astrebla pectinata	EA[3]	P	
Arundinella nepalensis	K[1]		
Avena sp.[b]	SE[4]	P	Sp
Botrichloa macra	SE[6]	P	Su/A
Brachyachne convergens	K[1,5]	A	Su
Brachiaria gilesii	C[7]	A	A
Brachiaria miliiformis	C[7]	A/P	Su/A
Cenchrus ciliaris [b]	C[7]	P	Su/A
Chloris truncata	SE[4]	A/P	Su
Chloris viragata [b]	K[1]	P	Su/A
Cymbopogon obtectus	C[7]	P	Sp/A
Cynodon dactylon	C[7] SE[4]	P	Su/A
Dactyloctenium radulans	EA[3]	A	Sp/Su
Danthonia caespitosa	SE[4]	P	Sp
***Danthonia* sp.**	SE[6]	P	
Dichanthium sericeum	C[7], EA[3]	P	Su
Dichanthium tenuiculum	C[7]	?	?
Digitaria brownii	C[7]	P	Sp–A
Digitaria sanguinalis[b]	SE[4]	A	Su/A
Echinochloa colonum[b]	C[2], K[5]	A	Su/A
Echinochloa crus-galli[b]	SE[4]	A	Su/A
Enneapogon avenaceous	EA[3]	A	Su/A
Enneapogon cylindricus	C[7], EA[3]	A/P	Su/A
Enneapogon oblongus	C[7]	P	?
Enneapogon pallidis	K[1]	P	Su/A
Enneapogon polyphyllus	C[7], EA[3]	A/P	Su/W
Enneapogon glaber	K[5]		
Eragrostis australis	WA[2]	P	Sp

Species	Location and source[a]	Growth habit[c]	Flowering[d]
Eragrostis eripoda	C[7],	P	A
Eriachne aristidea	EA[3]	A/P	Sp–A
Eriachne sp.	WA[2]		
Erichloa pseudo-acrotricha	SE[6]	P	Su/A
Iseilema membranaceum	C[2], K[2], EA[3]	A	Su/A
Neurachne michelliana	C[7]	P	W
Oryza rufipogon	K[5]		
Panicum decompositum	EA[3]	P	Su/A
Panicum prolutum	SE[4]	P	Sp
Panicum zymbirformae	K[2]		
Paspalidium sp.	C[7], K[2],	A/P	Su–W
Paspalidium criniforme	SE[6]	P	Su/A
Paspalidium dilatatum[b]	SE[4,6]	P	Sp–Au
Phalaris sp. [b]	SE[4]		
Poa annua [b]	SW[2], SE[4]	A/P	W/Sp
Sorgum stipoideum	K[5]	A	Su
Sporobolus australasicus	C[2], K[2]	A	?
Stipa densiflora	SE[6]	P	Sp
Stipa setacea	SE[6]	P	W/Sp
Themeda australis	C[7]	P	Sp/Su
Themeda triandra	K[5]		
Thyridolepsis michelliana	C[7]	P	Sp–A
Tragus australianus	C[7]	A	Su–Au
Trirapsis mollis	EA[3]	P	Sp–Au
Triodia basedowii	C[1]	P	Sp–W
Triodia wiseana	K[5]		
Tripogon loliiformis	C[7]	A	S/W
CHENOPODIACEAE			
Maireana scleroptera	C[1]	P	?
EUPHORBIACEAE			
Euphorbia drummondii	C[1]	A	?
POLYGONACAE			
Polygonum aviculare	SE[4]		

[a] Locations: C, Central Australia; K, Kimberley and western Northern Territory; EA, eastern arid zone, WA, western arid zone; SE, southeastern Australia; SW, southwestern Australia; Sources: 1 Immelmann 1962a; 2, Davies 1977a; 3, Morton and Davies 1983; 4, Zann and Straw 1984a; 5, Tidemann 1987; 6, Schöpfer 1989; 7, Zann *et al.* 1995.

[b] Introduced species.

[c] A, annual; P, perennial.

[d] Sp, spring; Su, summer; A, autumn; W, winter.

N.B. Animal matter found in the crops of Zebra Finches: termites (North 1909, Immelmann 1962a); dipterans (Immelmann 1962a); heteropterans (Morton and Davies 1983); aphids, ostracods, snails (Zann and Straw 1984a).

Appendix 3

Ethogram of Australian Zebra Finches (also see Figueredo *et al.* 1992).

Behaviour of individuals	Social behaviours
Locomotory	*Calls*
hopping	Tet Call
walking	Distance Call
pivoting	Stack Call
flying	Wsst Attack Call
taking-off	Thuk Call
landing	Distress Call
hovering	Kackle Call
	Ark Call
Recuperative	Whine Call
perching normal	Begging Call
perching fluffed	Long Tonal Call
squatting	
sleeked-bill open	*Agonistic*
carpel raising	horizontal-sleeked
sleep-bill buried	vertical-sleeked
sleep-head withdrawn	jabbing
	bill-fencing
Nutritive	supplanting
pecking	aerial combat
mandibulation	chasing
jumping	
pulling	*Sexual*
shovel-away	Head-tail twist
foot-clamping	Upright fluffed
drinking	allopreening
head-shake	clumping
head-flick	bill-wiping
defecating	hop-pivot
	hopping-to-and-fro
Exploratory	Undirected Song
monocular gaze	Directed Song
binocular gaze	Tail-quivering
neck-stretch	copulation

Behaviour of individuals	Social behaviours
Comfort	*Breeding*
self-preening	head-down tail-fan
bill-wiping	nest ceremony
head-wiping	pull-in building movement
yawning	push-up building movement
wing-leg-stretch	push-out-shake building movement
wing-arch	incubation
head-scratch	begging head-up
head-up body-shake	begging head-down
head-down body-shake (bathing)	food regurgitation
tail-flick	food transfer
wing-flick	nest-leading
sun-bathing sprawl	

Appendix 4

Handbook summary of major biological parameters. Numerical values are means or medians and sources may be found in the text. W: free-living Australian Zebra Finches, WC: wild-caught Australian Zebra Finches; D: domesticated Australian Zebra Finches; L: captive-bred Lesser Sundas Zebra Finches from Timor.

Classification

Family	Passeridae (Vigors, 1825)
Subfamily	Estrildinae (Blyth, 1829)
Tribe	Poephilini
Genus	*Taeniopygia* (Reichenbach, 1862)
Species	*guttata* (Vieillot, 1817)
Subspecies	*guttata* (Vieillot, 1817), Lesser Sundas
	castanotis (Gould, 1837), Australia

Captive breeding

Recommended density per pair	0.3–0.4 m^3 per pair
Minimum temperature	>12°C
Preferred temperature range	20–30°C
Breeding diet	dry seed, soaked seed, seeding grasses, greens, cuttle-bone, grit (siliceous and calcareous), egg food, vitamin supplements.
Seed preferences	1. panicum, 2. white millet, 3. Japanese millet, 4. canary seed

Captive breeding (continued)
Colour morph varieties
Hybrids (Immelmann, 1962a)

30 (D)
Double-barred Finch
Long-tailed Finch
Black-throated Finch
Masked Finch
Star Finch
Plum-headed Finch
Diamond Firetail
Chestnut-breasted Finch
Bengalese Finch
African Silverbill
Indian Silverbill
Grey-headed Silverbill
Tri-coloured Munia

Handling
Wild capture — mistnets around nests or water, walk-in traps

Marking
- metal bands (size) — 02—Lambournes, D—Hughes
- colour bands (size) — XCS—Hughes
- colour attractive to females — red (WC, D)
- colour unattractive to females — light green (WC, D)
- colour attractive to males — black (D)
- colour unattractive to males — light blue (D)

Anaesthesia
(50:50 mixture injected I. M.)
ketamine (mg/g)   0.03
xylazine (mg/g)   0.006

Morphology

Measurement	Sex	Values
Mass (g)	male	11.9–12.4 (W), 10.4–11.8 (L), 12.7 (D)
	female	12.2–12.7 (W), 10.0–10.3 (L), 13.0 (D)
Wing length (mm)	male	55 (W), 52 (L), 56–57 (D)
	female	55 (W), 52 (L), 56–57 (D)
Bill length (mm)	male	10.0 (W), 9.0 (L), 10.4 (D)
	female	10.0 (W), 9.0 (L), 10.6 (D)
Bill depth (mm)	males	8–9 (W), 7 (L), 6.5 (D)
	female	7–9 (W), 7 (L), 6.6 (D)
Head-bill length (mm)	males	23.3 (W)
	female	23.6 (W)
Rectrices length (mm)	male	35–38 (W), 34 (D)
	female	35–38 (W), 34 (D)
Remiges length numbers 1–9 (mm)	male	40–46 (W)
	female	41–46 (W)
Bill colour (Munsell scores)	male	8.75, 4/12 (W, D), 10, 4/12 (L)
	female	10, 5/12 (W, D), 10, 5/12 (L)
Iris colour (Munsell scores)	male	10, 3/10 (W), 10, 3/1, 12, 3/6 (D)
	female	10, 3/10 (W), 10, 3/1, 12, 3/6 (D)
Foot colour (Munsell scores)	male	10, 5/8 (W), 10, 5/8, 12.5, 5/10 (D)
	female	10, 5/8 (W), 10, 5/8, 12.5, 5/10 (D)
Back colour (Munsell scores)	male	20, 4/3 (W), 20, 4/1 (D)
	female	20, 4/3(W), 20, 4/1 (D)
Tarsus length (mm)	male	14.3–15.0 (W), 14.8 (D)
	female	14.3–15. 0 (W), 14.8 (D)
Primaries		9
Secondaries		9
Testis length (mm)	male	14–25 (W)

Morphology (continued)

Testis length (mm)—dehydrated	male	2–3 (WC)
Testis mass (mg)	male	1–51 (W)
Testis mass (mg)—dehydrated	male	12–41 (WC)
Ovary mass (mg)	female	20–170 (WC)
Ovary mass (mg)—dehydrated	female	2–37 (W)

Moult

Onset—sex plumage (age in days)	male	35–40 (W, D)
Completion—sex plumage (age)	male	55–70 (W, D)
Onset—primary moult (age)	male	80 (W)
	female	80 (W)
Primary moult cycle (days)	male	230–287 (W)
	female	240–287 (W)
Primary moult type		continuous step-wise/serially descendent

Nutrition

Seed (g) required per day at <15°C	5 (WC)
20–30°C	3–4 (WC, D)
>30°C	1–2 (D)
Seed (g) required per day to feed five fledglings	14 (D)
Energy intake (kJ per day) at 27°C	35.7 (D)
Standard metabolic rate (kJ per hour) at 23°C	0.80–0.88 (D)
Digestive efficiency (%)	78–88 (D)
Maximum seed intake (g per hour)	1–3.5 (WC), 3–6 (D)
Ingestion to egestion (minutes) at 23°C	87–106 (D)
Crop capacity (g)	0.3–0.6 (D)

Water intake (ml per day) at 23°C	3.1–4.6 (WC, D)
> 40°C	6–12 (WC, D)

Reproduction

Spermatogenesis—onset (days post-hatch)	60 (WC), 70 (D)
Ovulation—onset (days post-hatch)	60 (WC), 90 (D)
Pair formation—onset (days post-hatch)	50–60 (W, D)
Pair formation time (days)	2–10 (W, D)
Nest construction time (days)	5–13 (W), 1–5 (D)
Fertile period (days)	14–15 (D)
Onset of fertility (days before first egg)	11 (D)
Termination of fertility	day before last egg (D)
Fertility (% eggs hatched)	75–96 (W)
Extra-pair paternity (% young)	2.4 (W), 5.6 (D)
Extra-pair paternity (% broods)	8 (W), 11 (D)
Brood parasitism (% broods)	13–32 (W)
Laying time (diurnal)	first two hours of daylight (W, D)
Laying interval (h)	24 (W, D)
Clutch size (mode)	5 (W), 4–6 (D),
Laying type	indeterminate
Egg mass (g)	1.0 (W, D)
Egg volume (mm³)	927 (W), 948 (D)
Egg composition	
water content (% mass)	76.4 (W), 75.3 (D)
shell (% wet mass)	9.1 (W), 8.4 (D)
Calcium (mg)	18 (D)

Reproduction (continued)

yolk content (% wet mass)	26.4 (W), 24.2 (D)
lipid content (% wet mass)	6.9 (W), 5.8 (D)
Onset of incubation (laying day)	4 (W), 1–4 (D)
Incubation bout (minutes)	37–92 (W), 20 (D)
Incubation period (days)	11–14 (W), 11–14 (D)
Incubation temperature (°C)	36 (W), 35–37 (D)
Hatching interval (days)	1–2 (W), 2–5 (D)
Hatching mass (g)	0.6–0.9 (D)
Eyes open (days post-hatch)	6–7 (W, D)
Feathering complete (days post-hatch)	14 (W, D)
Nestling mortality (%)	9 (W), 13 (D)
Fledging mass (g)	10 (W), 10 (D)
Age at fledging (days post-hatch)	16–18 (W), 17–22 (D)
Inter-brood interval (days)	52 (W), 57 (D)
Nutritional independence (days post-hatch)	35 (W)
Roosting independence (days post-hatch)	50 (W)
Successive broods	>3 (W), 19 (WC), 14 (D)

Population

Colony size (nesting pairs)	7–47 (W)
Colony density (nests per hectare)	0.7–76.0 (W)
Population size	<200–18,000 (W)
Recapture rate (years to 10% threshold)	0.1–1.6
Mortality: annual adult (%)	72–96 (W)
nestling–independence (%)	67 (W)
nestling–sexual maturity (%)	77 (W)
first 12 months of life (%)	96 (W)
Life expectancy at hatching (days)	51 (W)
Maximum lifespan (years)	3.0–5.6 (W), 5–7 (D)

Sex ratio—adults (% males)	51–53 (W), 47–61(D)
—young (% males at 40 days)	44–52 (W)
Breeding philopatry (% natal pairs)	22 (W)
Nesting predation (% lost)	66 (W)
Rearing success (% nestlings fledged)	87–100 (W)
Breeding success (> 1 fledgling)	
(% breeding attempts)	35 (W)
(% incubated eggs)	41 (W)
(% clutches)	44–74 (W)
Age classes (days post-hatch)	
Nestling	1–17
Juvenile	18–35
Young immature	36–50
Late immature	51–80
Young adult	81–100
Adult	>100
Temperatures	
Diurnal cloacal temperature (°C)	38–44 (WC, D)
Critical temperature (°C)	45–46 (WC, D)
Thermal neutral zone (°C)	30–40 (WC, D)
Thermal conductance (j-g-h-°C)	5.2 (D)
Parasites	
ectoparasites	mites, mallophagan lice, hippoboscid flies
endoparasites	nematodes
Behaviour	
Subsong (days post-hatch)	28–50 (W), 24–50(D)
Plastic song (days post-hatch)	50–80 (W, D)
Full song (days post-hatch)	80 (W, D)

Behaviour (continued)

Number of distinct calls	12 (W, D), 4+ (L)
Main frequency of vocalisations (kHz)	2–5
Frequency of maximum emphasis (kHz)	3.6 (W, D), 4.7 (L)
Hearing sensitivity (kHz) range	1–6 (D)
best frequency	4 (D)
Temporal auditory acuity (ms)	2.5–7 (D)
Sensitive phase (days post-hatch)	
song stage I	18–35 (D)
stage II	36–65 (W, D)
imprinting stage I	15–40 (D)
stage II	>90 (D)

References

Abbott, I., Abbott, K. L. & Grant, P. R. 1977. Comparative ecology of Galápagos Ground Finches (*Geospiza* Gould): evaluation of the importance of floristic diversity and interspecific competition. Ecological Monographs 47: 151–184.

Adret, P. 1993. Operant conditioning, song learning and imprinting to taped song in the Zebra Finch. Animal Behaviour 46: 149–159.

Adkins-Regan, E. & Ascenzi, M. 1987. Social and sexual behaviour of male and female Zebra Finches treated with oestradiol during the nestling period. Animal Behaviour 35: 1100–1112.

Adkins-Regan, E. & Ascenzi, M. 1990. Sexual differentiation of behavior in the Zebra Finch: effect of early gonadectomy or androgen treatment. Hormones and Behavior 24: 114–127.

Amundson, T. & Slagsvold, T. 1991. Hatching asynchrony: facilitating adaptive or maladaptive brood reduction. Proceedings of XX International Ornithological Congress (Christchurch, 1990): 1707–1719.

Arnold, A. P. 1975a. The effects of castration and androgen replacement on song, courtship, and aggression in Zebra Finches (*Poephila guttata*). Journal of Experimental Zoology 191: 309–325.

Arnold, A. P. 1975b. The effects of castration on song development in Zebra Finches (*Poephila guttata*). Journal of Experimental Zoology 191: 261–277.

Arnold, A. P. 1980. Effects of androgens on volumes of sexually dimorphic brain regions in the Zebra Finch. Brain Research 185: 441–445.

Arnold, A. P. 1992. Developmental plasticity in neural circuits controlling birdsong: sexual differentiation and the neural basis of learning. Journal of Neurobiology 23: 1506–1528.

Arnold, A. P. 1993. Sexual differentiation of brain and behavior: the Zebra Finch is not just a flying rat. Brain Behavior and Evolution 42: 231–241.

Ashmole, N. P. 1962. The Black Noddy *Anous tenuirostris* on Ascension Island. Ibis 103b: 235–273.

Badman, F. J. 1979. Birds of the southern and western Lake Eyre drainage. South Australian Ornithologist 28: 29–55.

Badman, F. J. 1987. Boredrains and the birds of inland South Australia: a study of the relationships of boredrains to native bird populations in the far-north of South Australia. Nature Conservation Society of South Australia, Adelaide.

Baird, R. F. 1991. Avian fossils from the Quarternary of Australia. In Vickers-Rich, P., Monaghan, J. M., Baird, R. F. & Rich, T. C. (eds), Vertebrate Palaeontology of Australasia: 809–867. Monash University, Melbourne.

Baldwin, M. 1973. Divergent behaviour in the Zebra Finch. New South Wales Ornithologists Club 8: 4–7.

Baverstock, P. R., Christidis, L., Kreig, M. & Birrell, J. 1991. Immunological evolution of albumin in Estrildine finches (Aves: Passerformes) is far from clock-like. Australian Journal of Zoology 39: 417–425.

Beauchamp, G. & Kacelnik, A. 1991. Effects of the knowledge of partners on learning rates in Zebra Finches *Taeniopygia guttata*. Animal Behaviour 41: 247–253.

Beche, C. & Midtgård, U. 1981. Brain temperature and the rete mirabile ophthalmicum in the Zebra Finch (*Poephila guttata*). Journal of Comparative Physiology 145: 89–93.

Bernstein, M. H. 1971. Cutaneous water loss in small birds. Condor 73: 468–469.

Birkhead, T. R. 1987. Sperm-storage glands in a passerine: the Zebra Finch *Poephila guttata* (Estrildidae). Journal of the Zoological Society of London 212: 103–108.

Birkhead, T. R. 1991. Sperm depletion in the Bengalese Finch, *Lonchura striata*. Behavioural Ecology 2: 267–275.

Birkhead, T. R. & Biggins, J. D. 1987. Reproductive synchrony and extra pair copulation in birds. Ethology 74: 320–334.

Birkhead, T. R., Burke, T., Zann, R., Hunter, F. M. & Krupa, A. P. 1990. Extra-pair paternity and intraspecific brood parasitism in wild Zebra Finches *Taeniopygia guttata*, revealed by DNA fingerprinting. Behavioural Ecology and Sociobiology 27: 315–324.

Birkhead, T. R., Clarkson, K. & Zann, R. 1988a. Extra-pair courtship, copulation and mate-guarding in wild Zebra Finches *Taeniopygia guttata*. Animal Behaviour 36: 1853–1855.

Birkhead, T. R. & Fletcher, F. 1992. Sperm to spare? Sperm allocation in the Zebra Finch. Animal Behaviour 43: 1053–1055.

Birkhead, T. R., Hunter, F. M. & Pellat, J. E. 1989. Sperm competition in the Zebra Finch *Taeniopygia guttata*. Animal Behaviour 38: 935–950.

Birkhead, T. R. & Hunter, F. M. 1990. Numbers of sperm storage tubules in the Zebra Finch (*Poephila guttata*) and Bengalese Finch (*Lonchura striata*). Auk 107: 193–197.

Birkhead, T. R. & Møller, A. P. 1992. Sperm Competition in Birds. Academic Press, London.

Birkhead, T. R. & Møller, A. P. 1993. Why do male birds stop copulating while their partners are still fertile? Animal Behaviour 45: 105–118.

Birkhead, T. R., Pellat, J. E. & Fletcher, F. 1993. Selection and utilisation of spermatozoa in the reproductive tract of the female Zebra Finch *Taeniopygia guttata*. Journal of Reproduction and Fertility 99: 593–600.

Birkhead, T. R., Pellat, J. E. & Hunter, F. M. 1988b. Extra-pair copulation and sperm competition in the Zebra Finch. Nature (London) 334: 60–62.

Bischof, H.-J. 1980. Reaktionen von Zebrafinken- auf zweidimensionale Attrapen: Einfluss von Reizqualität und Prägung. Journal für Ornithologie 121: 288–290.

Bischof, H.-J. 1985. Der Anteil akustischer Komponenten an der Auslösung der Balz männlicher Zebrafinken (*Taeniopygia guttata castanotis*). Journal für Ornithologie 126: 273–279.

Bischof, H.-J., Böhner, J. & Sossinka, R 1981. Influence of external stimuli on the quality of song of the Zebra Finch (*Taeniopygia guttata castanotis* Gould). Zeitschrift für Tierpsychologie 57: 261–267.

Bischof, H.-J. & Clayton, N. S. 1991. Stabilization of sexual preferences by sexual experience in male Zebra Finches, *Taeniopygia guttata castanotis*. Behaviour 118: 144–155.

Bischof, H.-J. & Lassek, R. 1985. The gaping reaction and the development of fear in young Zebra Finches (*Taeniopygia guttata castanotis*). Zeitschrift für Tierpsychologie 69: 55–65.

Blakers, M., Davies, S. J. J. F. & Reilly, P. N. 1984. The Atlas of Australian Birds. Melbourne University Press, Melbourne.

Boag, P. 1987. Effects of nestling diet on growth and adult size of Zebra Finches (*Poephila guttata*). Auk 104: 155–166.

Böhner, J. 1983. Song learning in the Zebra Finch (*Taeniopygia guttata*): selectivity in choice of a tutor and accuracy of song copies. Animal Behaviour 31: 231–237.

Böhner, J. 1990. Early acquisition of song in the Zebra Finch, *Taeniopygia guttata*. Animal Behaviour 39: 369–374.

Böhner, J., Cooke, F. & Immelmann, K. 1984. Verhaltensbedingte Isolation zwischen den beiden Rassen des Zebrafinken (*Taeniopygia guttata*). Journal für Ornithologie 125: 473–477.

Boles, W. E. 1988. Comments on the subspecies of Australian native and introduced finches. Emu 88: 20–24.

Bottjer, S. W. & Arnold, A. P. 1984. The role of feedback from the vocal organ. I Maintenance of stereotypical vocalizations by adult Zebra Finches. Journal of Neurosciences 4: 2387–2396.

Bottjer, S. W. & Hewer, S. J. 1992. Castration and antisteroid treatment impair vocal learning in male Zebra Finches. Journal of Neurobiology 23: 337–353.

Bottjer, S. W., Miesner, E. A. & Arnold, A. P. 1984. Forebrain lesions disrupt development but not maintenance of song in passerine birds. Science (Washington, D. C.) 224: 901–903.

Bowler, J. M. 1982. Aridity in the Late Tertiary and Quarternary of Australia. In Barker, W. R. & Greenslade, P. J. M. (eds), Evolution of the Flora and Fauna of Arid Australia: 35–45. Peacock Publications, Adelaide.

Boyce, M. S. & Perrins, C. M. 1987. Optimization of great tit clutch size in a fluctuating environment. Ecology 68: 142–153.

Bridgewater, P. B. 1987. The present Australian environment—terrestrial and freshwater. In Dyne, G. R. & Walton, D. W. (eds), Fauna of Australia, Vol 1A: 69–100. Australian Government Publishing Service, Canberra.

Brooks, D. R. & McLennan, D. A. 1991. Phylogeny, Ecology, and Behavior: a research program in comparative biology. The University of Chicago Press, Chicago.

Burley, N. 1981a. Sex-ratio manipulation and selection for attractiveness. Science (Washington, D. C.) 211: 721–722.

Burley N. 1981b. The evolution of sexual indistinguishability. In Alexander, R. D. & Tinkle, D. W. (eds), Natural Selection and Social Behavior: Recent Research and New Theory: 121–137. Blackwells Scientific Publications, Oxford.

Burley, N. 1982. Reputed band attractiveness and sex manipulation in Zebra Finches (a reply). Science (Washington, D. C.) 215: 423–424.

Burley, N. 1985a. Leg-band color and mortality patterns of captive breeding populations of Zebra Finches. Auk 102: 647–651.

Burley, N. 1985b. The organization of behavior and the evolution of sexually selected traits. In Gowaty, P. A. & Mock, D. W. (eds), Avian Monogamy: 22–44. American Ornithologists Union: Washington, D. C.

Burley, N. 1986a. Comparison of band color preferences of two species of estrildid finches. Animal Behaviour 34: 1732–1741.

Burley, N. 1986b. Sexual selection for aesthetic traits in species with biparental care. American Naturalist 127: 415–445.

Burley, N. 1986c. Sex-ratio manipulation in color banded populations of Zebra Finches. Evolution 40: 1191–1206.

Burley, N. 1988a. The differential allocation hypothesis: an experimental test. American Naturalist 132: 613–628.

Burley, N. 1988b. Wild Zebra Finches have band-colour preferences. Animal Behaviour 36: 1235–1237.

Burley, N. & Bartels, P. J. 1990. Phenotypic similarities of sibling Zebra Finches. Animal Behaviour 39: 174–180.

Burley, N. & Coopersmith, C. 1987. Bill colour preferences of Zebra Finches. Ethology 76: 133–151.

Burley, N. T., Enstrom, D. A. & Chitwood, L. 1994. Extra-pair relations in Zebra Finches: differential male success results from female tactics. Animal Behaviour 48: 1031–1041.

Burley, N., Krantzberg, G. & Radman, P. 1982. Influence of colour banding on the conspecific preferences of Zebra Finches. Animal Behaviour 30: 444–455.

Burley, N., Minor, C. & Strachan, C. 1990. Social preferences of Zebra Finches for siblings, cousins and non-kin. Animal Behaviour 39: 775–784.

Burley, N. & Price, D. K. 1991. Extra-pair copulation and attractiveness in Zebra Finches. Proceedings of the XX International Ornithological Congress (Christchurch, 1991): 1367-1372.

Burley, N., Price, D. K. & Zann, R. A. 1992. Bill color, reproduction and condition effects in wild and domesticated Zebra Finches. Auk 109: 13–23.

Burley, N., Tidemann, S. C. & Kalupka, K. 1991. Bill colour and parasite loads of Zebra Finches. In Loye, J. E., & Zuk, M. (eds), Bird–parasite Interactions: ecology, evolution and behaviour: 359–376. Oxford University Press, Oxford.

Burley, N., Zann, R. A., Tidemann, S. C. & Male, E. B. 1989. Sex ratios of Zebra Finches. Emu 89: 83–92.

Butterfield, P. A. 1970. The pair bond in the Zebra Finch. In Crook, J. H. (ed.), Social Behaviour in Birds and Mammals: 249–278. Academic Press, London.

Cade, T. J., Tobin, C. A. & Gold, A. (1965). Water economy and metabolism of two estrildine finches. Physiological Zoology 38: 9–33.

Calder, W. A. 1964. Gaseous metabolism and water relations of the Zebra Finch, *Taeniopygia castanotis*. Physiological Zoology 37: 400–413.

Calder, W. A. & King, J. R. 1963. Evaporative cooling in the Zebra Finch. Experimentia 19: 603.

Carnaby, I. C. 1954. Nesting seasons of Western Australian birds. Western Australian Naturalist 4: 149–156.

Carr, R. A. 1982. The effects of domestication on the Distance Calls of the Zebra Finch (*Poephila guttata castanotis* Gould). Honours thesis, La Trobe University.

Carr, R. A. & Zann, R. A. 1986. The morphological identification of domesticated Zebra Finches, *Poephila guttata* (Passeriformes: Estrildidae), in Australia. Australian Journal of Zoology 34: 439–448.

Carter, T. 1889. Notes from Western Australia. Zoologist Series 3, 13: 267–268.

Carter, T. 1903. Birds occurring in the region of the North-West Cape. Emu 3: 89–96.

Caryl, P. G. 1975. Aggressive behaviour in the Zebra Finch *Taeniopygia guttata*. I. Fighting provoked by male and female social partners. Behaviour 52: 226–252.

Caryl, P. G. 1976. Sexual behaviour in the Zebra Finch *Taeniopygia guttata*: response to familiar and novel partners. Animal Behaviour 24: 93–107.

Caryl, P. G. 1981. The relationship between the motivation of directed and undirected song in the Zebra Finch. Zeitschrift für Tierpsychologie 57: 37–50.

Caughley, G. 1977. Analysis of Vertebrate Populations. Wiley, New York.

Cayley, N. 1932. Australian Finches in Bush and Aviary. Angus & Robertson, Sydney.

Cayley, N. 1959. What Bird is That? Angus & Robertson, Sydney.

Christidis, L. 1986a. Chromosomal evolution within the family Estrildidae (Aves). I. The Poephilae. Genetica 71: 81–97.

Christidis, L. 1986b. Chromosomal evolution within the family Estrildidae (Aves). II. The Lonchurae. Genetica 71: 99–113.

Christidis, L. 1987a. Biochemical systematics within Paleotropic finches (Aves: Estrildidae). Auk 104: 380–392.

Christidis, L. 1987b. Phylogeny and systematics of estrildine finches and their relationships to other seed-eating passerines. Emu 87: 119–123.

Christidis, L. & Boles, W. E. 1994. The taxonomy and Species of Birds of Australia and its Territories. Royal Australasian Ornithologists Union, Melbourne.

Clayton, N. S. 1987a. Song tutor choice in Zebra Finches. Animal Behaviour 35: 714–721.

Clayton, N. S. 1987b. Song learning in Bengalese Finches: a comparison with Zebra Finches. Ethology 76: 247–255.

Clayton, N. S. 1987c. Mate choice in male Zebra Finches: some effects of cross-fostering. Animal Behaviour 35: 596–622.

Clayton, N. S. 1987d. Song learning in cross-fostered Zebra Finches: a re-examination of the sensitive phase. Behaviour 102: 67–81.

Clayton, N. S. 1988a. Song discrimination learning in Zebra Finches. Animal Behaviour 36: 1016–1024.

Clayton, N. S. 1988b. Song learning and mate choice in estrildid finches raised by two species. Animal Behaviour 36: 1589–1600.

Clayton, N. S. 1988c. Song tutor choice in Zebra Finches and Bengalese Finches: the relative importance of visual and vocal cues. Behaviour 104: 281–299.

Clayton, N. S. 1989a. The effects of cross-fostering on selective learning in estrildid finches. Behaviour 109: 163–175.

Clayton, N. S. 1989b. Song, sex and sensitive phases in the behavioural development of birds. Trends in Ecology and Evolution 4: 82–84.

Clayton, N. S. 1990a. Mate choice and pair formation in Timor and Australian Mainland Zebra Finches. Animal Behaviour 39: 474–480.

Clayton, N. S. 1990b. The effects of cross-fostering on assortative mating between Zebra Finch subspecies. Animal Behaviour 40: 1102–1110.

Clayton, N. S. 1990c. Subspecies recognition and song learning in Zebra Finch subspecies. Animal Behaviour 40: 1009–1017.

Clayton, N. S. 1990d. Assortative mating in Zebra Finch subspecies, *Taeniopygia guttata guttata* and *T. g. castanotis*. Philosophical Transactions of the Royal Society of London Series B Biological Science 330: 351–370.

Clayton, N. S. 1994. Sensitive phases in avian brain and behavioural development. In Hogan, J. & Bolhuis, J. (eds), Causal Mechanisms of Behavioural Development: 98–115. Cambridge University Press, Cambridge.

Clayton, N. S. & T. R. Birkhead. 1989. Consistency in the scientific name of the Zebra Finch. Auk 106: 750.

Clayton, N. S., Hodson, D. & Zann, R. A. 1991. Geographic variation in Zebra Finch subspecies. Emu 91: 2–11.

Clayton, N. S. & Pröve, E. 1989. Song discrimination in female Zebra Finches and Bengalese Finches. Animal Behaviour 38: 352–362.

Cleland, J. B. 1931. Notes on the birds of central Australia between Alice Springs and Macdonald Downs. South Australian Ornithologist 11: 13–21.

Collins, S. A. 1994. Male displays: cause or effect of female preference? Animal Behaviour 48: 371–375.

Collins, S. A. 1995. The effect of recent experience on female choice in Zebra Finches. Animal Behaviour 49: 479–486

Collins, S. A., Hubbard, C. & Houtman, A. M. 1994. Female choice in the Zebra Finch—the effect of male beak colour and male song. Behavioural Ecology and Sociobiology 35: 21–25.

Condon, H. T. 1955. Aboriginal bird names of South Australia. Part 2. South Australian Ornithologist 21: 91–98.

Corbett, J. L. 1987. Zebra Finches: A Complete Introduction. T. F. H. Publications, Neptune City.

Corbett, L. H. & Newsome, A. E. 1987. The feeding ecology of the dingo. III. Dietary relationships with widely fluctuating prey populations in arid Australia: an hypothesis of alternation of predation. Oecologica 74: 215–227.

Cracraft, J. 1986. Origin and evolution of continental biotas: speciation and historical congruence within the Australian avifauna. Evolution 40: 977–996.

Cramp, S. 1980. Handbook of the Birds of Europe, and the Middle East and North Africa: the birds of the Western Palearctic. Volume 2. Oxford University Press, Oxford.

Crome, F. J. H. 1976. Breeding, moult and food of the Squatter Pigeon in north eastern Queensland. Australian Wildlife Research 3: 45–49.

Csicsáky, M. J. 1977. Body-gliding in the Zebra Finch. Fortschritte der Zoologie 24: 275–286.

Cynx, J. 1990. Experimental determination of a unit of song production in the Zebra Finch (*Taeniopygia guttata*). Journal of Comparative Psychology 104: 3–10.

Cynx, J. & Nottebohm, F. 1992a. Role of gender, season and familiarity in discrimination of conspecific song by Zebra Finches. Proceedings of the National Academy of Sciences, U. S. A. 89: 1368–1371.

Cynx, J. & Nottebohm, F. 1992b. Testosterone facilitates some conspecific song discriminations in castrated Zebra Finches (*Taeniopygia guttata*). Proceedings of the National Academy of Sciences, U. S. A. 89: 1376–1378.

Cynx, J., Williams, H. & . Nottebohm, F. 1990. Timbre discrimination in Zebra Finch (*Taeniopygia guttata*) song syllables. Journal of Comparative Psychology 104: 303–308.

Cynx, J., Williams, H. & Nottebohm, F. 1992. Hemispheric differences in avian song discrimination. Proceedings of the National Academy of Sciences, U. S. A. 89: 1372–1375.

Davidson, A. A. 1905. Notes on the birds in the journal of explorations in central Australia under the leadership of Allen A. Davidson. South Australian Parliamentary Paper No. 27, Adelaide.

Davies, S. J. J. F. 1971. Results of 40 hours' continuous watch at five water-points in an Australian desert. Emu 72: 8–12.

Davies, S. J. J. F. 1977a. The timing of breeding of the Zebra Finch *Taeniopygia guttata* at Mileura, Western Australia. Ibis 119: 369–372.

Davies, S. J. J. F. 1977b. Man's activities and birds' distribution in the arid zone. Emu 77: 169–172.

Davies, S. J. J. F. 1979. The breeding seasons of birds in southwestern Australia. Journal of the Royal Society of Western Australia 62: 53–64.

Davies, S. J. J. F. 1986. A biology of the desert fringe—Presidential address 1984. Journal of the Royal Society of Western Australia 68: 37–50.

Dawson, W. R. 1981. Adjustments of Australian birds to thermal conditions and water scarcity in arid zones. In Keast, A. (ed.), Ecological Biogeography of Australia: 1651–1674. Dr W. Junk, The Hague.

De Lope, F. & Møller, A. P. 1993. Female reproductive effort depends on the degree of ornamentation of their mates. Evolution 47: 1152–1160.

Delacour, J. 1943. A revision of the subfamily Estrildinae of the family Ploceidae. Zoologica 28: 69–86.

Delesalle, V. A. 1986. Division of parental care and reproductive success in the Zebra Finch (*Taeniopygia guttata*). Behavioral Processes 12: 1–22.

Dijkstra, C., Vuursteen, L., Daan, S. & Masman, D. 1982. Clutch-size and laying date in the Kestrel (*Falco tinnunculus* L.): the effect of supplementary food. Ibis 124: 210–213.

Dooling, R. J. 1982. Auditory perception in birds. In Kroodsma, D. E. & Miller, E. H. (eds), Acoustic communication in birds: 95–130. Academic Press, New York.

Dooling, R. J., Brown, S. D., Klump, G. M. & Okanoya, K. 1992. Auditory perception of conspecific and heterospecific vocalizations in birds: evidence for special processes. Journal of Comparative Psychology 106: 20–28.

Drent, R. H. 1975. Incubation. In Farner, D. S. & King, J. R. (eds), Avian Biology, Vol. 5: 333–420. Academic Press, New York.

Dunn, A. 1992. Unusual behaviour of a female Zebra Finch: possible egg–dumping. Corella 15: 150–152.

Dunn, A. 1994. The song of the Zebra Finch: context and possible functions. Ph.D thesis, La Trobe University.

Dunn, J. L. 1981. A study of courtship in the Zebra Finch *Taeniopygia guttata* (Vieillot). Ph.D. thesis, University of Newcastle upon Tyne.

Dunn, P. O. & Lifjeld, J. T. 1994. Can extra-pair copulations be used to predict extra-pair paternity in birds? Animal Behaviour 47: 983–985.

Eales, L. A. 1985. Song learning in Zebra Finches: some effects of song model availability on what is learnt and when. Animal Behaviour 33: 1293–1300.

Eales, L. A. 1987a. Do Zebra Finch males that have been raised by another species still tend to select a song tutor? Animal Behaviour 35: 1347–1355.

Eales, L. A. 1987b. Song learning in female-raised Zebra Finches; another look at the sensitive phase. Animal Behaviour 35: 1356–1365.

Eales, L. A. 1989. The influences of visual and vocal interaction on song learning in Zebra Finches. Animal Behaviour 37: 507–508.

El-Wailly, A. J. 1966. Energy requirements for egg laying and incubation in the Zebra Finch, *Taeniopygia guttata*. Condor 68: 582–594.

Evans, S. M. 1970. Aggressive and territorial behaviour in captive Zebra Finches. Bird Study 17: 28–35.

Evans, S. M. & Bougher, A. R. 1987. The abundance of estrildid finches at waterholes in the Kimberley (W. A.). Emu 87: 124–127.

Evans, S. M., Collins, J. A., Evans, R. & Miller, S. 1985. Patterns of drinking behaviour of some Australian estrildine finches. Ibis 127: 348–354.

Farner, D. S. 1967. The control of avian reproductive cycles. In Proceedings of the XIV International Ornithological Congress (Oxford, 1966): 106–133.

Farner, D. S. & Serventy, D. L. 1960. The timing of reproduction in birds in the arid regions of Australia. Anatomical Record 137: 354.

Fetherston, I. A. & Tyler Burley, N. 1990. Do Zebra Finches prefer to mate with close relatives? Behavioural Ecology and Sociobiology 27: 411–414.

Figueredo, A. J., Ross, D. M. & Petrinovich, L. 1992. The quantitative ethology of the Zebra Finch: a study in comparative psychometrics. Multivariate Behavioural Research 27: 435–458.

Finlayson, H. H. 1932. Heat in the interior of South Australia and in central Australia—holocaust of bird-life. South Australian Ornithologist 11: 158–160.

Finlayson, H. H. 1935. The Red Centre. Angus & Robertson, Sydney.

Fisher, C. D., Lingren, E. & Dawson, W. R. 1972. Drinking patterns and behavior of Australian desert birds in relation to their abundance and ecology. Condor 74: 111–136.

Ford, H. J. 1989. Ecology of Birds: An Australian perspective. Surrey Beatty & Sons, Chipping Norton.

Ford, J. & Sedgwick, E. H. 1967. Bird distribution in the Nullabor Plain and Great Victoria Desert region, Western Australia. Emu 67: 99–124.

Frakes, L. A., McGowran, B. & Bowler, J. M. (1987). The evolution of the Australian environments. In Dyne, G. R. & Walton, D. W. (eds), Fauna of Australia, Vol 1A: 1–16. Australian Government Publishing Service, Canberra.

Frith, H. J. 1973. Wildlife Conservation. Angus & Robertson, Melbourne.

Frith, H. J. & Tilt, R. A. 1959. Breeding of the Zebra Finch in the Murrumbidgee Irrigation Area, New South Wales. Emu 59: 289–295.

Galloway, R. W. & Kemp, E. M. 1981. Late Cainozoic environments in Australia. In Keast, A. (ed.), Ecological Biogeography of Australia Vol. 1: 52–80. Dr. W. Junk, The Hague.

Gard, R. & Gard, E. 1990. Canning Stock Route. Western Desert Guides, Wembley Downs.

Garson, P. J., Dunn, J. L., Walton, C. J. & Shaw, P. A. 1980. Stimuli eliciting courtship from domesticated Zebra Finches. Animal Behaviour 28: 1184–1187.

Gibbs, H. L., Grant, P. R. & Weiland, J. 1984. Breeding of Darwin's Finches at an unusually early age in an El Niño year. Auk 101: 872–874.

Ginn, H. B. & Melville, D. S. 1983. Moult in Birds. The British Trust for Ornithology, Tring.

Goodwin, D. 1982. Estrildid finches of the World. London: British Museum (Natural History). Oxford University Press, Oxford.

Gould, J. 1865. Handbook to the Birds of Australia. Volume 2. J. Gould, London.

Gowaty, P. A. 1993. Differential dispersal, local resource competition, and sex ratio variation in birds. American Naturalist 141: 263–280.

Greenewalt, C. H. 1968. Bird Song, Acoustics and Physiology. Smithsonian Institute Press, Washington, D. C.

Greenwood, P. J. 1980. Mating systems, philopatry and dispersal in birds and mammals. Animal Behaviour 28: 1140–1162.

Güttinger, H.-R. 1970. Zur Evolution von Verhaltenweisen und Lautäusserungen bei Prachtfinken (Estrildidae). Zeitschrift für Tierpsychologie 27: 1011–1075.

Güttinger, H.-R. & Nicolai, J. 1973. Struktur und Funktion der Rufe bei Prachtfinken (Estrildidae). Zeitschrift für Tierpsychologie 33: 319–334.

Hall, M. F. 1962. Evolutionary aspects of estrildid song. Symposium of the Zoological Society of London 8: 37–55.

Harding, C. F., Sheridan, K. & Walters, M. 1983. Hormonal specificity and activation of sexual behavior in male Zebra Finches. Hormones and Behaviour 17: 111–133.

Harrison, C. J. O. 1967. Apparent zoological dispersal patterns in two avian families. Bulletin of the British Ornithologists Club 87: 49–72.

Harrison, C. J. O. & Colston, P. R. 1969. Some records of breeding after late rains in the Hamersley region. The Western Australian Naturalist 11: 49–50.

Hashino, E. & Okanoya, K. 1989. Auditory sensitivity of the Zebra Finch (*Poephila guttata castanotis*). Journal of the Acoustical Society of Japan (E) 10: 51–52.

Haywood, S. 1993a. Sensory control of clutch size in the Zebra Finch (*Taeniopygia guttata*). Auk 110: 778–786.

Haywood, S. 1993b. Sensory and hormonal control of clutch size in birds. The Quarterly Review of Biology 68: 33–60.

Haywood, S. & Perrins, C. M. 1992. Is clutch size in birds affected by environmental conditions during growth? Proceedings of the Royal Society (London), Series B 249: 195–197.

Heidweiller, J. & Zweers, G. A. 1990. Drinking mechanisms in the Zebra Finch and the Bengalese Finch. Condor 92: 1–28.

Hindwood, K. 1951. Bird/insect relationship: with particular reference to a beetle (*Platydema pascoei*) inhabiting the nests of finches. Emu 50: 179–183.

Högstedt, G. 1981. Effects of additional food on reproductive success in the Magpie (*Pica pica*). Journal of Animal Ecology 50: 219–229.

Houston, D. C., Donnan, D. & Jones, P. J. (1995a). The source of nutrients required for egg production in Zebra Finches. Journal of Zoology 235: 469–483.

Houston, D. C., Donnan, D., Jones, P. J., Hamilton, I. & Osborne, D. (1995b). Changes in the muscle condition of female Zebra Finches *Poephila guttata* during egg laying and the role of protein storage in bird skeletal muscle. Ibis 137: 322–328.

Houtman, A. M. 1992. Female Zebra Finches choose extra–pair copulations with genetically attractive males. Proceedings of the Royal Society (London) Series B 249: 3–6.

Hoyt, D. F. 1979. Practical methods of estimating volume and fresh weight of bird eggs. Auk 96: 73–77.

Immelmann, K. 1959. Exerimentelle Untersuchungen über die biologische Bedeutung artsspezifischer Merkmale beim Zebrafinken (*Taeniopygia guttata* Gould). Zoologische Jahrbücher Abteilung für Systematik Okologie und Geographie der Tiere 86: 437–592.

Immelmann, K. 1962a. Beiträge zu einer vergleichenden Biologie australische Prachtfinken (Spermestidae). Zoologische Jahrbücher Abteilung für Systematik Okologie und Geographie der Tiere 90: 1–196.

Immelmann, K. 1962b. Biologische Bedeutung optischer und akustischer Merkmale bei Prachtfinken (Aves: Spermestidae). Verhandlungen der Deutschen Zoologischen Gesellschaft (Saarbrüchen 1961): 369–374.

Immelmann, K 1962c. Vergleichende Beobachtungen über Verhalten domestizierter Zebrafinken in Europa und ihrer wilden Stammform in Australien. Zeitschrift für Tierzuchtung und Zuchtungsbiologie 77: 198–216.

Immelmann, K. 1963a. Tierische Jahresperiodik in ökologischer Sicht: ein Beitrag zum Zeitgeberproblem, unter besonder Berücksichtigung der Brut- und Mauserzeiten australischer Vögel. Zoologische Jahrbücher Abteilung für Systematik Okologie und Geographie der Tiere 91: 91–200.

Immelmann, K. 1963b. Drought adaptations in Australian desert birds. Proceedings of the XIII International Ornithological Congress (Ithaca, 1962): 649–657.

Immelmann, K. 1965a. Australian Finches in Bush and Aviary. Angus and Robertson, Sydney.

Immelmann, K. 1965b. Versuch einer ökologischen Verbreitungsanalyse beim australischen Zebrafinken, *Taeniopygia guttata castanotis* (Gould). Journal für Ornithologie 106: 415–430.

Immelmann, K. 1965c. Prägungserscheinungen in der Gesangsentwicklung junger Zebrafinken. Naturwissenschaften 52: 169–170.

Immelmann, K. 1966. Ecology and behaviour of African and Australian grass finches—a comparison. The Ostrich Supplement 6: 371–379.

Immelmann, K. 1967. Zur ontogenetischen Gesangsentwicklung bei Prachtfinken. Verdhandlungen der Deutschen Zoologischen Gesellschaft (Göttingen 1966): 320–332.

Immelmann, K. 1968. Zur biologischen Bedeutung des Estrildidengesanges. Journal für Ornithologie 109: 284–299.

Immelmann, K. 1969. Song development in the Zebra Finch and other estrildid finches. In Hinde, R. A. (ed.), Bird Vocalizations: 64–74. Cambridge University. Press, Cambridge.

Immelmann, K. 1970. Der Zebrafink (*Taeniopygia guttata*). A. Ziemsen, Wittenberg Lutherstadt.

Immelmann, K. 1971. Ecological aspects of periodic reproduction. In Farner, D. S., King, J. R. & Parkes, K. C. (eds), Avian Biology. Vol. 1: 341–389. Academic Press, New York.

Immelmann, K. 1972a. The influences of early experience upon the development of social behaviour in estrildine finches. Proceedings of the XV International Ornithological Congress (Leiden 1970): 316–338.

Immelmann, K. 1972b. Sexual and other long-term aspects of imprinting in birds and other species. Advances in the Study of Animal Behaviour 4: 147–174.

Immelmann, K. 1975a. Ecological significance of imprinting and early learning. Annual Review of Ecology and Systematics 6: 15–37.

Immelmann, K. 1975b. The evolutionary significance of early experience. In Baerends, G., Beer, C. & Manning, A. (eds), Function and Evolution in Behaviour: 243–253. Clarendon Press, Oxford.

Immelmann, K. 1979. Genetical constraints on early learning: a perspective from sexual imprinting in birds. In Royce, J. R. & Mos L. P. (eds), Theoretical Advances in Behavior Genetics: 121–136. Sijthoff & Noordhoff, Germantown.

Immelmann, K. 1985. Sexual imprinting in Zebra Finches: mechanisms and biological significance. Proceedings of the XVIII International Ornithological Congress (Moscow 1982): 156–172.

Immelmann, K., Hailman, J. P. & Baylis, J. R. 1982. Reputed band attractiveness and sex manipulation in Zebra Finches. Science (Washington, D. C.) 215: 422.

Immelmann, K. & Immelmann, G. 1967. Verhaltensoekologischen Studieren an Afrikanischen und Australischen Estrildiden. Zoologische Jahrbücher Abteilung für Systematik Okologie und Geographie der Tiere 94: 609–686.

Immelmann, K. & Immelmann, G. 1968. Zur Fortpflanzungsbiologie einige Vögel in der Namib. Bonner Zoologische Beiträge 19: 329–339.

Immelmann, K., Kalberlah, H.-H., Rausch, P. & Stahnke, A. 1978. Sexuelle Prägung als möglicher Faktor innerartlicher Isolation beim Zebrafinken. Journal für Ornithologie 119: 197–212.

Immelmann, K., Piltz, A. & Sossinka, R. 1977. Experimentelle Untersuchungen zur Bedeutung der Rachenzeichnung junger Zebrafinken. Zeitschrift für Tierpsychologie 45: 210–218.

Immelmann, K., Pröve, R., Lassek, R. & Bischof, H.-J. 1991. Influence of adult courtship experience on the development of sexual preferences in Zebra Finch males. Animal Behaviour 42: 83–89.

Immelmann, K., Steinbacher, J. & Wolters, H. E. 1963. Vögel in Käfig und Voliere: Prachtfinken. Verlag Hans Limberg, Aachen.

Immelmann, K. & Suomi, S. 1981. Sensitive phases in development. In Immelmann, K., Barlow, G. W., Petrinovich, L. & Main, M. (eds), Behavioural Development: 395–431. Cambridge University Press, Cambridge.

Isaacs, J. 1980. Australian Dreaming: 40,000 years of Aboriginal History. Lansdown Press, Sydney.

Jones, A. E. & Slater, P. J. B. 1993. Do young Zebra Finches prefer to learn songs that are familiar to females with which they are housed? Animal Behaviour 46: 616–617.

Joseph, L. 1986. Seed-eating birds of southern Australia. In Ford, H. A. & Paton, D. C. (eds), The Dynamic Partnership: birds and plants in southern Australia: 85–93. Government Printer, South Australia.

Katz, L. C. & Gurney, M. E. 1981. Auditory responses in the Zebra Finch's motor system for song. Brain Research 211: 192–197.

Keast, A. 1958. Infra-specific variation in the Australian finches. Emu 58: 219–246.

Keast, A. 1961. Bird speciation on the Australian continent. Bulletin of the Museum of Comparative Zoology 123: 303–495.

Keast, A. 1968. Moult in birds of the Australian dry country relative to rainfall and breeding. Journal of the Zoological Society of London 155: 185–200.

Keast, A. 1981. The evolutionary biogeography of Australian birds. In Keast, A. (ed.), Ecological Biogeography of Australia Vol. 3: 1586–1635. Dr W. Junk, The Hague.

Keast, A. & Marshall, A. J. 1954. The influence of drought and rainfall on reproduction in Australian desert birds. Proceedings of the Zoological Society of London 124: 493–499.

Kemp, E. M. 1981. Tertiary palaeogeography and the evolution of Australian climate. In Keast, A. (ed.), Ecological Biogeography of Australia Vol. 1: 31–49. Dr W. Junk, The Hague.

Kendeigh, S. C., Dol'nik, V. R. & Gavrilov, V. M. 1977. Avian energetics. In Pinowski, J. & Kendeigh, S. C. (eds), Granivorous Birds in Ecosystems: 127–204. Cambridge University Press, Cambridge.

Kikkawa, J. 1980. Seasonally in the nesting season of Zebra Finches at Armidale, N.S.W. Emu 80: 13–20.

Kloren, H. 1982. Zebravinken. Bosch & Keuning, Baarn.

Kovach, J. K. 1975. The behavior of quail. In Hafez, E. S. E. (ed.), The Behavior of Domestic Animals: 437–453. Bailliere Tindall, London.

Krebs, J. R. & Davies, N. B. 1993. An Introduction to Behavioural Ecology. Third Edition. Blackwell Scientific Publications, Oxford.

Kruijt, J. P. & Meeuwissen, G. B. 1991. Sexual preferences of male Zebra Finches: effects of early experience. Animal Behaviour 42: 91–102.

Kruijt, J. P. & Meeuwissen, G. B. 1993. Consolidation and modification of sexual preferences in adult male Zebra Finches. Netherlands Journal of Zoology 43: 68–79.

Kruijt, J. P., ten Cate, C. J. & Meeuwissen, G. B. 1983. The influences of siblings on sexual preferences of male Zebra Finches. Developmental Psychobiology 16: 233–239.

Kunkel, P. 1959. Zum Verhalten einiger Prachtfinken (Estrildinae). Zeitschrift für Tierpsychologie 16: 302–350.

Kunkel, P. 1969. Die stammegeschichte der Prachtfinken (Estrildidae) im lichte des brutparasitismus der witwen (Viduinae). Ardea 57: 173–181.

Lack, D. 1968. Ecological Adaptations for Breeding in Birds. Metheun, London.

Lazarides, M. 1970. The Grasses of Central Australia. Australian National University Press, Canberra.

Lee, P. & Schmidt–Nielsen, K. 1971. Respiratory and cutaneous evaporation in the Zebra Finch: effect on water balance. American Journal of Physiology 220: 1598–1605.

Leeper, G. W. 1970. The Australian Environment. C.S.I.R.O. & Melbourne University Press, Melbourne.

Lemon, W. C. 1991. The fitness consequences of foraging behaviour in the Zebra Finch. Nature (London) 352: 153–155.

Lemon, W. C. 1993. The energetics of lifetime reproductive success in the Zebra Finch *Taeniopygia guttata*. Physiological Zoology 66: 946–963.

Lemon, W. C. & Barth, R. H. Jr. 1992. The effects of feeding rate on reproductive success in the Zebra Finch, *Taeniopygia guttata*. Animal Behaviour 44: 851–857.

Lidicker, W. Z. Jr & Stenseth, N. C. 1992. To disperse or not to disperse: who does it and why? In Stenseth, N. C. & Lidicker, W. Z. Jr (eds), Animal dispersal: small mammals as a model: 22–36. Chapman & Hall, London.

Lill, A. & Fell, P. 1990. Egg composition in some Australian birds. Emu 90: 33–39.

Lindsay, H. A. 1963. The Bushman's Handbook. Jacaranda Press, Brisbane.

Lofts, B. & Murton, R. K. 1968. Photoperiodic and physiological adaptations regulating avian breeding cycles and their ecological significance. Journal of the Zoological Society of London 155: 327–294.

Lombardi, C. M. & Curio, E. 1985a. Influence of environment on mobbing by Zebra Finches. Bird Behaviour 6: 28–33.

Lombardi, C. M. & Curio, E. 1985b. Social facilitation of mobbing in the Zebra Finch *Taeniopygia guttata*. Bird Behaviour 6: 34–40.

Long, J. L. 1981. Introduced Birds of the World. Reed, Adelaide.

Luine, V., Nottebohm, F., Harding, C. & McEwan, B. S. 1980. Androgen affects cholinergic enzymes in syringeal motor neurons and muscle. Brain Research 192: 89–107.

MacArthur, R. H. 1964. Environmental factors affecting species diversity. American Naturalist 98: 387–397.

MacGillivray, D. W. K. 1929. Through a drought-stricken land. Emu 29: 113–129.

Maclean, G. L. 1976. Rainfall and avian breeding seasons in northeastern New South Wales in spring and summer 1974–75. Emu 76: 139–142.

MacMillen, R. E. 1990. Water economy of granivorous birds: a predictive model. Condor 92: 379–392.

Mann, N. I., Slater, P. J. B., Eales, L. A. & Richards, C. 1991. The influence of visual stimuli on song tutor choice in the Zebra Finch, *Taeniopygia guttata*. Animal Behaviour 42: 285–293.

Mann, N. I. & Slater, P. J. B. 1994. What causes young male Zebra Finches, *Taeniopygia guttata*, to choose their father as a song tutor? Animal Behaviour 47: 671–677.

Mann, N. I. & Slater, P. J. B. 1995. Song tutor choice by Zebra Finches in aviaries. Animal Behaviour 49: 811–820.

Marler, P. & Nelson, D. A. 1993. Action-based learning: a new form of developmental plasticity in bird song. Netherlands Journal of Zoology 43: 91–103.

Marler, P. & Peters, S. 1982. Developmental overproduction and selective attrition: new processes in the epigenesis of bird song. Developmental Psychobiology 15: 369–378.

Marschall, U. & Prinzinger, R. 1991. Vergleichende Ökophysiologie von fünf Prachtfinkenarten (Estrildidae). Journal für Ornithologie 132: 319–323.

Marshall, A. J. 1949. Weather factors and spermatogenesis in birds. Proceedings of the Zoological Society of London 119: 711–716.

Marshall, A. J. & Serventy, D. L. 1958. The internal rhythm of reproduction of xerophilous birds under conditions of illumination and darkness. Journal of Experimental Biology 35: 666–670.

Martin, H.-J. 1985. Zebra Finches. Barrons, New York.

Mathews, G. M. 1913. A List of Birds of Australia. Weatherby, London.

Mayr, E. 1944a. Timor and the colonization of Australia by birds. Emu 44: 113–130.

Mayr, E. 1944b. The birds of Timor and Sumba. Bulletin of the American Museum of Natural History 83: 123–194.

Mayr, E. 1968. The sequence of genera in the Estrildidae (Aves). Brevoria 287: 1–14.

McGilp, J. N. 1923. Birds of Lake Frome district, South Australia. Emu 22: 274–287.

McGilp, J. N. 1932. Heat in the interior of South Australia. South Australian Ornithologist 11: 160–162.

McGilp, J. N. 1944. Bird life west of Oodnadatta, South Australia. South Australian Ornithologist 17: 1–9.

McIntosh, A. W. 1983. Individual recognition by the Distance Call in the Zebra Finch (*Poephila guttata castanotis* Gould). Honours thesis, La Trobe University.

Meggitt, M. J. 1971. Desert People: a study of the Walbiri aborigines of central Australia. University of Chicago Press, Chicago.

Meienberger, C. & Ziswiller, V. 1990. Die Artabhängigkeit der Umsatzquotienten nahverwandter Vogel—ein mathematisches Problem? Journal für Ornithologie 131: 267–277.

Menon, G. K., Baptista, L. F., Elias, P. M. & Bouvier, M. 1988. Fine structural basis of cutaneous water barrier in nestling Zebra Finches *Poephila guttata*. Ibis 130: 503–511.

Menon, G. K., Baptista, L. F., Brown, B. E. & Elias, P. M. 1989. Avian epidermal differentiation II. Adaptive response of permeability barrier to water deprivation and replenishment. Tissue & Cell 21: 83–92.

Miller, D. B. 1979a. The acoustic basis of mate recognition by female Zebra Finches (*Taeniopygia guttata*). Animal Behaviour 27: 376–380.

Miller, D. B. 1979b. Long–term recognition of father's song by female Zebra Finches. Nature (London) 280: 389–391.

Millington, R. W. & Winkworth, R. E. 1978. Climate. In Low, W. A. (ed.), The Physical and Biological Features of Kunoth Paddock in Central Australia: 16–22. Division of Land Resources Management Technical Paper No. 4.

Møller, A. P. 1990. Fluctuating asymmetry in male ornaments may reliably reveal male quality. Animal Behaviour 40: 1185–1187.

Møller, A. P. & Birkhead, T. R. 1991. Frequent copulations and mate guarding as alternative paternity guards in birds: a comparative study. Behaviour 118: 170–186.

Møller, A. P. & Birkhead, T. R. 1992. Validation of the heritability method to estimate extra–pair paternity in birds. Oikos 64: 485–488.

Møller, A. P. & Birkhead, T. R. 1993. Cuckoldry and sociality: a comparative study of birds. American Naturalist 142: 118–140.

Møller, A. P. & Petrie, M. 1991. Evolution of intraspecific variability in bird's eggs: is intraspecific nest parasitism the selective agent? Proceedings of the XX International Ornithological Congress (Christchurch, 1990): 1041–1048.

Morley, R. J. & Flenley, J. R. 1987. Late Cainozoic vegetational and environmental changes in the Malay archipelago. In Whitemore, T. C. (ed.), Biogeographical Evolution of the Malay archipelago: 50–59. Clarendon Press, Oxford.

Morris, D. 1954. The reproductive behaviour of the Zebra Finch (*Poephila guttata*) with special reference to pseudofemale behaviour and displacement activities. Behaviour 7: 1–31.

Morris, D. 1955. The seed preferences of certain finches under controlled conditions. Avicultural Magazine 61: 271–287.

Morris, D. 1957. The reproductive behaviour of the Bronze Mannikin, *Lonchura cullata*. Behaviour 11: 156–201.

Morris, D. 1958. The comparative ethology of grassfinches (Erythrurae) and mannikins (Amadinae). Proceedings of the Zoological Society of London 131: 389–439.

Morrison, R. G. & Nottebohm, F. 1993. Role of a telencephalic nucleus in the delayed song learning of socially isolated Zebra Finches. Journal of Neurobiology 24: 1045–1064.

Morton, S. R. 1985. Granivory in arid regions: comparison of Australia with north and south America. Ecology 66: 1859–1866.

Morton, S. R. & Davies P. H. 1983. Food of the Zebra Finch (*Poephila guttata*), and an examination of granivory in birds of the Australia arid zone. Australian Journal of Ecology 8: 235–243.

Mott, J. J. 1972. Germination studies of some annual species from an arid region of Western Australia. Journal of Ecology 60: 293–304.

Mountford, C. P. 1976. Nomads of the Australian deserts. Rigby, Melbourne.

Muller, R. F. & Smith, D. G. 1978. Parent–offspring interactions in Zebra Finches. Auk 95: 485–495.

Nelson, T. W. A. 1993. The development of feeding behaviour in juvenile Zebra Finches *Poephila guttata*. Honours thesis, Monash University.

Neunzig, K. 1965. Fremdländischen Stubenvögel. Asher, Amsterdam.

Newton, I. 1967. The adaptive radiation and feeding ecology of some British finches. Ibis 108: 41–67.

Newton, I. & Marquiss, M. 1981. Effect of additional food on laying dates and clutch-size of sparrowhawks. Ornis Scandia 12: 224–229.

Newton, I. 1989. Synthesis. In Newton, I. (ed.), Lifetime Reproduction in Birds: 441–469. Academic Press, London.

Nicolai, J. 1964. Der Brutparasitismus der Viduinae als ethologische Problem. Zeitschrift für Tierpsychologie 21: 129–204.

Nielsen, L. 1959. Yellow–tailed Thornbill and Zebra Finch nesting record. Emu 59: 274.

Nix, H. A. 1976. Environmental control of breeding, post–breeding dispersal and migration of birds in the Australian region. Proceedings of the XVI International Ornithological Congress (Canberra 1974): 272–305.

Nix, H. A. 1982. Environmental determinants of biogeography and evolution in Terra Australis. In Barker, W. R. & Greenslade, P. J. M. (eds), Evolution of the Flora and Fauna of Arid Australia: 47–66. Peacock Publications, Adelaide.

Nordeen, K. W. & Nordeen, E. J. 1988. Projection neurons within a vocal motor pathway are born during song learning in Zebra Finches. Nature (London) 334: 149–151.

Nordeen, K. W. & Nordeen, E. J. 1992. Auditory feedback is necessary for the maintenance of stereotyped song in adult Zebra Finches. Behavioral and Neural Biology 57: 58–66.

North, A. J. 1909. Nests and eggs of birds found breeding in Australia and Tasmania. Volume II. Australian Museum, Sydney.

Nottebohm, F. 1980. Brain pathways for vocal learning in birds: a review of the first 10 years. Progress in Psychobiology and Physiological Psychology 9: 85–124.

Nottebohm, F. 1993. The search for neural mechanisms that define the sensitive period for song learning in birds. Netherlands Journal of Zoology 43: 193–234.

Nottebohm, F, Alvarez-Buylla, A., Cynx, J., Kirn, J., Ling, C., Nottebohm, M., Suter, R., Tolles, A., & Williams, H. 1990. Song learning in birds: the relation between perception and production. Philosophical Transactions of the Royal Society (London), Series B 329: 115–124.

Okanoya, K. & Dooling, R. J. 1987. Hearing in passerine and psittacine birds: a comparative study of absolute and masked thresholds. Journal of Comparative Psychology 101: 7–15.

Okanoya, K. & Dooling, R. 1988. Obtaining acoustic similarity measures from animals: a method for species comparisons. Journal of the Acoustical Society of America 83: 1690–1693.

Okanoya, K. & Dooling, R. 1990. Detection of gaps in noise in Budgerigars (*Melopsittacus undulatus*) and Zebra Finches (*Poephila guttata*). Hearing Research 50: 185–192.

Okanoya, K. & Dooling, R. 1991a. Perception of Distance Calls by Budgerigars (*Melopsittacus undulatus*) and Zebra Finches (*Poephila guttata*): assessing species–specific advantages. Journal of Comparative Psychology 105: 60–72.

Okanoya, K. & Dooling, R. 1991b. Detection of species–specific calls in noise by Zebra Finches *Poephila guttata* and Budgerigars *Melopsittacus undulatus*: time or frequency domain? Bioacoustics 3: 163–172.

Okanoya, K., Yoneda, T. & Kimura, T. 1993. Acoustical variations in sexually dimorphic features of distance calls in domesticated Zebra Finches (*Taeniopygia guttata castanotis*). Journal of Ethology 11: 29–36.

Oksche, A., Farner, D. S., Serventy, D. L., Wolff, F. & Nicholls, C. A. 1963. The hypothalamo-hypophysial neurosecretory system of the Zebra Finch, *Taeniopygia castanotis*. Zeitschrift für Zellforschung 58: 846–914.

Palmeros, V. L. 1983. Seed preferences in Zebra Finches. M. Sc thesis, University of Sussex.

Parsons, F. E. 1968. Pterlyography. Libraries Board of South Australia, Adelaide.

Pellat, E. J. & Birkhead, T. R. 1994. Ejaculate size in Zebra Finches *Taeniopygia guttata* and a method for obtaining ejaculates from passerine birds. Ibis 136: 97–106.

Pohl-Apel, G. & Sossinka, R. 1982. III. The effect of pre- and postnatally administered oestrogen. Verhandlungen der Deutschen Zoologischen Gesellschaft 1982: 365.

Pohl-Apel, G. & Sossinka, R. 1984. Hormonal determination of song plasticity in females of the Zebra Finch: critical phase of treatment. Zeitschrift für Tierpsychologie 64: 330–336.

Powell, C. McA., Johnson, B. D. & Veevers, J. J. 1981. The early Cretaceous break-up of Eastern Gonwanaland, the separation of Australia and India, and their interaction with southeast Asia. In Keast, A. (ed.), Ecological Biogeography of Australia: 16–29. Dr W. Junk, The Hague.

Price, D. K. & Burley, N. T. 1993. Constraints on the evolution of attractive traits: genetic (co)variance of Zebra Finch bill colour. Hereditary 71: 405–412.

Price, P. H. 1979. Developmental determinants of structure in Zebra Finch song. Journal of Comparative Physiological Psychology 93: 260–277.

Priedkalns, J., Oksche, A., Vleck, C. & Bennett, B. K. 1984. The response of the hypothalamo–gonadal system to environmental factors in the Zebra Finch, *Poephila guttata castanotis*. Cell and Tissue Research 238: 23–35.

Pröve, E. 1974. Der Einfluss von Kastration und Testosteronsubstitution auf das Sexualverhalten männlicher Zebrafinken (*Taeniopygia guttata castanotis* Gould). Journal für Ornithologie 115: 338–347.

Pröve, E. 1978. Quantitative Untersuchungen zu Wechselbeziehungen zwischen Balzaktivität und Testosterontitern bei männlichen Zebrafinken (*Taeniopygia guttata castanotis* Gould. Zeitschrift für Tierpsychologie 48: 46–67.

Pröve, E. 1983. Hormonal correlates of behavioural development in male Zebra Finches. In Balthazart, J., Pröve, E. & Gilles, R. (eds), Hormones and Behaviour in Higher Vertebrates: 368–374. Springer-Verlag, Berlin, Heidelberg.

Pröve, E. & Immelmann, K. 1982. Behavioral and hormonal responses of male Zebra Finches to antiandrogens. Hormones and Behavior 16: 121–131.

Queensland Finch Society. 1987. Finch Breeders Handbook: Volume 1—The Australians. Queensland Finch Society Inc., Brisbane.

Rabinowitch, V. E. 1969. The role of experience in the development and retention of seed preferences in Zebra Finches. Behaviour 33: 222–236.

Rahn, H., Sotherland, P. R. & Paganelli, C. V. 1985. Interrelationships between egg mass and adult body mass and metabolism among passerine birds. Journal für Ornithologie 126: 263–271.

Ratcliffe, L. M. & P. T. Boag. 1987. Effects of color bands on male competition and sexual attractiveness in Zebra Finches (*Poephila guttata*). Canadian Journal of Zoology 65: 333–338.

Reid, J. & Fleming, M. 1992. The conservation status of birds in arid Australia. Rangelands Journal 14: 65–91.

Rix, C. E. 1943. A review of the birds between Mt Lofty Ranges and the River Murray—a site for a real sanctuary. South Australian Ornithologist 16: 57–78.

Rix, C. E. 1970. Birds of the Northern Territory. South Australian Ornithologist 25: 147–190.

Rogers, C. H. 1986. The World of Zebra Finches. Nimrod Press, London.

Roper, A. 1993. The structure of the Zebra Finch (*Taeniopygia guttata*) Begging Call as a function of sex, age and motivation. Honours thesis, La Trobe University.

Rowe, L., Ludwig, D. & Schluter, D. 1994. Time, condition, and the seasonal decline of avian clutch size. American Naturalist 143: 698–722.

Rowley, I. 1975. Bird Life. Collins, Sydney.

Rowley, I. 1983. Re–mating in birds. In Bateson, P. (ed.), Mate Choice: 331–360. Cambridge University Press, Cambridge.

Rowley, I. & Russell, E. 1991. Demography of passerines in the temperate southern hemisphere. In Perrins, C. M., Lebreton, J.-D. & Hirons, G. J. M. (eds), Bird Population Studies: 22–44. Oxford University Press, Oxford.

Sargent, T. D. 1965. The role of experience in the nest building of the Zebra Finch. Auk 82: 48–61.

Schleucher, E. 1993. Life in extreme dryness and heat: a telemetric study of the behaviour of the Diamond Dove *Geopelia cuneata* in its natural habitat. Emu 93: 251–258.

Schlinger, B. A. & Arnold, A. P. 1992. Plasma sex steroids and tissue aromatization in hatchling Zebra Finches: implications for sexual differentiation of singing behaviour. Endocrinology 130: 289–299.

Schodde, R. 1975. Interim List of Australian Song Birds. Royal Australasian Ornithologists Union, Melbourne.

Schodde, R. 1982. Origin, adaptation and evolution of birds in arid Australia. In Barker, W. R. & Greenslade, P. J. M. (eds), Evolution of the Flora and Fauna of Arid Australia: 191–224. Peacock Publications, Adelaide.

Schodde, R. & Callaby, J. H. 1972. The biogeography of the Australo-Papuan bird and mammal faunas in relation to Torres Strait. In Walker, D. (ed.), The Natural and Cultural History of Torres Strait: 257–300. Australian National University Press, Canberra.

Schodde, R. & McKean, J. 1976. The relationships of some monotypic genera of Australian oscines. Proceedings of the XVI Ornithological Congress (Canberra 1974): 531–541.

Schoepfer, M. 1989. Moult strategies of four species of grassfinches living in the same area. Emu 89: 102–111.

Schöpfer, M. 1989. Feeding ecology of five sympatric species of grassfinches in southeastern Australia. Ph. D thesis, La Trobe University, Bundoora.

Schubert, C. A., Ratcliffe, L. M. & Boag, P. T. 1989. A test of inbreeding avoidance in the Zebra Finch. Ethology 82: 265–274.

Schwab, C. 1986. Comparison of songs between related individuals of the Zebra Finch *Poephila guttata*. Honours thesis, Queen's University, Kingston.

Seller, T. J. 1979. Unilateral nervous control of the syrinx in Java Sparrows. Journal of Comparative Physiology (A) 129: 281–288.

Serventy, D. L. 1971. Biology of desert birds. In Farner, D. S., King, J. R. & Parkes, K. C. (eds), Avian Biology. Vol. 1: 287–339. Academic Press, New York.

Serventy, D. L. & Marshall, A. J. 1957. Breeding periodicity in Western Australian birds: with an account of unseasonal nesting in 1953 and 1955. Emu 57: 99–126.

Serventy, D. L., Nicholls, C. A. & Farner, D. S. 1967. Pneumatization of the cranium of the Zebra Finch *Taeniopygia guttata*. Ibis 109: 570–578.

Serventy, D. L. & Whittell, H. M. 1976. Birds of Western Australia. Fifth Edition. University of Western Australia, Perth.

Shephard, M. 1989. Aviculture in Australia. Black Cockatoo Press, Melbourne.

Sheridan, K 1985. Mate selection in the Zebra Finch, *Poephila guttata*: the relationship of specific hormone dependent male behaviors to female choices. Ph.D. thesis. City University of New York.

Sibley, C. G. & Ahlquist, J. E. 1985. The phylogeny and classification of birds of the Australo-Papuan passerine birds. Emu 85: 1–14.

Silcox, A. P. 1979. Pair bonding in the Zebra Finch. Ph.D. thesis. University of Newcastle Upon Tyne.

Silcox, A. P. & Evans, S. M. 1982. Factors affecting the formation and maintenance of pair bonds in the Zebra Finch, *Taeniopygia guttata*. Animal Behaviour 30: 1237–1243.

Simpson, H. 1932. Bird notes. South Australian Ornithologist 11:195.

Simpson, H. B. & Vicario, D. S. 1990. Brain pathways for learned and unlearned vocalizations differ in Zebra Finches. Journal of Neuroscience 10: 1541–1556.

Simpson, H. B. & Vicario, D. S. 1991a. Early oestrogen treatment of female Zebra Finches masculinizes the brain pathway for learned vocalizations. Journal of Neurobiology 22: 777–793.

Simpson, H. B. & Vicario, D. S. 1991b. Early estrogen treatment alone causes female Zebra Finches to produce learned, male-like vocalizations. Journal of Neurobiology 22: 755–776.

Skadhauge, E. 1981. Osmoregulation in Birds. Springer-Verlag, Berlin.

Skadhauge, E. & Bradshaw, D. S. 1974. Drinking and cloacal excretion of salt and water in the Zebra Finch. American Journal of Physiology 227: 1263–1267.

Skagen, S. K. 1988. Asynchronous hatching and food limitation: a test of Lack's hypothesis. Auk 105: 78–88.

Slater, P. J. B. & Clayton, N. S. 1991. Domestication and song learning in Zebra Finches *Taeniopygia guttata*. Emu 91: 126–128.

Slater, P. J. B. & Clements, F. A. 1981. Incestuous mating in Zebra Finches. Zeitschrift für Tierpsychologie 57: 201–208.

Slater, P. J. B., Eales, L. A. & Clayton, N. S. 1988. Song learning in Zebra Finches (*Taeniopygia guttata*): progress and prospects. Advances in the Study of Behaviour 18: 1–34.

Slater, P. J. B., Jones, A. E. & ten Cate, C. J. 1993. Can lack of experience delay the end of the sensitive phase for song learning? Netherlands Journal of Zoology 40: 80–90.

Slater, P. J. B. & Jones, A. E. 1995. The timing of song and distance call learning in Zebra Finches. Animal Behaviour 49: 548–550.

Slater, P. J. B. & Mann, N. S. 1990. Do male Zebra Finches learn their father's songs? Trends in Ecology and Evolution 5: 415–417.

Slater, P. J. B. & Mann, N. S. 1991. Early experience and song learning in Zebra Finches *Taeniopygia guttata*. Proceedings of XX International Ornithological Congress, (Christchurch, 1990): 1074–1080.

Slater, P. J. B. & Richards, C. 1990. Renesting and song learning in the Zebra Finch, *Taeniopygia guttata*. Animal Behaviour 40: 1191–1192.

Slater, P. J. B., Richards, C. & Mann, N. I. 1991. Song learning in Zebra Finches exposed to a series of tutors during the sensitive phase. Ethology 88: 163–171.

Slatyer, R. O. 1962. Climate of the Alice Springs area. In Perry, R. A (ed.), General Report on the Lands of Alice Springs Area, Northern Territory, 1956–57: 109–128. CSIRO Land Research Series , Melbourne.

Sonnemann, P. & Sjölander, P. 1977. Effects of cross-fostering on sexual imprinting of the female Zebra Finch, *Taeniopygia guttata*. Zeitschrift für Tierpsychologie 45: 337–348.

Sossinka, R. 1970. Domestikationserscheinungen beim Zebrafinken *Taeniopygia guttata castanotis* (GOULD). Jahrbücher Abteilung für Systematik Okologie und Geographie der Tiere 97: 455–521.

Sossinka, R. 1972a. Besonderheiten in der sexuellen Entwicklung des Zebrafinken *Taeniopygia guttata castanotis* (Gould). Journal für Ornithologie 113: 29–36.

Sossinka, R. 1972b. Langfristiges Durstvermögen wilder und domestizieter Zebra Finken (*Taeniopygia guttata castanotis*). Journal für Ornithologie 113: 418–426.

Sossinka, R. 1974. Der Einfluss von Durstperioden auf die Schilddrüsen und Gonadenaktivität und ihre Bedeutung für die Brutperiodik des Zebrafinken (*Taeniopygia castanotis* Gould). Journal für Ornithologie 115: 128–140.

Sossinka, R. 1975. Quantitative Untersuchungen zur sexuellen Reifung des Zebrafinken, *Taeniopygia castanotis* (GOULD). Deutschen Zoologischen Gesellschaft 1974: 344–347.

Sossinka, R. 1980a. Ovarian development in an opportunistic breeder, the Zebra Finch *Poephila guttata*. The Journal of Experimental Zoology 211: 225–230.

Sossinka, R. 1980b. Reproductive strategies of estrildid finches in different climatic zones of the tropics: gonadal maturation. Proceedings of XVII International Ornithological Congress, (Berlin, 1978): 1074–1080.

Sossinka, R. 1982. Domestication in birds. In Farner, D. S., King, J. R. and Parks, K. C. (eds), Avian Biology Vol. 6: 373–403. Academic Press, New York.

Sossinka, R., and J. Böhner. 1980. Song types in the Zebra Finch (*Poephila guttata castanotis*). Zeitschrift für Tierpsychologie 53: 123–132.

Specht, R. L. 1981. Major vegetation formations in Australia. In Keast, A. (ed.), Ecological Biogeography of Australia Vol. 1: 163–298. Dr. W. Junk, The Hague.

Stafford Smith, D. M. & Morton, S. R. 1990. A framework for the ecology of arid Australia. Journal of Arid Environments 18: 255–278.

Stearns, S. C. 1992. The Evolution of Life Histories. Oxford University Press, Oxford.

Steiner, H. 1955. Das Brutverhalten der Prachtfinken, Spermestidae, als Ausdruck ihres selbständigen Familiencharakters. Proceedings of the XI International Ornithological Congress: 350–355.

Steiner, H. 1960. Klassifikation der Prachtfinken, Spermestidae, auf Grund der Rachenzeichnungen ihrer Nestlinge. Journal für Ornithologie 101: 92–112.

Sullivan, M. S. 1994. Discrimination among males by female Zebra Finches based on past as well as current phenotype. Ethology 96: 97–104.

Swaddle, J. P. & Cuthill, I. C. 1994a. Preference for symmetric males by female Zebra Finches. Nature (London) 367: 165–166.

Swaddle, J. P. & Cuthill, I. C. 1994b. Female Zebra Finches prefer mates with symmetric chest plumage. Philosophical Transactions of the Royal Society of London, Series B Biological Sciences 258: 267-271.

ten Cate, C. J. 1982. Behavioural differences between Zebra Finch and Bengalese Finch (foster)parents raising Zebra Finch offspring. Behaviour 81: 152–172.

ten Cate, C. J. 1984. The influence of social relations on the development of species recognition in Zebra Finch males. Behaviour 91: 263–285.

ten Cate, C. J. 1985a. On sex differences in imprinting. Animal Behaviour 33: 1310–1317.

ten Cate, C. J. 1985b. Directed song of male Zebra Finches as a predictor of subsequent intra- and interspecific social behaviour and pair formation. Behavioral Processes 10: 369–374.

ten Cate, C. J. 1986a. Listening behaviour and song learning in Zebra Finches. Animal Behaviour 34: 1267–1269.

ten Cate, C. J. 1986b. Sexual preferences in Zebra Finch males raised by two species: I: a case of double-imprinting. Journal of Comparative Psychology 100: 248–252.

ten Cate, C. J. 1987. Sexual preferences in Zebra Finch males raised by two species: II: the internal representation resulting from double imprinting. Animal Behaviour 35: 321–330.

ten Cate, C. J. 1991a. Population lateralization in Zebra Finch courtship: a re–assessment. Animal Behaviour 41: 900–901.

ten Cate, C. J. 1991b. Behaviour-contingent exposure to taped song and Zebra Finch song learning. Animal Behaviour 42: 857–859.

ten Cate, C. J., Baauw, A., Ballintun, M., Majoor, B. & van der Horst, I. 1990. Lateralization of orientation in sexually active Zebra Finches: eye use asymmetry or locomotor bias. Animal Behaviour 39: 992–994.

ten Cate, C. J. & Bateson, P. 1988. Sexual selection of conspicuous characteristics in birds by means of imprinting. Evolution 42: 1355–1358.

ten Cate, C. J., Kruijt, J. P. & Meeuwissen, G. G. 1989. The influence of testing conditions on sexual preferences in double-imprinted Zebra Finch males. Animal Behaviour 37: 694–696.

ten Cate, C. J., Los, L. & Schilperoord, L. 1984. The influence of differences in social experience on the development of species recognition in Zebra Finch males. Animal Behaviour 32: 852–860.

ten Cate, C. J. & Mug, G. 1984. The development of mate choice in Zebra Finches. Behaviour 90: 125–150.

ten Cate, C. J. & Slater, P. J. B. 1991. Song learning in Zebra Finches: how are elements from two tutors integrated? Animal Behaviour 42: 150–152.

ten Cate, C. J., Vos, D. R. & Mann, N. 1993. Sexual imprinting and song learning: two of one kind? Netherlands Journal of Zoology 43: 34–45.

Thissen, D. & Martin, E. 1982. Reputed band attractiveness and sex manipulation in Zebra Finches. Science (Washington, D. C.) 215: 423.

Tidemann, S. C. 1987. Gouldian finches in the wild. Bird Keeping in Australia 30: 145–153.

Tidemann, S. C. & Woinarski, J. C. Z. 1994. Moult characteristics and breeding seasons of Gouldian *Erythrura gouldiae*, Masked *Poephila personata* and Long-tailed Finches *P. acuticauda* in savannah woodland in the Northern Territory. Emu 94: 46–52.

Ueda, K. 1985. Juvenile female breeding of the Fan-tailed Warbler *Cisticola juncidis* : occurrence of two generations per year. Ibis 127: 111–116.

van Tets, G. F. 1974. Fossil birds (Aves) from Weeke's Cave, Nullabor Plain, South Australia. Transactions of the Royal Society of South Australia 94: 229–230.

Verheijen, J. A. J. 1964. Breeding season on the island of Flores, Indonesia. Ardea 52: 194–201.

Verheijen, J. A. J. 1976. Some data on the avifauna of the island of Roti, Lesser Sunda Islands, Indonesia. Zoologische Mededelingen 50: 1–21.

Vleck, C. M. 1981. Energetic cost of incubation in the Zebra Finch. Condor 83: 229–237.

Vleck, C. M. & Priedkalns, J. 1985. Reproduction in Zebra Finches: hormone levels and effect of dehydration. Condor 87: 37–46.

Vos, D. R. 1994. Sex recognition in Zebra Finch males results from early experience. Behaviour 128: 1–14.

Vos, D. R., Prijs, J. & ten Cate, C. 1993. Sexual imprinting in Zebra Finch males: a differential effect of successive and simultaneous experience with two colour morphs. Behaviour 126: 137–154.

Vriends, M. M. 1980. Handbook of Zebra Finches. T. F. H. Publications, Neptune City.

Wallace, A. R. 1864. A list of birds inhabiting the islands of Timor, Flores, and Lombock, with descriptions of the new species. Proceedings of the Zoological Society of London 1863: 18–32.

Walter, H., Harnickell, E. & Mueller-Dombois, D. 1975. Climate-diagram maps of the individual continents and the ecological climatic regions of the earth. Springer–Verlag, Berlin.

Walter, M. J. 1973. Effects of parental colouration on the mate preference of offspring in the Zebra Finch *Taeniopygia guttata castanotis* GOULD. Behaviour 46: 154-173.

Walters, M. J., Collado, D. & Harding, C. F. 1991. Oestrogenic modulation in male Zebra Finches: differential effects on directed and undirected songs. Animal Behaviour 42: 445–452.

Webb, D. R. 1987. Thermal tolerance of avian embryos: a review. Condor 89: 874–898.

Webster, M. D. 1991. Behavioral and physiological adaptations of birds to hot desert climates. Proceedings of XX International Ornithological Congress (Christchurch, 1990): 1765–1766.

Weisman, R., Shackelton, S. Ratcliffe, L., Weary, D. & Boag, P. 1994. Sexual preferences of female Zebra Finches: imprinting on bill colour. Behaviour 128: 15–24.

White, C. M. N. & Bruce, M. D. 1986. The Birds of Wallacia. British Ornithologists' Union, London.

White, F. N., Bartholomew, G. A. & Howell, T. R. (1975). The thermal significance of the nest of the sociable weaver, *Philetairus socius*: winter observations. Ibis 117: 171–179.

White, M. E. 1986. The Greening of Gonwana: the 400 million year of Australia's Plants. Reid, Sydney.

White, S. R. 1946. Notes on the bird life of Australia's heaviest rainfall region. Emu 46: 81–122.

Whitlock, F. L. 1948. Abnormal nests and eggs of Western Australian birds. Western Australian Naturalist 1: 75–78

Wiens, J. A. & Rotenberry, J. T. 1979. Diet niche relationships among North American grassland and shrubsteppe birds. Oecologica 42: 253–292.

Wiens, J. A. & Johnston, R. F. 1977. Adaptive correlates of granivory in birds. In Pinowski, J. & Kendeigh, S. C. (eds), Granivorous Birds in Ecosystems: 301–340. Cambridge University Press, Cambridge.

Williams, H. 1985. Sexual dimorphism of auditory activity in the Zebra Finch song system. Behavioral and Neural Biology 44: 470–484.

Williams, H. 1989. Multiple representations and auditory–motor interactions in the avian song system. Annals of the New York Academy of Science 563: 148–164.

Williams, H. 1990. Models for song learning in the Zebra Finch: fathers or others? Animal Behaviour 39: 747–757.

Williams, H., Crane, L. A., Hale, T. K., Esposito, M. A. & Nottebohm, F. 1992. Right-side dominance for song control in the Zebra Finch. Journal of Neurobiology 23: 1006–1020.

Williams, H., Cynx, J., & Nottebohm, F. 1989. Timbre control in Zebra Finch (*Taeniopygia guttata*) song syllables. Journal of Comparative Psychology 103: 366–380.

Williams, H., Kilander, K. & Sotanski, M. L. 1993. Untutored song, reproductive success and song learning. Animal Behaviour 45: 695–705.

Williams, H. & McKibben. J. R. 1992. Changes in stereotyped central motor patterns controlling vocalizations are induced by peripheral nerve injury. Behavioral and Neural Biology 57: 67–68.

Williams, H. & Nottebohm, F. 1985. Auditory responses in avian vocal motor neurons: a motor theory for song perception in birds. Science (Washington, D. C.) 229: 279–282.

Williams, H. & Staples, K. 1992. Syllable chunking in Zebra Finch (*Taeniopygia guttata*) song. Journal of Comparative Psychology 106: 278–286.

Willson, M. F. 1971. Seed selection in some North American finches. Condor 73: 415–429.

Winfield, C. 1982. Bush tucker: a guide to, and resources on traditional aboriginal foods of the north west of S. A. and central Australia. Mimili Aboriginal School, North West South Australia.

Woinarski, J. C. Z. & Tidemann, S. C. 1992. Survivorship and some population parameters for the endangered Gouldian finch *Erythrura gouldiae* and two other finch species at two sites in tropical northern Australia. Emu 92: 33–38.

Wolters, H. E. 1957. Die Klassifikation der Webefinken (Estrildidae). Bonner Zoologische Beiträge 8: 90–129.

Workman, L. & Andrew, R. J. 1986. Asymmetries of eye use in birds. Animal Behaviour 34: 1582–1584.

Wyndham, E. 1980. Environment and food of the Budgerigar *Melopsittacus undulatus*. Australian Journal of Ecology 5: 47–61.

Wyndham, E. 1981. Moult of Budgerigars *Melopsittacus undulatus*. Ibis 123: 145–157.

Wyndham, E. 1986. Length of birds' breeding seasons. American Naturalist 128, 155–164.

Wynn, S. E. & Price, T. 1993. Male and female choice in Zebra Finches. Auk 110: 635–638.

Yoneda, T. & Okanoya, K. 1991. Ontogeny of sexually dimorphic distance calls in Bengalese Finches. Journal of Ethology 9: 41–46.

Zann, R. 1972. Ethology of *Poephila* grassfinches (Estrildidae). Ph.D. thesis, University of Queensland, Brisbane.

Zann, R. 1975. Inter- and intraspecific variation in the calls of three species of grassfinches of the subgenus *Poephila* (Gould) (Estrildidae). Zeitschrift für Tierpsychologie 39: 85–115.

Zann, R. 1976a. Inter- and intraspecific variation in the courtship of three species of grassfinches of the subgenus *Poephila* (Gould) (Estrildidae). Zeitschrift für Tierpsychologie 41: 409–433.

Zann, R. 1976b. Variation in the songs of three species of Estrildine grassfinches. Emu 76: 97–108.

Zann, R. 1977. Pairbond and bonding behaviour in three species of grassfinches of the genus *Poephila* (Gould). Emu 97: 97–106.

Zann, R. 1984. Structural variation in the Zebra Finch distance call. Zeitschrift für Tierpsychologie 66: 328–345.

Zann, R. 1985a. Slow continuous wing-moult of Zebra Finches *Poephila guttata* from southeast Australia. Ibis 127: 187–196.

Zann, R. 1985b. Ontogeny of the Zebra Finch distance call: I. Effects of cross-fostering to Bengalese finches. Zeitschrift für Tierpsychologie 68: 1–23.

Zann, R. 1990. Song and call learning in wild Zebra Finches in south-east Australia. Animal Behaviour 40: 811–828.

Zann, R. 1993a. Structure, sequence and evolution of song elements in wild Australian Zebra Finches. Auk 110: 702–715.

Zann, R. 1993b. Variation in song structure within and among populations of Australian Zebra Finches. Auk 110: 716–726.

Zann, R. 1994a. Reproduction in a Zebra Finch colony in southeastern Australia: the significance of monogamy, precocial breeding and multiple broods in a highly mobile species. Emu 94: 285–299.

Zann, R. 1994b. Effects of band color on survivorship, body condition and reproductive effort of free-living Australian Zebra Finches. Auk 111: 131–142.

Zann, R. A., Morton, S. R., Jones, K. R. & Burley, N. 1995. The timing of breeding of Zebra Finches in relation to rainfall at Alice Springs, central Australia. Emu 95: 208–222.

Zann, R. & Rossetto, M. 1991. Zebra Finch incubation: brood patch, egg temperature and thermal properties of the nest. Emu 91: 107–120.

Zann, R. & Runciman, D. 1994. Survivorship, dispersal and sex ratios of Zebra Finches *Taeniopygia guttata* in southeast Australia. Ibis 136: 136–146.

Zann, R. & Straw, B. 1984a. Feeding ecology and breeding of Zebra Finches in farmland in northern Victoria. Australian Wildlife Research 11: 533–552.

Zann, R. & Straw, B. 1984b. A non-destructive method to determine the diet of seed-eating birds. Emu 84: 40–41.

Ziswiller, V. 1959. Besonderheiten in der Ontogenese der Prachtfinken (Spermestidae). Vierteljahrsschrift Naturforschenden Gesellschaft (in Zürich) 104: 222–226.

Ziswiller, V., Güttinger, H. R. & Bregulla, H. 1972. Monographie der Gattung *Erythrura* Swainson 1873 (Aves, Passeres, Estrildidae). Bonner Zoologische Monographien No. 2.

Author index

Subject index